**Exploring the Human
Plasma Proteome**

Edited by
Gilbert S. Omenn

Related Titles

Jungblut, P. R.,
Hecker, M. (Eds.)

Proteomics of Microbial Pathogens

2006
ISBN-13: 978-3-527-31759-2
ISBN-10: 3-527-31759-7

Liebler, D. C., Petricoin, E. F.,
Liotta, L. A. (Eds.)

Proteomics in Cancer Research

2005
ISBN-13: 978-0-471-44476-3
ISBN-10: 0-471-44476-6

Lion, N., Rossier, J. S.,
Girault, H. (Eds.)

Microfluidic Applications in Biology

From Technologies to Systems Biology

2006
ISBN-13: 978-3-527-31761-5
ISBN-10: 3-527-31761-9

Sanchez, J.-C., Corthals, G. L., Hochstrasser, D. F. (Eds.)

Biomedical Applications of Proteomics

2004
ISBN-13: 978-3-527-30807-1
ISBN-10: 3-527-30807-5

Hamacher, M., Marcus, K., Stühler, K.,
van Hall, A., Warscheid, B., Meyer, H.E.
(Eds.)

Proteomics in Drug Research

2006
ISBN-13: 978-3-527-31226-9
ISBN-10: 3-527-31226-9

Exploring the Human Plasma Proteome

Edited by
Gilbert S. Omenn

WILEY-VCH Verlag GmbH & Co. KGaA

The Editor

Prof. Dr. Gilbert S. Omenn
University of Michigan
A520 MSRB I Bldg
1150 West Medical Center Dr.
Ann Arbor,
MI 48109–0626
USA

■ All books published by Wiley-VCH are carefully produced. Nevertheless, authors, editors, and publisher do not warrant the information contained in these books, including this book, to be free of errors. Readers are advised to keep in mind that statements, data, illustrations, procedural details or other items may inadvertently be inaccurate.

Library of Congress Card No.:
applied for

British Library Cataloguing-in-Publication Data
A catalogue record for this book is available from the British Library.

Bibliographic information published by Die Deutsche Bibliothek
Die Deutsche Bibliothek lists this publication in the Deutsche Nationalbibliografie; detailed bibliographic data is available in the Internet at <http://dnb.ddb.de>.

© 2006 WILEY-VCH Verlag GmbH & Co. KGaA, Weinheim

All rights reserved (including those of translation into other languages). No part of this book may be reproduced in any form – by photoprinting, microfilm, or any other means – nor transmitted or translated into a machine language without written permission from the publishers. Registered names, trademarks, etc. used in this book, even when not specifically marked as such, are not to be considered unprotected by law.

Printed in the Federal Republic of Germany
Printed on acid-free paper

Typesetting X Con Media AG, Bonn
Printing Strauss GmbH, Mörlenbach
Binding Litges & Dopf Buchbinderei GmbH, Heppenheim

ISBN: 3-527-31757-0
ISBN 13: 978-3-527-31757-8

Table of Contents

1		Overview of the HUPO Plasma Proteome Project: Results from the pilot phase with 35 collaborating laboratories and multiple analytical groups, generating a core dataset of 3020 proteins and a publicly-available database 1

Gilbert S. Omenn, David J. States, Marcin Adamski, Thomas W. Blackwell, Rajasree Menon, Henning Hermjakob, Rolf Apweiler, Brian B. Haab, Richard J. Simpson, James S. Eddes, Eugene A. Kapp, Robert L. Moritz, Daniel W. Chan, Alex J. Rai, Arie Admon, Ruedi Aebersold, Jimmy Eng, William S. Hancock, Stanley A. Hefta, Helmut Meyer, Young-Ki Paik, Jong-Shin Yoo, Peipei Ping, Joel Pounds, Joshua Adkins, Xiaohong Qian, Rong Wang, Valerie Wasinger, Chi Yue Wu, Xiaohang Zhao, Rong Zeng, Alexander Archakov, Akira Tsugita, Ilan Beer, Akhilesh Pandey, Michael Pisano, Philip Andrews, Harald Tammen, David W. Speicher and Samir M. Hanash

1.1	Introduction 2	
1.2	PPP reference specimens 4	
1.3	Bioinformatics and technology platforms 5	
1.3.1	Constructing a PPP database for human plasma and serum proteins 5	
1.3.2	Analysis of confidence of protein identifications 14	
1.3.3	Quantitation of protein concentrations 15	
1.4	Comparing the specimens 17	
1.4.1	Choice of specimen and collection and handling variables 17	
1.4.2	Depletion of abundant proteins followed by fractionation of intact proteins 19	
1.4.3	Comparing technology platforms 22	
1.4.4	Alternative search algorithms for peptide and protein identification 23	
1.4.5	Independent analyses of raw spectra or peaklists 24	
1.4.6	Comparisons with published reports 25	
1.4.7	Direct MS (SELDI) analyses 27	
1.4.8	Annotation of the HUPO PPP core dataset(s) 27	
1.4.9	Identification of novel peptides using whole genome ORF search 30	
1.4.10	Identification of microbial proteins in the circulation 30	

Exploring the Human Plasma Proteome. Edited by Gilbert S. Omenn
Copyright © 2006 WILEY-VCH Verlag GmbH & Co. KGaA, Weinheim
ISBN: 3-527-31757-0

1.5	Discussion 31
1.6	References 33

2	**Data management and preliminary data analysis in the pilot phase of the HUPO Plasma Proteome Project** 37
	Marcin Adamski, Thomas Blackwell, Rajasree Menon, Lennart Martens, Henning Hermjakob, Chris Taylor, Gilbert S. Omenn and David J. States
2.1	Introduction 37
2.2	Materials and methods 39
2.2.1	Development of the data model 39
2.2.1.1	Laboratory 39
2.2.1.2	Experimental protocol 39
2.2.1.3	Protein identification data set 39
2.2.1.4	Peak list 41
2.2.1.5	Summary of technologies and resources 41
2.2.1.6	MS/MS spectra 41
2.2.1.7	SELDI peak list 42
2.2.2	Data submission process 42
2.2.3	Design of the data repository 42
2.2.4	Receipt of the data 43
2.3	Inference from peptide level to protein level 44
2.4	Summary of contributed data 46
2.4.1	Cross-laboratory comparison, confidence of the identifications 49
2.5	False-positive identifications 51
2.6	Data dissemination 56
2.7	Discussion 57
2.8	Concluding remarks 58
2.9	Computer technologies applied 60
2.10	References 61

3	**HUPO Plasma Proteome Project specimen collection and handling: Towards the standardization of parameters for plasma proteome samples** 63
	Alex J. Rai, Craig A. Gelfand, Bruce C. Haywood, David J. Warunek, Jizu Yi, Mark D. Schuchard, Richard J. Mehigh, Steven L. Cockrill, Graham B. I. Scott, Harald Tammen, Peter Schulz-Knappe, David W. Speicher, Frank Vitzthum, Brian B. Haab, Gerard Siest and Daniel W. Chan
3.1	Introduction 63
3.2	Materials and methods 65
3.2.1	HUPO reference sample collection protocol 65
3.2.2	Differential peptide display 66
3.2.3	Stability studies and SELDI analysis 66
3.2.4	SDS-PAGE analysis for stability studies 67

3.2.5	2-DE for stability studies 67	
3.2.6	SELDI-TOF analysis for protease inhibitor studies 67	
3.2.7	2-DE for plasma protease inhibition studies 68	
3.2.8	Tryptic digestion and protein identification for protease inhibition studies 69	
3.2.9	Antibody microarray analysis using two-color rolling circle amplification 69	
3.3	Results 69	
3.3.1	Comparisons of specimen types 71	
3.3.1.1	Analysis of serum 71	
3.3.1.2	Analysis of plasma 71	
3.3.2	Evaluation of storage and handling conditions 71	
3.3.3	Evaluations of the use of protease inhibitors 73	
3.3.3.1	Analysis with SELDI-TOF MS of "time zero" effects of protease inhibitors in plasma 73	
3.3.3.2	Analysis by 2-DE 73	
3.3.3.3	Analysis with antibody arrays 76	
3.4	Discussion 77	
3.4.1	Other pre-analytical variables and control considerations 83	
3.4.2	Reference materials 84	
3.5	Concluding remarks 87	
3.6	References 88	

4 **Immunoassay and antibody microarray analysis of the HUPO Plasma Proteome Project reference specimens: Systematic variation between sample types and calibration of mass spectrometry data** 91
Brian B. Haab, Bernhard H. Geierstanger, George Michailidis, Frank Vitzthum, Sara Forrester, Ryan Okon, Petri Saviranta, Achim Brinker, Martin Sorette, Lorah Perlee, Shubha Suresh, Garry Drwal, Joshua N. Adkins and Gilbert S. Omenn

4.1	Introduction 92	
4.2	Materials and methods 93	
4.2.1	Reference specimens 93	
4.2.2	DB immunoassays 93	
4.2.3	Antibody arrays at GNF 94	
4.2.3.1	Antibodies, reagents, microarray printing, and platform 94	
4.2.3.2	Microarray layout and processing 94	
4.2.3.3	Array imaging and data analysis 95	
4.2.4	Antibody microarrays at MSI 95	
4.2.4.1	Chip manufacture 95	
4.2.4.2	Rolling circle amplification (RCA) immunoassay 96	
4.2.4.3	Conversion of mean fluorescent intensity to concentration 96	

4.2.5	Antibody microarrays at VARI 96
4.2.5.1	Fabrication of antibody microarrays 96
4.2.5.2	Serum labeling 97
4.2.5.3	Processing of antibody microarrays 97
4.2.5.4	Analysis 97
4.2.6	Retrieval and matching of IPI numbers for the analytes 97
4.3	Results 98
4.3.1	Antibody-based measurements of the HUPO reference specimens 98
4.3.2	Systematic variation between the preparation methods of the PPP reference specimens 100
4.3.3	Consistent alterations in specific protein abundances 107
4.3.4	Linkage of MS data and antibody-based measurements 108
4.4	Discussion 110
4.5	References 113

5	**Depletion of multiple high-abundance proteins improves protein profiling capacities of human serum and plasma** 115
	Lynn A. Echan, Hsin-Yao Tang, Nadeem Ali-Khan, KiBeom Lee and David W. Speicher
5.1	Introduction 116
5.2	Materials and methods 117
5.2.1	Serum/plasma collection 117
5.2.2	MARS 118
5.2.3	Multiple affinity removal spin cartridge 118
5.2.4	Microscale solution IEF (MicroSol IEF) (ZOOM™-IEF) fractionation 118
5.2.5	2-DE 119
5.2.6	LC-MS/MS 119
5.3	Results 120
5.3.1	Depletion of major proteins to enhance detection of lower abundance proteins 120
5.3.2	Evaluation of high-abundance protein removal using 2-DE 121
5.3.3	Specificity of major protein depletion 123
5.3.4	Impact of Top-6 protein depletion on detection of lower abundance proteins using 2-D gels 125
5.3.5	Combining Top-6 protein depletion with microSol IEF prefractionation and narrow pH range gels 125
5.3.6	Analysis of Top-6 depleted serum and plasma using protein array pixelation 128
5.4	Discussion 130
5.5	References 134

6	**A novel four-dimensional strategy combining protein and peptide separation methods enables detection of low-abundance proteins in human plasma and serum proteomes** *135*
	Hsin-Yao Tang, Nadeem Ali-Khan, Lynn A. Echan, Natasha Levenkova, John J. Rux and David W. Speicher
6.1	Introduction *135*
6.2	Materials and methods *138*
6.2.1	Materials *138*
6.2.2	Top six protein depletion *138*
6.2.3	MicroSol-IEF fractionation *139*
6.2.4	Protein array pixelation *139*
6.2.5	LC-ESI-MS/MS methods *140*
6.2.6	Data analysis *140*
6.3	Results and discussion *141*
6.3.1	Protein array pixelation strategy *141*
6.3.2	Optimization of protein array pixelation *143*
6.3.3	Total analysis time for protein array pixelation of human plasma proteome *146*
6.3.4	Systematic protein array pixelation of the human plasma proteome *147*
6.3.5	Systematic protein array pixelation of the human serum proteome *150*
6.3.6	Analyses of human plasma and serum proteomes using HUPO filter criteria *153*
6.4	Concluding remarks *157*
6.5	References *157*
7	**A study of glycoproteins in human serum and plasma reference standards (HUPO) using multilectin affinity chromatography coupled with RPLC-MS/MS** *159*
	Ziping Yang, William S. Hancock, Tori Richmond Chew and Leo Bonilla
7.1	Introduction *159*
7.2	Materials and methods *160*
7.2.1	Materials *160*
7.2.2	Isolating glycoproteins using multilectin affinity columns *161*
7.2.3	Analysis of glycoproteins on LC-LCQ MS *161*
7.2.4	Analysis of glycoproteins on LC-LTQ MS *162*
7.2.5	Protein database search *162*
7.3	Results and discussion *162*
7.3.1	Protein IDs from the plasma and serum samples *162*
7.3.2	Comparison between serum and plasma glycoproteomes *179*
7.3.3	Comparison of the glycoproteins present in the samples collected from three ethnic groups *179*

7.4	Concluding remarks 182
7.5	References 183

8	**Evaluation of prefractionation methods as a preparatory step for multidimensional based chromatography of serum proteins** 185
	Eilon Barnea, Raya Sorkin, Tamar Ziv, Ilan Beer and Arie Admon
8.1	Introduction 185
8.1.1	The HUPO Plasma Proteome Project (PPP) goals and the serum as a complex sample 185
8.1.2	The scope of this manuscript 187
8.2	Materials and methods 187
8.2.1	Depletion from serum albumin and antibodies 187
8.2.2	MudPIT and mass segmentation 187
8.2.3	Protein separation by SDS-PAGE 188
8.2.4	SCX separation of intact proteins followed by MudPIT 188
8.2.5	Liquid-phase IEF followed by MudPIT 188
8.2.6	Capillary RP-LC-MS/MS 189
8.2.7	MS data processing and peptide/protein identifications 189
8.3	Results 189
8.3.1	Comparisons between the prefractionation methods 190
8.3.2	Identification of different protein subsets 191
8.3.3	Proteins identified by only one prefractionation method 193
8.3.4	Different methods resulted in diverse peptide coverage 193
8.4	Discussion 196
8.4.1	Giving every peptide a chance 196
8.4.2	How to identify more of the marginal proteins 197
8.4.3	Clustering and comparing raw data 197
8.4.4	High throughput and ruggedness versus high sensitivity 197
8.4.5	The cost effectiveness of the different methods 198
8.5	Concluding remarks 198
8.6	References 199

9	**Efficient prefractionation of low-abundance proteins in human plasma and construction of a two-dimensional map** 201
	Sang Yun Cho, Eun-Young Lee, Joon Seok Lee, Hye-Young Kim, Jae Myun Park, Min-Seok Kwon, Young-Kew Park, Hyoung-Joo Lee,, Min-Jung Kang, Jin Young Kim, Jong Shin Yoo, Sung Jin Park, Jin Won Cho, Hyon-Suk Kim and Young-Ki Paik
9.1	Introduction 202
9.2	Materials and methods 203
9.2.1	Plasma sample preparation 203
9.2.2	Depletion of major abundance proteins with an immunoaffinity column 203

9.2.3	2-DE	204
9.2.4	Identification of proteins by MS	204
9.2.5	Fractionation of the plasma samples by FFE	204
9.2.6	LC-MS/MS	205
9.2.7	Bioinformatics	206
9.3	Results and discussion	206
9.3.1	2-DE map of human plasma devoid of high-abundance proteins	206
9.3.2	Expression of different anticoagulant-treated plasma	214
9.3.3	FFE/1-DE/nanoLC-MS/MS and 2-DE/MALDI-TOF	215
9.4	Concluding remarks	219
9.5	References	219

10 Comparison of alternative analytical techniques for the characterisation of the human serum proteome in HUPO Plasma Proteome Project 221
Xiaohai Li, Yan Gong, Ying Wang, Songfeng Wu, Yun Cai, Ping He, Zhuang Lu, Wantao Ying, Yangjun Zhang, Liyan Jiao, Hongzhi He, Zisen Zhang, Fuchu He, Xiaohang Zhao and Xiaohong Qian

10.1	Introduction	222
10.2	Materials and methods	223
10.2.1	Materials	223
10.2.2	Human serum samples	223
10.2.3	Integrated strategy for characterising analytical approaches	223
10.2.4	Depletion of the highly abundant serum proteins by MARS	224
10.2.5	Desalting and concentrating the flow-through fractions by centrifugal ultrafiltration	224
10.2.6	Fractionation of depleted serum samples by anion-exchange HPLC	225
10.2.7	Protein fractionation by 2-D HPLC with nonporous RP-HPLC	225
10.2.8	The 2-DE strategy for the analysis of serum proteins	226
10.2.8.1	2-DE	226
10.2.8.2	In-gel digestion *via* automated workstation	227
10.2.8.3	Protein spot identification by MALDI-TOF-MS/MS	227
10.2.9	Shotgun strategy for the analysis of serum proteins	228
10.2.9.1	Trypsin digestion of serum proteins	228
10.2.9.2	Protein identification by micro2-D LC-ESI-MS/MS	228
10.2.9.3	Data processing	229
10.2.10	Protein fractionation strategy for the analysis of serum proteins	229
10.2.10.1	2-D LC fractionation of serum proteins	229
10.2.10.2	Digestion of the 2-D LC separated fractions	229
10.2.10.3	1-D microRP-HPLC-ESI-MS/MS identification of digested serum proteins	230
10.2.11	Offline shotgun strategy for the analysis of serum proteins	230

10.2.11.1	Offline SCX for first-dimension chromatographic separation of peptides *230*	
10.2.11.2	1-D capillary RP-HPLC/microESI-IT-MS/MS analysis for the SCX-separated peptide fractions *231*	
10.2.12	Optimised nanoRP-HPLC-nanoESI IT-MS/MS for the reanalysis of offline SCX-separated peptides (offline-nanospray strategy) *231*	
10.3	Integrated analysis of the whole data sets *231*	
10.3.1	Protein grouping analysis *231*	
10.3.2	Sequence clustering *232*	
10.4	Results and discussion *233*	
10.4.1	Depletion of the highly abundant serum proteins *233*	
10.4.2	The 2-DE strategy for the analysis of serum proteins *233*	
10.4.3	2-D HPLC fractionation for the analysis of serum proteins *234*	
10.4.4	Shotgun strategy for the analysis of serum proteins with online SCX *237*	
10.4.5	Shotgun strategy for the analysis of serum proteins with offline SCX *237*	
10.4.6	Offline SCX shotgun-nanospray strategy for the analysis of serum proteins *239*	
10.4.7	Comparison of the five strategies for the analysis of the human serum proteome *241*	
10.5	Concluding remarks *246*	
10.6	References *246*	
11	**A proteomic study of the HUPO Plasma Proteome Project's pilot samples using an accurate mass and time tag strategy** *249*	
	Joshua N. Adkins, Matthew E. Monroe, Kenneth J. Auberry, Yufeng Shen, Jon M. Jacobs, David G. Camp II, Frank Vitzthum, Karin D. Rodland, Richard, C. Zangar, Richard D. Smith and Joel G. Pounds	
11.1	Introduction *250*	
11.2	Materials and methods *251*	
11.2.1	Human blood serum and plasma *251*	
11.2.2	Depletion of Igs and trypsin digestion *252*	
11.2.3	Peptide cleanup *252*	
11.2.4	Capillary RP-LC *253*	
11.2.5	IT-MS *254*	
11.2.6	SEQUEST identification of peptides *254*	
11.2.7	Putative mass and time tag database from SEQUEST results *254*	
11.2.8	FT-ICR-MS *255*	
11.2.9	cLC-FT-ICR MS data analysis *255*	
11.2.10	OmniViz cluster and visual analysis *257*	
11.3	Results *257*	
11.3.1	PuMT tag database *257*	

11.3.2	Summary of peptide/protein identifications by AMT tags	258
11.3.3	Protein concentration estimates from ion current	260
11.3.4	Global protein analysis	261
11.4	Discussion	264
11.4.1	Application of FT-ICR MS as a proteomic technology bridge	264
11.4.2	Confidence in any MS-based proteomic approach	266
11.4.3	Peptide/protein redundancy	267
11.4.4	Identification sensitivity versus specificity	267
11.4.5	Throughput and differential analysis	269
11.5	References	270

12 Analysis of Human Proteome Organization Plasma Proteome Project (HUPO PPP) reference specimens using surface enhanced laser desorption/ ionization-time of flight (SELDI-TOF) mass spectrometry: Multi-institution correlation of spectra and identification of biomarkers 273

Alex J. Rai, Paul M. Stemmer, Zhen Zhang, Bao-ling Adam, William T. Morgan, Rebecca E. Caffrey, Vladimir N. Podust, Manisha Patel, Lih-yin Lim, Natalia V. Shipulina, Daniel W. Chan, O. John Semmes and Hon-chiu Eastwood Leung

12.1	Introduction	273
12.2	Materials and methods	275
12.2.1	Sample preparation	275
12.2.2	Sample preprocessing	275
12.2.3	Target (CM10) chip preparation and sample incubation	275
12.2.4	Scanning protocol	276
12.2.5	Data processing	276
12.2.6	Bioinformatics analysis of data and correlation coefficient matrix	276
12.2.7	Protein purification, SDS-PAGE analysis, and extraction of proteins	276
12.2.8	Peptide mass fingerprinting (PMF)	277
12.2.9	MS/MS analysis	277
12.2.10	Western blot analysis	277
12.3	Results	278
12.4	Discussion	283
12.5	References	286

13 An evaluation, comparison, and accurate benchmarking of several publicly available MS/MS search algorithms: Sensitivity and specificity analysis 289

Eugene A. Kapp, Frédéric Schütz, Lisa M. Connolly, John A. Chakel, Jose E. Meza, Christine A. Miller, David Fenyö, Jimmy K. Eng, Joshua N. Adkins, Gilbert S. Omenn and Richard J. Simpson

13.1	Introduction	289
13.1.1	Heuristic algorithms	291

13.1.2	Probabilistic algorithms	292
13.2	Materials and methods	292
13.2.1	HUPO-PPP reference specimens	292
13.2.2	Sample preparation and MS analysis	293
13.2.3	Protein sequence databases	293
13.2.4	MS/MS database search strategy	293
13.2.4.1	SEQUEST and MASCOT workflow performed by the JPSL research group	294
13.2.4.2	SEQUEST and PeptideProphet workflow performed by the ISB research group	294
13.2.4.3	Spectrum Mill workflow performed by the Agilent group	295
13.2.4.4	Sonar and X!Tandem workflow performed by David Fenyö	295
13.2.5	Web interface for data validation, integration, and cross annotation	295
13.2.6	ROC curve generation	297
13.3	Results and discussion	298
13.3.1	Comparison of MS/MS search algorithms	299
13.3.1.1	Sensitivity and concordance between MS/MS search algorithms	299
13.3.1.2	Specificity and discriminatory power of the primary score statistic for the different MS/MS search algorithms: Distribution of scores and ROC plots	301
13.3.1.3	Calculation of score thresholds based on specified FP identification error rates	304
13.3.1.4	Benchmarking of the different MS/MS search algorithms at 1% FP error rate	310
13.3.1.5	Effect of database size and search strategy	311
13.3.1.6	Utility of reversed sequence searches	311
13.3.1.7	Consensus scoring between MS/MS search algorithms	312
13.4	Concluding remarks	313
13.5	References	314
14	**Human Plasma PeptideAtlas**	**317**
	Eric W. Deutsch, Jimmy K. Eng, Hui Zhang, Nichole L. King, Alexey I. Nesvizhskii, Biaoyang Lin, Hookeun Lee, Eugene C. Yi, Reto Ossola and Ruedi Aebersold	
14.1	References	322
15	**Do we want our data raw? Including binary mass spectrometry data in public proteomics data repositories**	**323**
	Lennart Martens, Alexey I. Nesvizhskii, Henning Hermjakob, Marcin Adamski, Gilbert S. Omenn, Joël Vandekerckhove and Kris Gevaert	
15.1	References	328

16	A functional annotation of subproteomes in human plasma *329*
	Peipei Ping, Thomas M. Vondriska, Chad J. Creighton, TKB Gandhi, Ziping Yang, Rajasree Menon, Min-Seok Kwon, Sang Yun Cho, Garry Drwal, Markus Kellmann, Suraj Peri, Shubha Suresh, Mads Gronborg, Henrik Molina, Raghothama Chaerkady, B. Rekha, Arun S. Shet, Robert E. Gerszten, Haifeng Wu,, Mark Raftery, Valerie Wasinger, Peter Schulz-Knappe, Samir M. Hanash, Young-ki Paik, William S. Hancock, David J. States, Gilbert S. Omenn and Akhilesh Pandey
16.1	Introduction *330*
16.2	Materials and methods *330*
16.2.1	Coagulation pathway and protein interaction network analysis *331*
16.2.2	Gene ontology annotations *331*
16.2.3	Analysis of MS-derived data for identification of proteolytic events and post-translational modifications *331*
16.3	Results and discussion *331*
16.3.1	Bioinformatic analyses of the functional subproteomes *332*
16.3.1.1	An interaction map of human plasma proteins *332*
16.3.1.2	Gene Ontology annotation of protein function *334*
16.3.2	Proteins involved in the blood coagulation pathway *335*
16.3.3	Proteins potentially derived from mononuclear phagocytes *337*
16.3.4	Proteins involved in inflammation *338*
16.3.5	Analyzing the peptide subproteome of human plasma *339*
16.3.6	Liver related plasma proteins *339*
16.3.7	Cardiovascular system related plasma proteins *341*
16.3.8	Glycoproteins *342*
16.3.9	DNA-binding proteins *342*
16.3.9.1	Histones *343*
16.3.9.2	Helicases *344*
16.3.9.3	Zinc finger proteins *345*
16.3.10	Annotation through reanalysis of mass spectrometry data *345*
16.3.10.1	Cleavage of signal peptides and transmembrane domains *346*
16.3.10.2	Identification of PTMs *347*
16.4	Concluding remarks *348*
16.5	References *349*

17	Cardiovascular-related proteins identified in human plasma by the HUPO Plasma Proteome Project Pilot Phase *353*
	Beniam T. Berhane, Chenggong Zong, David A. Liem, Aaron Huang, Steven Le, Ricky D. Edmondson, Richard C. Jones, Xin Qiao, Julian P. Whitelegge, Peipei Ping and Thomas M. Vondriska
17.1	Introduction *353*
17.1.1	HUPO Plasma Proteome Project pilot phase *354*

17.1.2	Need for novel insights into cardiovascular disease	*354*
17.2	Materials and methods	*355*
17.3	Groups of cardiovascular-related proteins	*356*
17.3.1	Markers of inflammation and CVD	*356*
17.3.2	Vascular and coagulation proteins	*357*
17.3.3	Signaling proteins	*359*
17.3.4	Growth- and differentiation-associated proteins	*360*
17.3.5	Cytoskeletal proteins	*360*
17.3.6	Transcription factors	*361*
17.3.7	Channel and receptor proteins	*363*
17.3.8	Heart failure- and remodeling-related proteins	*364*
17.4	Functional analyses and implications	*365*
17.4.1	Organ specific cardiovascular-related proteins in plasma	*365*
17.4.2	Novel cardiovascular-related proteins identified in plasma	*366*
17.5	Methodology considerations	*368*
17.6	Conclusions and future directions	*368*
17.7	References	*370*

Preface

Plasma and serum are the preferred specimens for non-invasive sampling of normal individuals, at-risk groups, and patients for protein biomarkers discovered and validated to reflect physiological, pathological, and pharmacological phenotypes. These specimens present enormous challenges due to extreme complexity, representing potentially all proteins in the body and their isoforms; at least ten orders of magnitude range in protein concentrations; intra-individual and inter-individual variation from genetics, diet, smoking, hormones, and many other sources; and especially non-standardized methods of sample processing. Furthermore, the inherent limitations of incomplete sampling of peptides by mass spectrometry and high error rates of peptide identifications and protein assignments with various search algorithms and databases lead to low concordance of protein identifications even with repeat analyses of the same sample. These features complicate diagnostic comparisons of specimens.

The Human Proteome Organization (HUPO) has launched several major initiatives to explore the proteomes of liver, brain, and plasma and to generate informatics standards and large-scale antibody production. This book presents the major findings from the pilot phase of the Plasma Proteome Project (PPP). The 17 chapters embrace a combination of collaborative analyses of HUPO PPP reference specimens and several lab-specific projects, both experimental and analytical. The investigators compared PPP reference specimens of human serum and EDTA, heparin, and citrate-anti-coagulated plasma; EDTA-plasma was determined to be the preferred specimen. Together these chapters examine many features of specimen handling, depletion of abundant proteins, fractionation of intact proteins, fractionation of tryptic digest peptides, and analysis of those peptides with various MS/MS instruments. Combinations of technologies gave the most resolution. The subsequent step of matching spectra to peptide sequences with a variety algorithms has numerous, often unspecified parameters. The alignment of peptide sequences with proteins via protein or gene databases likewise is laden with uncertainties and redundancies. Especially for longitudinal and collaborative studies, the periodic issuance of modified versions of the databases creates a moving target for protein identification and annotation, let alone comparison of results from different studies. These challenges are explored in depth. As in the special issue of Proteomics (August 2005) with a total of 28 papers, the authors here provide a revealing snapshot of the output from a variety of proteomics technology platforms across laboratories.

Exploring the Human Plasma Proteome. Edited by Gilbert S. Omenn
Copyright © 2006 WILEY-VCH Verlag GmbH & Co. KGaA, Weinheim
ISBN: 3-527-31757-0

The extensive annotations show that present methods already are capable of detecting in plasma large numbers of low-abundance proteins of great biological interest from essentially all cellular compartments. Studies focusing on sub-proteomes based on glycoprotein enrichment or molecular weight yielded additional findings. As more powerful technologies are applied, we can expect ever more extensive identification, as well as quantitation, of proteins and their isoforms. The high proportion of genes which generate detectable splice isoforms further complicates protein identifications, yet helps to clarify the basis on which humans can have such complex phenotypes with a surprisingly small complement of genes (latest Human Genome Project estimate is about 22,000 protein-encoding genes).

The PPP Core Dataset has 5102 proteins identified with 2 or more peptides, of which 3020 remain after application of our integration algorithm for protein matches which cannot be distinguished with the available peptides. A special feature of the PPP is the set of independent analyses from the raw spectra or peaklists across the multiple laboratories. These independent analyses eliminate the high variability from lab-specific search algorithms, different databases, and investigators' judgments, though each independent analysis has its own peculiar attributes. We also provide comparisons with several published datasets. Meta-analysis of separate studies has similar challenges to those experienced in the integration of datasets from the collaborating PPP laboratories.

Numerous other "cuts" of the data can be made. The primary data are available for such additional analyses at the European Bioinformatics Institute (www.ebi.,a-c.uk/pride); the University of Michigan (www.bioinformatics.med.umich.edu/hupo/ppp); and the Institute for Systems Biology (www.peptideatlas.org). We are keen to encourage such further analyses. Two examples have already appeared, introducing adjustments for protein length and multiple comparisons testing [1] and enhancing the characterization of the human genome from these proteomics data and gene mapping [2]. This publication presents the foundation for planning the next phases of the Plasma Proteome Project, with Young-Ki Paik, Matthias Mann, and myself as co-chairs. We will:

1. develop standardized operating procedures for specimens, protein and peptide fractionation, and analyses, with attention to replicability of results, to make proteomics practicable for clinical chemistry;
2. select priority PPP proteins for the HUPO Antibody Production Initiative, to generate reagents for biomarker and pathways studies and plasma/organ proteome comparisons;
3. collaborate on informatics, databases, annotations, and error estimation for plasma and serum studies, both HUPO-initiated and published by others;
4. stimulate proteomics technology advances, with special attention to high-resolution/higher-throughput methods and to quantitation of proteins and characterization of modified proteins (primarily glycoproteins and phosphoproteins); and
5. assure paired analyses of plasma and tissue specimens in organ-based and disease-focused proteomics initiatives.

The spirit of collaboration in the Plasma Proteome Project has been splendid. The substantial commitment of so many investigators and sponsors to this pilot phase has been admirable. As a work-in-progress the PPP has generated productive discussions at many scientific meetings. On behalf of the Executive Committee and Technical Committees, I thank everyone involved.

Gil Omenn
Gilbert S. Omenn
University of Michigan, Ann Arbor
August 2006

1. States, D. J., Omenn, G. S., Blackwell, T. W., Fermin, D., Eng, J., Speicher, D. W., Hanash, S. M. *Challenges in deriving high-confidence protein identifications from data gathered by HUPO plasma proteome collaborative study. Nature Biotech* 2006, *24*, 333–338.
2. Fermin, D., Allen, B. B., Blackwell, T. W., Menon, R., Adamski, M., Xy, Y., Ulintz, P., Omenn, G. S., States, D. J. *Novel gene and gene model detection using a whole genome open reading frame analysis in proteomics. Genome Biology* 2006, *7:R35*, Published online: http://genomebiology.com/2006/7/4/R35.

List of Contributors

Dr. Gilbert S. Omenn
Internal Medicine,
University of Michigan, MSRB 1,
1150 W. Medical Center Dr.
Ann Arbor,
MI 48109-0656, USA

Dr. David J. States
University of Michigan,
2017 Palmer Commons,
100 Washtenaw Avenue,
Ann Arbor,
MI 48109-2218, USA

Dr. Alex J. Rai
Assistant Professor and Director
of General Chemistry,
The Johns Hopkins University
School of Medicine,
Department of Pathology,
600 N. Wolfe St.,
Meyer B-121, Baltimore,
MD 21287-7065, USA

Brian B. Haab
Ph.D.,
The Van Andel Research Institute,
333 Bostwick NE,
Grand Rapids,
MI 49503, USA

Dr. David W. Speicher
The Wistar Institute,
3601 Spruce St. Rm. 151,
Philadelphia,
PA 19104, USA

Dr. David W. Speicher
The Wistar Institute,
3601 Spruce Street,
Philadelphia,
PA 19104, USA

Professor William S. Hancock
Barnett Institute and Department
of Chemistry and Chemical Biology,
Northeastern University,
Boston,
MA 02115, USA

Professor Arie Admon
Department of Biology,
Technion,
Haifa 3200, Israel

Professor Young-Ki Paik
Yonsei Proteome Research Center
and Biomedical Proteome
Research Center,
Yonsei University,
134 Shinchon-dong,
Sudaemoon-ku,
Seoul 120-749, Korea

Exploring the Human Plasma Proteome. Edited by Gilbert S. Omenn
Copyright © 2006 WILEY-VCH Verlag GmbH & Co. KGaA, Weinheim
ISBN: 3-527-31757-0

Dr. Xiaohong Qian
Ph.D.,
Beijing Institute of Radiation
Medicine,
27 Taiping Road,
Beijing 100850,
China

Dr. Joel G. Pounds
Biological Sciences Division,
Pacific Northwest National
Laboratory,
Box 999 MSIN: P7-58,
Richland,
WA 99352, USA

Dr. Alex J. Rai
Assistant Professor and Director
of General Chemistry,
Department of Pathology,
Division of Clinical Chemistry,
Johns Hopkins University
School of Medicine,
600 N. Wolfe St., Meyer B-121,
Baltimore,
MD 21287-7065, USA

Professor Richard J. Simpson
Joint ProteomicS Laboratory,
Ludwig Institute for Cancer
Research,
P.O. Box 2008,
Royal Melbourne Hospital,
Parkville, Victoria 3050,
Australia

Dr. Eric W. Deutsch
Institute for Systems Biology,
1441 N 34th Street,
Seattle,
WA 98103, USA

Dr. Lennart Martens
Department of Biochemistry,
Faculty of Medicine and Health
Sciences,
Ghent University,
A. Baertsoenkaai 3,
B-9000 Ghent, Belgium

Dr. Akhilesh Pandey
McKusick-Nathans Institute of
Genetic Medicine,
733 N. Broadway, BRB 569,
Johns Hopkins University,
Baltimore,
MD 21205, USA

Thomas M. Vondriska
Cardiovascular Research
Laboratories,
Departments of Physiology
and Medicine,
Division of Cardiology,
David Geffen School of Medicine
at UCLA,
Room 1619, MRL Building,
Los Angeles,
CA 90095, USA

1
Overview of the HUPO Plasma Proteome Project: Results from the pilot phase with 35 collaborating laboratories and multiple analytical groups, generating a core dataset of 3020 proteins and a publicly-available database*

Gilbert S. Omenn, David J. States, Marcin Adamski, Thomas W. Blackwell, Rajasree Menon, Henning Hermjakob, Rolf Apweiler, Brian B. Haab, Richard J. Simpson, James S. Eddes, Eugene A. Kapp, Robert L. Moritz, Daniel W. Chan, Alex J. Rai, Arie Admon, Ruedi Aebersold, Jimmy Eng, William S. Hancock, Stanley A. Hefta, Helmut Meyer, Young-Ki Paik, Jong-Shin Yoo, Peipei Ping, Joel Pounds, Joshua Adkins, Xiaohong Qian, Rong Wang, Valerie Wasinger, Chi Yue Wu, Xiaohang Zhao, Rong Zeng, Alexander Archakov, Akira Tsugita, Ilan Beer, Akhilesh Pandey, Michael Pisano, Philip Andrews, Harald Tammen, David W. Speicher and Samir M. Hanash

HUPO initiated the Plasma Proteome Project (PPP) in 2002. Its pilot phase has (1) evaluated advantages and limitations of many depletion, fractionation, and MS technology platforms; (2) compared PPP reference specimens of human serum and EDTA, heparin, and citrate-anti-coagulated plasma; and (3) created a publicly-available knowledge base (www.bioinformatics.med.umich.edu/hupo/ppp; www.ebi.ac.uk/pride). Thirty-five participating laboratories in 13 countries submitted datasets. Working groups addressed (a) specimen stability and protein concentrations; (b) protein identifications from 18 MS/MS datasets; (c) independent analyses from raw MS-MS spectra; (d) search engine performance, subproteome analyses, and biological insights; (e) antibody arrays; and (f) direct MS/SELDI analyses. MS-MS datasets had 15 710 different International Protein Index (IPI) protein IDs; our integration algorithm applied to multiple matches of peptide sequences yielded 9504 IPI proteins identified with one or more peptides and 3020 proteins identified with two or more peptides (the Core Dataset). These proteins have been characterized with Gene Ontology, InterPro, Novartis Atlas, OMIM, and immunoassay-based concentration determinations. The database permits examination of many other subsets, such as 1274 proteins identified with three or more peptides. Reverse protein to DNA matching identified proteins for 118 previously unidentified ORFs.

---

* Originally published in Proteomics 2005, 13, 3226–3245

Exploring the Human Plasma Proteome. Edited by Gilbert S. Omenn
Copyright © 2006 WILEY-VCH Verlag GmbH & Co. KGaA, Weinheim
ISBN: 3-527-31757-0

We recommend use of plasma instead of serum, with EDTA (or citrate) for anticoagulation. To improve resolution, sensitivity and reproducibility of peptide identifications and protein matches, we recommend combinations of depletion, fractionation, and MS/MS technologies, with explicit criteria for evaluation of spectra, use of search algorithms, and integration of homologous protein matches.

This Special Issue of PROTEOMICS presents papers integral to the collaborative analysis plus many reports of supplementary work on various aspects of the PPP workplan. These PPP results on complexity, dynamic range, incomplete sampling, false-positive matches, and integration of diverse datasets for plasma and serum proteins lay a foundation for development and validation of circulating protein biomarkers in health and disease.

1.1
Introduction

A comprehensive, systematic characterization of circulating proteins in health and disease will greatly facilitate development of biomarkers for prevention, diagnosis, and therapy of cancers and other diseases [1]. Proteomics technologies now permit extensive fractionation of proteins in complex specimens, analysis of peptides by MS, and matching of peptide sequences to protein "hits" through gene and protein databases generated directly and indirectly from the sequencing of the human genome [2, 3], as well as other methods for identifying proteins.

The HUPO, formed in 2001, aims to accelerate the development of the field of proteomics and to stimulate and organize international collaborations in research and education [4]. HUPO has launched major initiatives focused on the plasma, liver, and brain proteomes, proteomics standards and databases, and large-scale antibody production. The plasma proteome is linked with these other initiatives (see Fig. 1).

The long-term scientific goals of the HUPO Plasma Proteome Project (PPP) are (1) comprehensive analysis of the protein constituents of human plasma and serum; (2) identification of biological sources of variation within individuals over time due to physiology (age, sex, menstrual cycle, exercise, stress), pathology (various diseases, special cohorts), and treatments (common medications); and (3) determination of the extent of variation across individuals within populations and across populations due to genetic, nutritional and other factors. The pilot phase aims to (1) compare advantages and limitations of many technology platforms; (2) contrast reference specimens of human plasma (EDTA, heparin, or citrate-anticoagulated) and serum in terms of numbers of proteins identified and any interferences with various technology platforms; and (3) create a global, open-source knowledge base/data repository.

The collaborative nature of this Project permitted exploration of many variables and adoption during the study phase of emerging technologies. Planning proceeded expeditiously from the organizing meeting of HUPO in Bethesda in

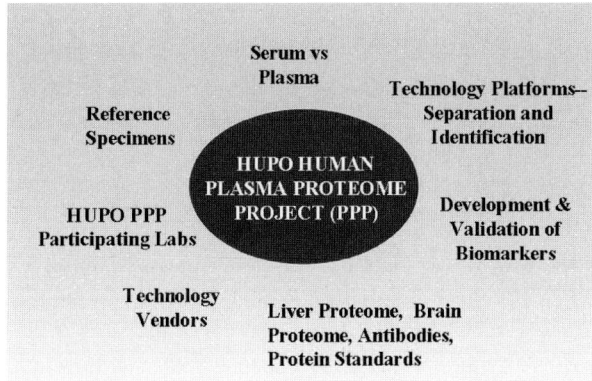

Fig. 1 Schema showing relationship of HUPO Plasma Proteome Project (PPP) to other HUPO initiatives and components of the PPP.

April 2002, to the first PPP meeting in Ann Arbor in September 2002, the expression of interest by numerous investigators at the 1st HUPO World Congress on Proteomics in Versailles in November 2002, and then the PPP Workshop for Technical Committees and participating laboratories in Bethesda in July 2003 to launch the pilot phase. PPP reference specimens were prepared and distributed, beginning in September 2003, and first data were submitted, analyzed, and presented at a workshop at the 2nd HUPO World Congress in Montreal in November 2003. An intensive 4 day Jamboree Workshop was organized for Ann Arbor in June 2004, at which numerous work groups pursued cross-laboratory analyses and proposed further work. Investigators were advised to adopt more stringent criteria for high confidence peptide and protein identifications, and a commitment was made to collect raw spectra from the 18 laboratories that had submitted MS/MS or FT-ICR/MS datasets for independent analyses by three different groups. The datasets were moving targets, as some, but not all, labs submitted expanded or updated analyses, and about 15 laboratories completed "special projects" stimulated by HUPO PPP with a competition for small grants following the Montreal workshop.

The PPP provided participating laboratories with 1.0 mL of reference specimens of serum and plasma by three different methods of anticoagulation for plasma (EDTA, citrate, heparin) from specific donor pools. Investigators utilized their established and emerging technologies for fractionation and analysis of proteins. Investigators were encouraged to "push the limits" of their methods to detect and identify low abundance proteins. Comparisons of findings across laboratories provide a special opportunity for confirmation of protein identifications. Results were submitted to centralized bioinformatics functions at the University of Michigan and the European Bioinformatics Institute to create an integrated data repository from which PPP and other investigators could initiate further analyses and annotations. The approaches and core results have been presented at the US HUPO inaugural meeting in March 2005, the HUPO World Congress in Munich in August 2005, and at other meetings.

Here we present a comprehensive account of the major findings from the pilot phase of the Human Plasma Proteome Project, including the many associated special projects.

1.2
PPP reference specimens

The primary specimens were sets of four reference specimens prepared under the direction of the HUPO PPP Specimens Committee by BD Diagnostics for each of three ethnic groups: Caucasian-American (B1), African-American (B2), and Asian-American (B3). Each pool consisted of 400 mL of blood each from one male and one post-menopausal female healthy, fasting donor, collected into 10 mL tubes in a prescribed sequence (see Supplementary Protocol) after informed consent. Very large pools were rejected as requiring too prolonged specimen handling and processing unlike the collection of individual specimens; even a protocol for two males and two females proved to require more than the 2 h limit we set. Equal numbers of tubes and aliquots were generated with appropriate concentrations of K_2-EDTA, lithium heparin, or sodium citrate for plasma or permitted to clot at room temperature for 30 min to yield serum (with micronized silica as clot activator). The additives were dry-sprayed on the inner walls of the tubes, except for 1.0 mL of 0.105 M buffered sodium citrate, which gave a final ratio of 9:1 for blood to citrate in a 10 mL final volume, causing an 11% dilution of the blood. No protease inhibitor cocktails were used. This procedure required 2 h, mostly at 2 to 6°C. After centrifugation, volumes from the male and female donors in each donor pair for each specimen type were pooled and then aliquoted into numerous 250 µL portions in vials which were frozen and stored at $-70°C$. The centrifugation conditions with citrate consistently produced platelet-poor plasma (platelet count $<10^3/\mu L$). Aliquots tested negative for HIV, HBV, HCV, HTLV-1, and syphilis. We supplied four × 250 µL aliquots for each of the four plasma/serum specimens in each set. These vials were shipped on dry ice *via* courier in early May 2003 (and later to additional laboratories which petitioned to join the project, some of which could no longer be supplied the B1 set). No reshipping was permitted.

The Chinese Academy of Medical Sciences (CAMS) used a variant of the BD protocol to generate similar reference serum and plasma specimens, as described by Li *et al.* [5] and He *et al.* [6]. Pools were prepared after review by the CAMS Ethics Committee and informed consent by ten male and ten female donors in Beijing. Donors were fasting and avoided taking medicines or drinking alcohol for the 12 h before sampling. A subsequent pooling of 20 mL from each of the male and female serum or plasma specimens created the C1-CAMS PPP reference specimens which were sent to the 15 laboratories requesting these specimens after storage at $-80°C$. They were shipped on dry ice using the same courier in September 2003. C1-CAMS specimens were centrifuged originally, and then again upon thawing, at 4°C [6].

Finally, the UK National Institute of Biological Standards and Control (NIBSC) made available to the PPP their lyophilized citrated plasma standard prepared for hemostasis and thrombosis studies from a pool of 25 donors [1].

A standard questionnaire was sent to all laboratories expressing interest. Of 55 laboratories that originally committed to participate, 41 received the BD B1 specimens, 27 the B2 and B3 specimens, 15 the CAMS specimens, and 45 the NIBSC specimens. Laboratories varied on how many of the specimens they actually analyzed.

1.3 Bioinformatics and technology platforms

As intended, laboratories used a wide variety of methods, including multiple LC-MS/MS instruments, MALDI-MS, and FT-ICR-MS; depletion of abundant proteins; fractionation of intact proteins on 2-D gels or with LC or IEF methods; protein enrichment or labeling methods; immunoassays or antibody arrays; and direct (SELDI) MS. They also varied on choice of search algorithm and database, and criteria for declaring high or lower confidence identification of peptide sequences and matching proteins (Tab. 1). In general, the numbers of proteins reported individually by the labs do not have the integration feature which was applied to the whole PPP dataset. In several cases, much more extensive analyses were reported. Thus, many of the individual papers in this special issue have additional protein identifications not included in the project-wide dataset(s).

1.3.1 Constructing a PPP database for human plasma and serum proteins

Data management for this project included guidance and protocols for data collection, then centralized integration, analysis, and dissemination of findings worldwide *via* a communications infrastructure. As described in great detail by Adamski *et al.* [7, 8], key challenges were integration of heterogeneous datasets, reduction of redundant information to minimal identification sets, and data annotation. Multiple factors had to be balanced, including when to "freeze" on a particular release of the ever-changing database selected for the PPP and how to deal with "lower confidence" peptide identifications. Freezing of the database was essential to conduct extensive comparisons of complex datasets and annotations of the dataset as a whole. However, it complicates the work of linking findings of the current study to evolving knowledge of the human genome and its annotation. Many of the entries in the protein sequence database(s) available at the initiation of the project or even the analytical phase were revised, replaced, or withdrawn over the course of the project, and continue to be revised. Our policies and practices anticipated the guidelines issued recently by Carr *et al.* [9], as documented by Adamski *et al.* [7].

The 18 participating laboratories using MS/MS or FT-ICR-MS submitted a total of 42 306 protein identifications using various search engines and databases to handle spectra and generate peptide sequence lists from the specimens analyzed.

Tab. 1 Protein identifications by lab, by specimen, and by methods

Lab ID	Specimen	Deple-tion	Protein separation	Reduction/ alkylation	Peptide separation	Mass spectrum	Search software	3020 High confidence	3020 Lower confidence	Single peptide
1	b1-cit	aig	none	iam	rp/scx/rp	esi-ms/ms_decaxp	PepMiner	61	39	12
1	b1-edta	aig	none	iam	rp/scx/rp	esi-ms/ms_decaxp	PepMiner	35	30	14
1	b1-hep	aig	none	iam	rp/scx/rp	esi-ms/ms_decaxp	PepMiner	50	38	13
1	b1-serum	aig	none	iam	rp/scx/rp	esi-ms/ms_decaxp	PepMiner	21	6	5
1	b2-cit	aig	none	iam	rp/scx/rp	esi-ms/ms_decaxp	PepMiner	57	37	12
1	b2-hep	aig	none	iam	rp/scx/rp	esi-ms/ms_decaxp	PepMiner	58	30	12
1	b2-serum	aig	none	iam	rp/scx/rp	esi-ms/ms_decaxp	PepMiner	59	31	12
1	b3-serum	aig	none	iam	rp/scx/rp	esi-ms/ms_decaxp	PepMiner	17	6	7
2	b1-cit	none	cho affinity	iam	scx/rp	esi-ms/ms_qtof	SEQUEST	165	79	94
2	b1-serum	none	cho affinity	iam	scx/rp	esi-ms/ms_qtof	SEQUEST	136	48	38
2	nibsc	none	cho affinity	iam	scx/rp	esi-ms/ms_qtof	SEQUEST	171	121	85
11	b1-cit	none	cho affinity	iam	rp	esi-ms/ms_decaxp	SEQUEST	59	4	9
11	b1-edta	none	cho affinity	iam	rp	esi-ms/ms_decaxp	SEQUEST	64	6	4
11	b1-hep	none	cho affinity	iam	rp	esi-ms/ms_decaxp	SEQUEST	62	9	15
11	b1-serum	none	cho affinity	iam	rp	esi-ms/ms_decaxp	SEQUEST	64	3	16
12	b1-cit	aig	none	iam	rp/scx/rp	esi-ms/ms_deca	SEQUEST	111	0	113
12	b1-edta	aig	none	iam	rp/scx/rp	esi-ms/ms_deca	SEQUEST	111	0	101
12	b1-hep	aig	none	iam	rp/scx/rp	esi-ms/ms_deca	SEQUEST	127	0	130
12	b1-serum	aig	none	iam	rp/scx/rp	esi-ms/ms_deca	SEQUEST	123	0	111
17	b1-serum	aig	1s sds	iam	rp	esi-ms/ms_lcq	SEQUEST	50	19	7
21	b1-cit	top6	rotofor-ief/rp/1d-sds	iam	rp	esi-ms/ms_qtof	MASCOT	40	0	1
21	b1-cit	top6	rotofor-ief/rp/1d-sds	none	none	maldi-ms/ms[a]bi4700	MASCOT	51	0	3
21	b1-cit	top6	rotofor-ief/rp/1d-sds	none	rp	esi-ms/ms_qtof	MASCOT	39	0	1
21	b1-edta	top6	rotofor-ief/rp/1d-sds	iam	rp	esi-ms/ms_qtof	MASCOT	40	0	1

1.3 Bioinformatics and technology platforms

Tab. 1 Continued

Lab ID	Specimen	Deple-tion	Protein separation	Reduction/ alkylation	Peptide separation	Mass spectrum	Search software	3020 High confidence	3020 Lower confidence	Single peptide
21	b1-edta	top6	rotofor-ief/rp/1d-sds	none	none	maldi-ms/ms^{a}bi4700	MASCOT	51	0	3
21	b1-edta	top6	rotofor-ief/rp/1d-sds	none	rp	esi-ms/ms_qtof	MASCOT	39	0	1
21	b1-serum	top6	rotofor-ief/rp/1d-sds	iam	rp	esi-ms/ms_qtof	MASCOT	40	0	1
21	b1-serum	top6	rotofor-ief/rp/1d-sds	none	none	maldi-ms/ms^{a}bi4700	MASCOT	51	0	3
21	b1-serum	top6	rotofor-ief/rp/1d-sds	none	rp	esi-ms/ms_qtof	MASCOT	39	0	1
21	b1-serum	top6	1s sds	iam	rp/scx/rp	esi-ms/ms_decaxp	SEQUEST	277	0	161
24	b1-serum	a	rp	iam	rp	esi-ms/ms_qtrap	MASCOT	7	12	1
24	b1-serum	none	rp	iam	rp	esi-ms/ms_qtrap	MASCOT	17	21	3
26	b2-cit	none	rotofor-ief/1d-sds	iam	rp	esi-ms/ms_qtof	MASCOT	160	44	12
28	b1-cit	ig	none	none	rp	esi-fticr	VIPER	218	45	208
28	b1-serum	ig	none	none	rp	esi-fticr	VIPER	223	50	239
28	b2-cit	ig	none	none	rp	esi-fticr	VIPER	255	140	346
28	b2-serum	ig	none	none	rp	esi-fticr	VIPER	244	181	405
28	b3-cit	ig	none	none	rp	esi-fticr	VIPER	214	188	359
28	b3-serum	ig	none	none	rp	esi-fticr	VIPER	218	193	384
29	b1-cit	top6	none	iam	scx/rp	esi-ms/ms_decaxp	SEQUEST	19	129	136
29	b1-cit	top6	none	iam	scx/rp/2mz	esi-ms/ms_decaxp	SEQUEST	51	160	181
29	b1-edta	top6	none	iam	scx/rp	esi-ms/ms_decaxp	SEQUEST	50	199	264
29	b1-edta	top6	none	iam	scx/rp/2mz	esi-ms/ms_decaxp	SEQUEST	82	491	557
29	b1-hep	top6	none	iam	scx/rp	esi-ms/ms_decaxp	SEQUEST	26	97	122
29	b1-serum	top6	none	iam	scx/rp	esi-ms/ms_decaxp	SEQUEST	90	338	432
29	c1-cit	top6	none	iam	scx/rp/2mz	esi-ms/ms_decaxp	SEQUEST	82	449	517
29	c1-edta	top6	none	iam	scx/rp/2mz	esi-ms/ms_decaxp	SEQUEST	72	555	610
29	c1-hep	top6	none	iam	scx/rp/2mz	esi-ms/ms_decaxp	SEQUEST	82	227	283

Tab. 1 Continued

Lab ID	Specimen	Deple-tion	Protein separation	Reduction/alkylation	Peptide separation	Mass spectrum	Search software	3020 High confidence	3020 Lower confidence	Single peptide
29	c1-serum	top6	none	iam	scx/rp/2mz	esi-ms/ms_decaxp	SEQUEST	97	519	570
29	nibsc	top6	none	iam	scx/rp/2mz	esi-ms/ms_decaxp	SEQUEST	82	371	432
33	nibsc	top6	ffe/rp	none	rp/ziptip	maldi-ms/ms_qstar	Digger	54	0	0
33	nibsc	top6	ffe/rp	none	rp/ziptip	maldi-ms/ms_qstar	MASCOT	58	0	3
34	b1-hep	top6	zoom-ief/1d-sds	iam	rp	esi-ms/ms_decaxp	SEQUEST	123	148	146
34	b1-serum	top6	zoom-ief/1d-sds	iam	rp	esi-ms/ms_ltq	SEQUEST	427	741	1172
40	b1-hep	none	aig affinity/rp	iam	scx/rp	esi-ms/ms_lcq	Sonar	160	253	185
41	b1-cit	none	gradiflow/tca	none	scx/rp	esi-ms/ms_qstar	SEQUEST	72	0	34
41	b1-edta	none	gradiflow/tca	none	scx/rp	esi-ms/ms_qstar	SEQUEST	62	0	16
41	b1-hep	none	gradiflow/tca	none	scx/rp	esi-ms/ms_qstar	SEQUEST	51	0	7
41	b1-serum	none	gradiflow/tca	none	scx/rp	esi-ms/ms_qstar	SEQUEST	76	0	27
41	nibsc	none	gradiflow/tca	none	scx/rp	esi-ms/ms_qstar	SEQUEST	53	0	1
43	b1-cit	aig	none	iam	rp	esi-ms/ms_qtof	MASCOT	26	0	0
43	b1-edta	aig	none	iam	rp	esi-ms/ms_qtof	MASCOT	31	0	0
43	b1-hep	aig	none	iam	rp	esi-ms/ms_qtof	MASCOT	37	0	0
43	b1-hep	aig	none	iam	rp	maldi-ms/ms[a]bi4700	MASCOT	26	0	0
43	b1-serum	aig	none	iam	rp	esi-ms/ms_qtof	MASCOT	24	0	0
43	nibsc	aig	none	iam	rp	esi-ms/ms_qtof	MASCOT	21	0	0
46	c1-serum	top6	none	iam	rp	esi-ms/ms_ltq	SEQUEST	185	522	571
55	b1-cit	none	sax	iam	rp	esi-ms/ms_ltq	SEQUEST	216	48	73

High and lower confidence
1. PepMiner results: score $>80/100$
2. ProteinProphet: high $p \geq 0.95$; lower $0.95 > p \geq 0.2$
11. $X_{corr} \geq 1.5/2.0/2.5$ for charge states $+1/+2/+3$. Tryptic cleavage rules. High confidence: two or more peptide ids or single peptide ID manually inspected; spectrum must show high signal and top 3 ions must be assigned either b or y. Otherwise, lower confidence
12. PeptideProphet high confidence $p \geq 0.35$. All IDs reported as high confidence.
17. SEQUEST results: no-enzyme searches, acceptance criteria not stated. (For the automatic interpretation of fragment ion spectra the SEQUEST algorithm is used screening the NCBI protein database (weekly updated version)). The chosen parameters are: aver
21. MASCOT result; high confidence only: probability $\geq 98\%$, numerous isoforms identified
22. SEQUEST result: $X_{corr} \geq 1.9/2.5/3.75$ for charge states $+1/+2/+3$, no manual inspection, no other criteria used
24. MASCOT result. High confidence: if two or more peptides, each of them has to have MASCOT score ≥ 20; if single peptide, it has to have MASCOT score ≥ 30.
26. High confidence fully bryptic peptides: MASCOT individual peptides score ≥ 21 or total score ≥ 80; if single peptide hit, score ≥ 60; if lower scores, manually inspected to check fragment ions and mass error.
28. Confidence is based on reproducibility of identification in triplicate analyses of a sample. High confidence = identification of AMT peptides for a given ORF in two or three of triplicate FT-ICR analyses. Lower confidence = identification of AMT peptides in only one of three FT-ICR analyses. VIPER and Q-Rollup software were used to match FT-ICR accurate masses to the AMT database
29. High confidence: $X_{corr} \geq 1.9/2.2/3.75$ (for charges $+1/+2/+3$), deltaCn ≥ 0.1, and $R_{sp} \leq 4$. Lower confidence: $X_{corr} \geq 1.5/2.0/2.5$ (for charges $+1/+2/+3$), deltaCn ≥ 0.1
33. High confidence: Digger nxc ≥ 0.3; MASCOT score ≥ 15
34. High and lower confidence both used PPP stringent segment parameters of $X_{corr} \geq 1.9$, 2.2 and 3.15; deltaCN ≥ 0.1; Rsp ≤ 4; high-two or more peptides; lower-one peptide.
40. Sonar results. High confidence: protein expect value < 1; lower confidence: protein expect value ≥ 1
41. DTA Select results, criteria not stated, manually inspected
43. MASCOT results: protein p-value ≤ 0.05 and at least one peptide with MASCOT score ≥ 20.
46. High confidence: $X_{corr} \geq 1.9/2.2/3.75$ (for charges $+1/+2/+3$), deltaCn ≥ 0.1, and $R_{sp} \leq 4$; lower confidence: $X_{corr} > 1.5/2.0/2.5$.
55. Identical sets of .dta files were searched using SEQUEST, Sonar and X!Tandem. SEQUEST criteria: $X_{corr} > 1.8/2.0/2.5$ for charge states $+1/+2/+3$, deltaCn ≥ 0.1, Sp ≤ 200. X!Tandem criteria: expectation value ≤ 0

These reports matched to 15 710 non-redundant entries (of which 15 519 were based on peptides with six or more amino acids) in the International Protein Index, which had been chosen as the standard reference database for this Project (IPI version 2.21, July 2003) [9]. We designed an integration algorithm which selected one representative protein among multiple proteins (homologs and isoforms) to which identified peptides gave 100% sequence matches. This integration process resulted in 9504 proteins in the IPI v2.21 database identified with one or more peptides. From this point of view, the PPP database is conservative, counting homologous proteins and all isoforms of particular proteins (and their corresponding genes) just once, unless the sequences actually differentiated any additional matches. We included at this stage proteins identified by matches to one or more peptide sequences of "high" or "lower" confidence according to cutpoints utilized with the various search engines used by different MS/MS instruments. Tab. 1 shows the details of the cutpoints or filters used by each investigator and the numbers of "high" and "lower" confidence protein IDs. All laboratories utilizing SEQUEST were asked to reanalyze their results using the PPP specified filters of X_{corr} values $\geq$ 1.9, 2.2, and 3.75 for singly, doubly, and triply charged ions, with deltaCN value $\geq$ 0.1 and $R_{sp} \geq 4$ for fully tryptic peptides for "high confidence" identifications; most did so. No equivalency rules were applied across all the search algorithms for all the cutpoints.

However, Kapp et al. [11] provide such a cross-algorithm analysis for three specified false-positive rates using one laboratory dataset. Since the approaches and analytical instruments used by the various laboratories (Tab. 1) were far too diverse to utilize a standardized set of mass spec/search engine criteria, we created a relatively stringent defined set of protein IDs from the 9504 above by requiring that the same protein be identified with at least a second peptide. In a peptide chromatography run for MS, not all peaks are selected for MS/MS analysis, and the identification of peptide fragment ions is a low-percentage sampling process. Thus, additional analyses in the same lab and in other labs would be expected to enhance the yield of peptide IDs. Consequently, MS data from the individual laboratories were combined to increase the probability of peptide and protein identification. The use of different instrumentation with proprietary software and different search engines for identification made it unfeasible to apply a standard set of parameters to peptide sequences. Therefore, we required a minimum of two distinct peptides to be inferred from mass spectra and matched 100% to the database protein sequence, as a uniform criterion for a given protein to be considered identified.

Of this total of 9504 protein IDs, 6484 were based on one peptide, while 3020 were based on two or more peptides (Tab. 2). That process generated the list of 3020 proteins (5102 before integration) which is utilized as our Core Protein Dataset for the HUPO PPP knowledge base. Full details with unique IPI accession numbers for each protein are accessible for examination and re-analysis at http://www.bioinformatics.med.umich.edu/hupo/ppp and www.ebi.ac.uk/pride. Fig. 2 shows the numbers of proteins identified with $\geq$ n peptides with the percentage of those IDs confirmed in a second laboratory. Of these peptides, the vast majority were ten or more amino acids in length, with a median of 12.9 and a minimum of six amino acids in this dataset; the distribution of lengths is shifted to the right compared with the theoretical tryptic peptides from the total IPI database. The 3020 proteins represent a very broad sampling of the IPI proteins in terms of characterization by pI and by molecular weight of the transcription product (often a "precursor" protein).

Tab. 2 Protein identifications by lab and specimen, based on two or more peptides for each protein match, generating the PPP 3020 protein core dataset

Lab Id	nibsc	b1-cit	b1-edta	b1-hep	b1-serum	b2-cit	b2-hep	b2-serum	b3-cit	b3-serum	c1-cit	c1-edta	c1-hep	c1-serum	plasma	serum	both
1	0	100	65	88	27	94	88	90	0	23	0	0	0	0	197	108	220
2	292	244	0	0	184	0	0	0	0	0	0	0	0	0	399	184	469
11	0	63	70	71	67	0	0	0	0	0	0	0	0	0	102	67	120
12	0	111	111	127	123	0	0	0	0	0	0	0	0	0	277	123	348
17	0	0	0	0	69	0	0	0	0	0	0	0	0	0	0	69	69
21	0	78	78	0	78	0	0	0	0	0	0	0	0	0	78	78	78
22	0	0	0	0	277	0	0	0	0	0	0	0	0	0	0	277	277
24	0	0	0	0	51	0	0	0	0	0	0	0	0	0	0	51	51
26	0	0	0	0	0	204	0	0	0	0	0	0	0	0	204	0	204
28	0	263	0	0	273	395	0	425	0	0	0	0	0	0	565	572	693
29	453	323	724	123	428	0	0	0	402	411	531	627	309	616	1576	867	1839
33	60	0	0	0	0	0	0	0	0	0	0	0	0	0	60	0	60
34	0	0	0	271	1168	0	0	0	0	0	0	0	0	0	271	1168	1251
40	0	0	0	413	0	0	0	0	0	0	0	0	0	0	413	0	413
41	53	72	62	51	76	0	0	0	0	0	0	0	0	0	113	76	137
43	21	26	31	43	24	0	0	0	0	0	0	0	0	0	51	24	52
46	0	0	0	0	0	0	0	0	0	0	0	0	0	707	0	707	707
55	0	264	0	0	0	0	0	0	0	0	0	0	0	0	264	0	264
nibsc	b1-cit	b1-edta	b1-hep	b1-serum	b2-cit	b2-hep	b2-serum	b3-cit	b3-serum	c1-cit	c1-edta	c1-hep	c1-serum	plasma	serum	both
679	1016	876	838	1749	568	88	470	402	419	531	627	309	1124	2580	2353	3020

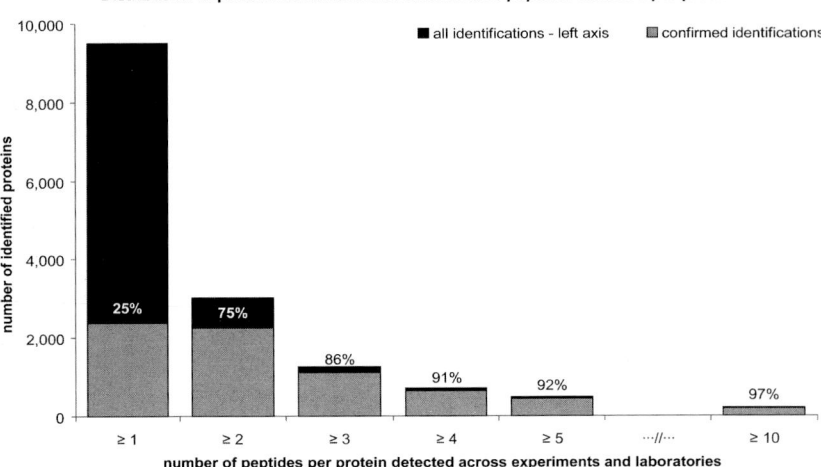

Fig. 2 Number of proteins identified as a function of number of peptides matched.

The PPP database permits future users to choose their own cut-points for sub-analyses, including 2857 proteins identified at least once with "high confidence" criteria; 1555 proteins based on two or more peptides, at least one of which was reported as high confidence (from the intersection of the 3020 and the 2857); and 1274 proteins based on matching to three or more peptides.

Fig. 3 shows the methods used and the log of the number of proteins identified by the various laboratories. At the top of the figure are results with MALDI-MS. Four labs reported MALDI-MS without MS/MS for certain specimens. For example, Lab 22 analyzed all four samples of each of the B1, B2, and B3 specimens by MALDI-MS, and then used in-depth ESI-MS/MS Deca-xp for B1 serum only. Altogether there were 367 distinct protein IDs by MALDI-MS, of which 226 were confirmed by MS/MS or FT-ICR/MS in the core dataset of 3020 IPI proteins, while 141 were not so confirmed. The mean and median numbers of peptides for the confirmed proteins were significantly higher than for those not confirmed. The MALDI-MS data were not used in identifying the 3020 protein dataset or creating Fig. 2.

The capillary LC-FT-ICR-MS results (Lab 28) were included. This method (Adkins et al. [12]) depends upon previous ion-trap MS/MS studies to generate a database of highly accurate mass and normalized elution time parameters for each peptide. Proteins in new specimens cannot be recognized if those proteins were not already detected and characterized in creating (and updating) the AMT database. Only 22% of 722 proteins identified across the six PPP specimens had more than one peptide match; ProteinProphet clustered these 722 into 377 non-redundant proteins. The LC-MS/AMT method has the potential to expedite analysis of large numbers of specimens once the mass tolerance is tightened, the elution times are made highly reproducible, and the AMT parameters are known for a very substantial number of true-positive peptides. Even then, however, samples of dif-

1.3 Bioinformatics and technology platforms

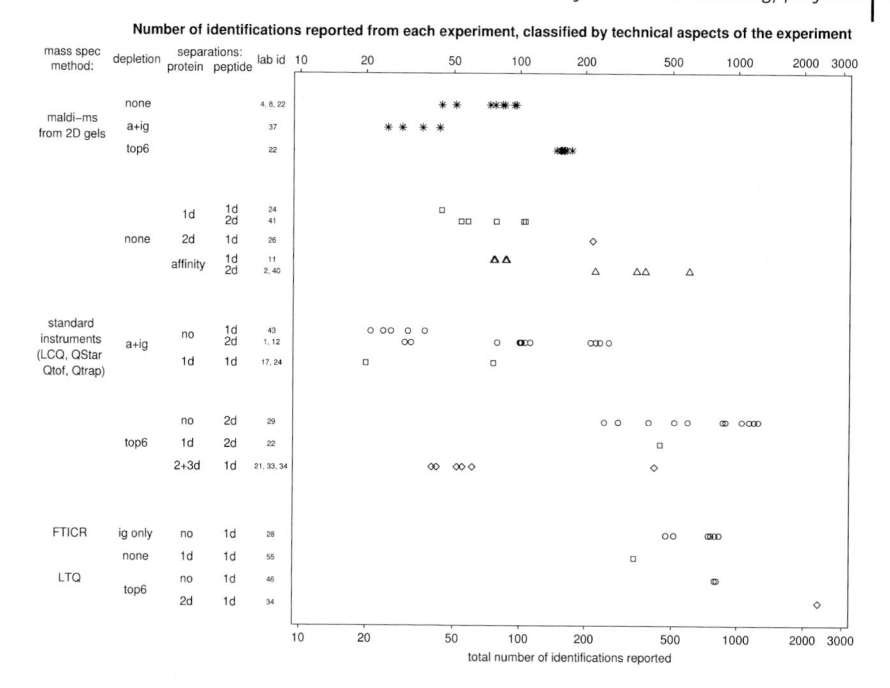

Fig. 3 Categorization of depletion, fractionation, and MS methods and yield of proteins identified (log scale).

fering origin and complexity may have different PTMs and different elution times, limiting the usefulness of the AMT tags. At present, peptide coverage seems to be quite limited. However, powerful MS-FT-ICR-MS (MS3) combinations are being introduced [13]. Lab 28 contributed valuable data on serum/plasma comparisons. Adkins et al. [12] also demonstrated that their approach gives a rough quantitative estimation of protein concentrations based on average ion current for all the peptides identified for 18 particular proteins, correlated in log-log plots with nephelometric immunoassay results.

The most striking difference in MS was the comparison of LCQ-Deca XP+ ion trap (IT) and LTQ linear IT MS/MS instruments by Lab 34. The analyses were of two different specimens from the BD B1 set, using similar depletion, protein array pixelation prefractionation, and tryptic peptide fractionation (Tang et al. [14] this issue). LCQ analysis of B1-heparin-plasma yielded 575 IDs, while LTQ analysis of B1-serum yielded 2890 protein IDs, both with the PPP high-stringency SEQUEST filters. Many low abundance proteins in the low ng/mL to pg/mL range were identified. The comparison is complicated, however, by the fact that the protein identifications used different amounts of starting material. Depletion was applied to 193 μL (14.5 mg) of plasma and 415 μL (35.3 mg) of serum. After the fractionation steps, fractions equivalent to 0.6 μL (45 μg) of the plasma and 2.4 μL (204 μg) of the serum were analyzed in the LC-MS/MS. Thus, some or possibly most of the difference in yield may be attributable to a larger volume analyzed. There were some other differences, as well including use of protease inhibitors with the depletion buffer, higher DTT concentration, fewer Micro-Sol-IEF fractions, and data-dependent MS/MS scans of the three most abundant ions with the LCQ instead of ten ions in the LTQ B1-serum experiment. There were also some differences in the searching of databases with one (serum) *versus* two (plasma) missed cleavage sites permitted. Tang *et al.* [14] describe extensive sensitivity analyses of experimental parameters that affect the tradeoff between numbers of high confidence protein IDs and analysis time. For example, gas phase fractionation to analyze different segments of the m/z range in each run was judged to be inefficient.

Labs 46 and 55 also employed LTQ instruments and obtained large numbers of identifications for reference specimens C1-serum and B1-citrate-plasma, respectively (Tables 1 and 2, Fig. 3).

1.3.2
Analysis of confidence of protein identifications

High false-positive rates are acknowledged to be a major problem in protein identification. Estimates can be generated, at least in relatively homogeneous datasets, by probabilistic methods using PeptideProphet and ProteinProphet, by matching to reversed-sequence databases [15–20]. The alternative of careful manual inspection of the spectra becomes a huge task and is subjective. The spectrum may represent a mixture of different peptides with almost equal parent masses and elution times. The biological specimen may have allelic variants or a contaminant not recorded in the database. Even if the sequence is correct, PTMs may take the sequence outside the scope of the match. However, true positives may be a problem, too, especially when the database sequence is simply not the same as that of the biological specimen analyzed.

To estimate the confidence of protein identifications across our heterogeneous database, we compared the observed data on number of peptide matches *per* identification to a model in which identifications are randomly distributed. False-positive and true positive peptide identifications should show opposite behavior when numbers of identifications become large. We expect false-positive IDs to accumulate roughly proportional to the total, so that the chance of two or more false-positive identifications coinciding on the same database entry should be the product of their random probabilities. In contrast, a protein which is present in detectable concentration will produce many tryptic peptides in nearly stoichiometric quantities. Increased sampling, therefore, should increase the number of distinct peptides mapping to the same (correct) database entry. This model results in a Poisson distribution of number of peptides matched *per* sequence. Two parameters are needed to specify the model, the total number of proteins (Ndb) and the expected proportion of false peptide matches *per* database entry (lambda, ranging in this case from 0.211 to 0.146). The IPI 2.21 database contains 49 924 sequences after adjustment for redundancy. The upper bound for lambda corresponds to the assumption that every identified protein has at least one false-positive matching peptide; this bound eliminates all single-peptide hits. The lower bound accepts as

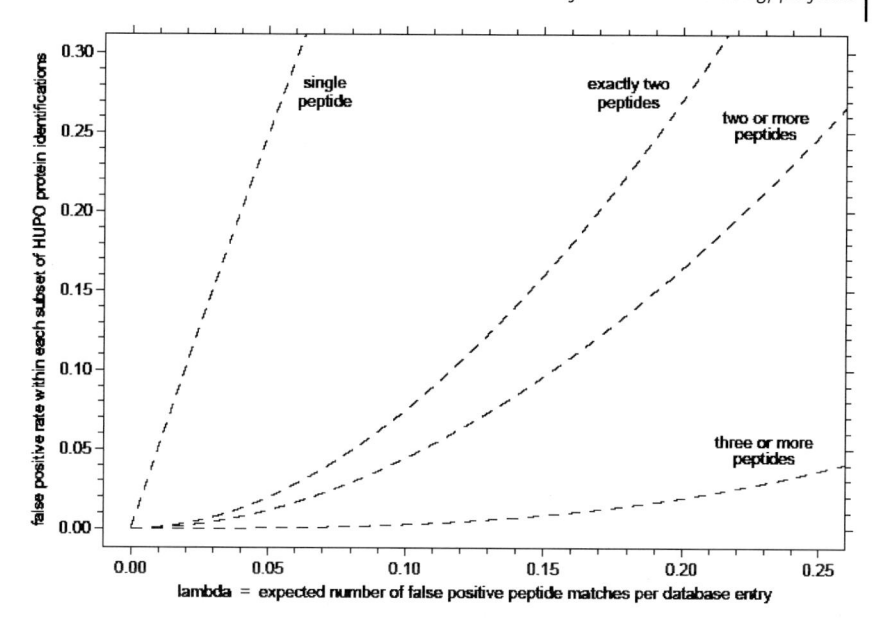

Fig. 4 Plot of estimated error rate for subsets of PPP proteins based on one, two, or three or more peptides, Poisson model.

correct all 1956 protein identifications based on a high confidence single peptide report, but treats all the 4528 lower confidence single peptide identifications as false. Throughout this range of values of lambda, proteins with four or more supporting peptides are predicted to be correct with better than 0.99 confidence; with exactly three peptides, 0.95–0.98; and with exactly two peptides 0.70 to 0.85 (Fig. 4). We based our annotations on the 3020 identifications made with two or more peptides project-wide to avoid a bias toward highly abundant proteins, if we had limited annotation to proteins based on three or more peptides. Furthermore, a substantial majority of protein IDs based on exactly two peptides is probably correct. Independent conclusions from manual review of a large number of spectra led one of our investigators to estimate at least 20% of one-peptide hits appear to be true positives. In addition, MacCoss et al. [21] concluded that the chance that multi-peptide proteins are false-positives declines exponentially with the number of peptides identified.

1.3.3
Quantitation of protein concentrations

A critical parameter for detection and identification of proteins is the abundance or concentration of the protein and its isoforms. We generated a calibration curve for a set of sentinel proteins for which quantitative immunoassays were available. Four different immunoassay and antibody microarray methods were performed by four independent laboratories (DadeBehring, Genomics Institute of Novartis Foundation, Molecular Staging, and Van Andel Research Institute). A total of 323 assays measured 237 unique analytes (Haab et al. [22]). In the cases of multiple assays, we cannot be certain that the same epitopes were targeted. This approach permits assessment of systematic variation in concentration of proteins associated with blood preparation methods (serum and the three anticoagulation methods for plasma in each specimen set) and, after matching to IPI identifiers, facilitates an analysis of dependence on concentration for MS-based protein identifications using the HUPO PPP specimens. Some proteins were at such low concentrations that they were even undetectable with immunoassay or microarray methods. After extensive curation, we matched 76 IPI proteins among the 9504 dataset (based on one or more peptides) and 49 proteins among the 3020 protein dataset (based on two or more peptides) to quantitative analytes. Fig. 1 in Haab et al. [22] shows four parameters used to determine the sensitivity of detection of these proteins as a function of immunoreactive concentration: number of labs reporting that protein, number of peptides on which protein IDs were based, percent coverage of the protein sequence, and score. The correlation coefficient for the total number of peptides matching that protein is $r = 0.86$ for the 3020 dataset and $r = 0.90$ for the 9504 protein dataset:

$\log10(N) = 0.365*\log10(conc) - 0.711$.

As expected, the most abundant proteins are the most readily detected, with essentially 100% agreement; with much less abundant proteins, only the laboratories with protocols and instruments capable of much more sensitive detection identified these proteins. Among the 49 proteins matched to the 3020 protein dataset, 12 are biologically interesting proteins identified with measured concentrations from 200 pg/mL to 20 ng/mL (Tab. 3).

Tab. 3 Least abundant proteins identified with two or more peptides (included in core dataset) with measured concentrations in the range of 200 to 20 000 pg/mL serum or plasma

Protein	Concentraion (pg/mL)
Alpha fetoprotein	2.9E+02
TNF-R-8	3.3E+02
TNF-ligand-6	1.5E+03
PDGF-R alpha	4.6E+03
Leukemia inhibitory factor receptor	5.0E+03
MMP-2/gelatinase	8.8E+03
EGFR	1.1E+04
TIMP-1	1.4E+04
IGFBP-2	1.5E+04
Activated leukocyte adhesion mol	1.6E+04
Selectin L	1.7E+04

1.4
Comparing the specimens

1.4.1
Choice of specimen and collection and handling variables

Pre-analytical variables can alter the analysis of blood-derived samples. Publications and protocols are generally deficient in this regard. Besides preparing the reference specimens of serum and plasma for direct comparisons, we undertook special studies on choice of sample type, stability during storage, use of protease inhibitors, and criteria for clinical standardization. The Specimens Committee concluded (Rai et al. [23]) that plasma is preferable to serum, due to less degradation ex vivo (as shown specifically by Tammen et al. [24] and Misek et al. [25]). Nevertheless, there is a view that standardization of proteomics assays with serum may be desirable, since archived specimens are so frequently sera.

They concluded that platelet-depletion of plasma may be desirable to avoid platelet activation with release of proteins, especially if there is a 4C step in the preparation. BD explained that 4C was chosen for centrifugation and holding of the tubes prior to aliquoting to aid in stabilizing labile biomarkers. For investigators concerned about platelet contamination, options include filtration of the plasma through a 0.2 µ low protein binding filter; double centrifugation of the specimen; and use of additives that minimize platelet activation, such as CTAD, a mixture of citrate, theophylline, adenosine, and dipyridamole. Samples should be aliquoted and stored frozen with minimization of thaw/re-freeze cycles, preferably in liquid nitrogen, though $-80°C$ seems to be very nearly as good. Protease inhibitors would be desirable, but present cocktails introduce complications due to peptide inhibitors that may interfere in the MS and small molecule inhibitors that form covalent bonds with proteins, shifting the isoform pattern. The Committee recommends diligent tracking of pre-analytical variables, and development and use of certified reference materials for quality control and quality assurance.

Haab et al. [22] extensively analyzed the concentrations of assayable proteins in the PPP specimen sets. They noted a systematic 15% lower value for many proteins in citrate-plasma, compared with other specimens; it turns out that this can be attributed to dilution and osmotic effect with the citrate solution, without any impairment in detection of proteins compared with the other specimens. However, David Warunek and Bruce Haywood of BD advised us that results with citrate-anticoagulated plasma can be quite sensitive to the blood:additive ratio and the subject's hematocrit. EDTA, meanwhile, is a much better chelator of calcium and more effective at platelet inactivation.

The sets of four specimens from a given donor pool yielded rather similar numbers of proteins when analyzed by the same lab and same techniques (see Tab. 2). Naturally, the agreement on identification of specific proteins was greater for higher abundance proteins. Since the laboratories exercised considerable discretion in deciding how many and which of the reference specimens to request and how many to actually analyze, as well as how intensively to analyze them, comparisons across the specimen results is of limited validity in this exploratory phase of the PPP. However, comparisons within several laboratories (1, 2, 11, 12, 28, 29, 41, 43 in Tab. 2) show quite close values for numbers of proteins identified, with deficiencies for B1-serum and B3-serum in Lab 1, and B1-heparin and possibly B3-heparin in Lab 29). It is curious that several laboratories chose citrate-plasma if they analyzed only one plasma specimen (Tables 1 and 2). Lab 28 shows greater similarity within each of the three donor pools for three citrate-plasma *versus* serum comparisons, than for citrate-plasma or sera across the three pools. The values for total number of proteins within each pair were quite close, whereas the B1 specimens yielded significantly fewer identifications than the B2 and B3 pairs. For B2 serum and B2 citrate plasma, they reported 365 proteins in common, of 542 and 572 identified in each. Ion current estimation of concentrations put 275 of the 365 within ±2-fold; 59 proteins had plasma/serum values >2X and 31 had P/S values <0.5X (Adkins *et al.* [12], this issue). Lab 34 is a special case, because different instruments were used for B1-heparin (LCQ) and B1-serum (LTQ), as noted above (Section 3.1).

Tab. 2 summarizes the protein IDs by lab and specimen. As noted above, the numbers of proteins identified in the consolidated database may be different from those in the individual papers in this special issue due to the integration procedure applied to the Core Dataset and the expanded analyses for these papers. The most analyzed specimen, B1-serum (Caucasian American) had 1749 IDs among the 3020. The three anticoagulated B1 specimens yielded a total of 1904 unduplicated IDs, of which 1023 were in common with the proteins identified in the B1-serum. The total number of unique IDs in the four B1 specimens that meet the two or more peptides criterion in either plasma or serum is 2630. A similar analysis of the combined C1 (Chinese) pooled specimens in just two labs yielded 1693 proteins, of which 1416 were identified in the B1 pool. With the exception of Labs 26 and 28, no very extensive analyses of the B2/B3 African-American and Asian American specimens were submitted. Combining all datasets, including the lyophilized NIBSC citrate-plasma specimen, we reached the 3020 protein dataset.

Tammen *et al.* [24] focused on the "peptidome" with mass <15 kDa. Peptides may be fragments of higher M_r proteins, or hormones, growth factors, and cytokines with specific biological functions. Their findings are not included in the Core Dataset since they used differential peptide display, plotting m/z ratios against retention time, with RP-HPLC-MALDI-TOF-MS. They do use nESQ-qTOF-MS/MS or MALDI-Tof-Tof-MS to confirm some peptide identifications. They did not actually attempt to identify proteins from the peptides. However, they made observations highly relevant to specimen processing. A large number of peptides, including many abundant peptides, are present only in serum, presumably due to the multi-protease events of clotting (AP-FXIII), enzyme activities (kallikrein), or peptides derived from cellular components, especially platelets, or the clot itself (thymosin beta-4, zyxin). In fact, at least 40% of the peptides detected in serum were serum-specific. Clotting is unpredictable due to influences of temperature, time, and medications, which are hard to standardize. These observations with serum may be highly relevant to the interpretation of SELDI results. They reported altered elution behavior of peptides in the presence of heparin, due to the polyanion nature of polydisperse low M_r heparin. Heparin acts through activation of antithrombin III, while citrate and EDTA inhibit coagulation and other enzymatic processes by chelate formation with ion-dependent enzymes. They recommend platelet-depleted EDTA or citrate-plasma, which gave consistent and similar results. They do not recommend addition of protease inhibitors, especially aprotinin, which requires µg/mL concentrations that interfere with analysis.

1.4.2
Depletion of abundant proteins followed by fractionation of intact proteins

Reducing the complexity of protein mixtures by depletion and fractionation of intact proteins greatly simplifies the task for MS/MS analysis. There are essentially three patterns of depletion in Tab. 2 and Fig. 3: no depletion of the most abundant proteins, depletion only of albumin or Ig or both, and depletion of the top-6 proteins, which are albumin, IgG, IgA, haptoglobin, alpha-1 anti-trypsin, and transferrin (Agilent column). There is clear evidence from the main database and from a series of special project studies by PPP investigators that depletion makes it significantly more feasible to visualize, detect, and then identify lower abundance proteins (Echan et al. [26], Li et al. [5], Zolotarjova et al. [27], Huang et al. [28], Tang et al. [14], Misek et al. [25], Yang et al. [29], Barnea et al. [30], Moritz et al. [31], Cho et al. [32], Kim et al. [33]). However, when only 2-DE is employed, the many "new" spots detected after depletion are unmasked isoforms of medium-abundance proteins, rather than lower abundance proteins [5, 26]. There is a counterbalancing problem, namely non-target or inadvertent removal of other proteins [6], which could be due to peptides and proteins bound to the target proteins, especially albumin; cross-reactivity with the bound antibodies; or non-specific binding to the column or resin or dye. Details of the protocols, proprietary buffers, column capacity, and previous use of the columns may be important variables. With older and much less expensive albumin-removal agents, such as Cibacron Blue dye, there is thought to be binding to the dye (as well as any binding to the albumin).

Moritz et al. [31] provide a preliminary report using free-flow electrophoresis (FFE-IEF) and rapid (6 min) RP-HPLC to fractionate citrate-plasma (Lab 33). They analyzed both bound and flow-through fractions from immunoaffinity depletion of the top-6 proteins. From 15 of 96 FFE fractions, with 72 780 MS/MS spectra analyzed with MASCOT and Digger and subjected to manual validation, they obtained 55 proteins based on two or more peptides and 23 more based on one peptide, across a mass range of from 4 to 190 kDa; these included several with estimated concentrations of 0.5–1 ng/mL. They highlight the identification in the bound fraction of a 35-amino acid serine protease protection peptide (CRISPP) that is cleaved from the C-terminus of alpha-1 anti-trypsin, non-covalently complexed with alpha-1 anti-trypsin, and not included in the IPI 2.21 database. They detected protein complexes by using non-denaturing, non-reducing buffers. They enhanced their yield by building a data-dependent exclusion list to prevent re-identifying abundant peptides.

Tang et al. [14] investigated many experimental parameters of depletion, fractionation, and such MS variables as gas phase fractionation. They combined solution isoelectrofocusing and 1-D SDS gel electrophoresis to generate "pixels" of proteins with defined pI and M_r ranges, then fractionated tryptic digests with 2-D LC, followed by LCQ-Deca-XP+ or LTQ-linear IT-MS/MS for B1-heparin-plasma and B1-serum reference specimens, respectively. These methods yielded 575 and 2890 high-confidence protein identifications (see Section 3.1) using the stringent HUPO PPP SEQUEST parameters; they did not remove potential homologous database entries; 319 of the 575 plasma proteins were identified in the serum specimen. Of these 319, half are single-peptide proteins in plasma, but many more are multiple-peptide proteins in serum, with the LTQ instrument, and have rich MS/MS fragmentation patterns. They estimated that proteins in the low ng/mL range were detected from 45 µg of plasma protein using the LCQ-Deca XP+, whereas proteins in the low pg/mL range were detected from 204 µg of serum using the LTQ. They uniquely utilized a SEQUEST Sf score, which combines X_{corr}, deltaCn, Sp, R_{sp}, and ions scores using a neural network to reflect the strength of peptide assignment on a scale of 0 to 1; scores ≥ 0.7 were considered to have a high probability of being correct, regardless of other parameters; when Sf scores replaced $R_{sp} \geq 4$, they obtained 744 and 4377 non-redundant protein identifications from the plasma and serum specimens, respectively.

Misek et al. [25] identified many isoforms and compared relative abundance of proteins in serum, EDTA-plasma, and citrate-plasma labeled, respectively, with the fluorescent dyes Cy3, Cy5, and Cy2 after top-6 immunoaffinity depletion. The three labeled, depleted samples were subjected to three-dimensional protein fractionation by pI, hydrophobicity, and M_r. About 3000 bands on 1-D SDS gels with $\pm >$ two-fold differences in intensity of fluorescence in dye pairs were excised and analyzed by MS/MS, yielding a total of only 82 non-redundant proteins; 28 proteins were identified in ten or more different fractions. Complement C3 and clusterin are presented as examples of proteins whose biologically significant cleavage products can be identified with this method. Not surprisingly, the yield in MS/MS was greater for proteins with higher intensity (abundance). Multiple isoforms reduce the concentration of a protein in any particular spot or fraction and may react very differently with antibodies used to quantify the proteins or detect the proteins, as on microarrays.

Subfractionation of the complex mixtures that are plasma and serum can be performed chemically or with capture agents. A very good example is the glycoprotein subproteome. Labs 2 and 11 (Tables 1 and 2) utilized hydrazide chemistry and binding with three lectins, respectively, to enrich for glycoproteins. The chemical method, which captures N-linked glycoproteins subsequently treated with PNGase F, was published by Zhang et al. in 2003 [34]. Yang et al. [29] used wheat germ agglutinin, Jacalin lectin, and Con A together on agarose to isolate and characterize approximately 150 glycoproteins in PPP serum and plasma reference specimens after analysis by LCQ-MS/MS, with confirmation in some cases using a linear IT LTQ instrument. There was close similarity for the composition of the glycoproteome across the plasma and serum specimen sets, except for fibrinogen, which was absent from serum (after clotting). Samples from the individuals from three different ethnic groups showed only a few individual differences. Together the two laboratories identified 254 glycoproteins, of which 164 were identified by other laboratories in this collaboration. That means that 90 were found only in the glycoprotein-enriched studies. Glycoprotein has an important incidental benefit in that the non-glycosylated albumin protein should be excluded; in fact, some albumin remains, given its very high abundance and its tendency to bind glycoproteins.

Cho et al. [32] combined immunoaffinity depletion of the top-6 proteins with free-flow electrophoresis or 2-DE of fractions, and MALD-TOF-MS PMF; they found only minor differences across the donor and specimen preparation variables. With 2-DE they found few non-target proteins in the immunoaffinity bound fraction.

Kim et al. [33] sought to identify and eliminate false-positive peptide identifications and subsequent protein matches by analyzing molecular weight on 1-D SDS gels after immunoaffinity depletion. Of 494 proteins identified with 2-D-LC/ESI-MS/MS of 28 1-D fractions, using SEQUEST with stringent PPP filters, 202 were excluded as single-peptide hits as well as estimated M_r too deviating from theoretical M_r, but 166 one-peptide matches were retained based on good M_r match. This approach requires careful curation for biologically cleaved proteins. Their method actually increased the number of accepted proteins, since only 128 (26% of 494) were based on two or more peptides among the total of 292 protein identifications claimed for the B1-serum specimen.

Echan et al. [26] compared the immunoafffinity top-6 depletion column and corresponding spin cartridge from Agilent with a prototype ProteoPrep dual anti-albumin/anti-IgG antibody column from Sigma Aldrich, with five commercially available kits using Cibacron Blue for albumin and/or Protein A or G for immunoglobulin depletion, and with no depletion. These variables correspond to the categories depicted in Fig. 3. The polyclonal antibody column gave nearly complete depletion, showed low non-specific binding, based on 2-DE profiles, and permitted many new spots to be visualized. However, the number of new proteins was quite small, due to the emergence of newly visualized spots representing numerous isoforms of the now-most abundant remaining proteins. They estimated that silver staining on 2-D gels should have been able to detect proteins originally present in the serum or plasma at 40 ng/mL or higher, while the protein identified with lowest known concentration is at about 30 µg/mL, before accounting for heterogeneity of isoforms. The two-protein column had more capacity for albumin and IgG removal, but also removed many non-target proteins, which may be improved with optimized buffers. Apparently, buffer variables are very influential with all of the antibody columns. Given published reports of up to 63 proteins bound to albumin [35], secondary binding conditions can introduce major variability in results. Clearly, more potent technology combinations are required to adequately evaluate the non-target binding of proteins during immunoaffinity depletion, as well as to reach down to the ng/mL to pg/mL concentration range. Echan et al. [26] point out that the inexpensive and convenient dye and protein A/G methods can be used for fractionation rather than depletion. They also note the potential to specifically deplete many more proteins with expanded immunoaffinity columns.

Additional papers by Zolotarjova et al. [27] and by Huang et al. [28], scientists at Agilent and at GenWay Biotech, respectively, present laboratory results with their immunoaffinity products. The polyclonal rabbit antibody column from Agilent removes albumin, IgG, IgA, haptoglobin, transferrin, and alpha-1 anti-trypsin. The polyclonal chicken IgY antibodies on microbeads from GenWay remove six (albumin, IgG, IgA, IgM, transferrin, and fibrinongen) or 12 (also alpha-1 anti-trypsin, alpha-2 macroglobulin, haptoglobin, apolipoproteins A-I and A-II, and orosomucoid/alpha-1 acid glycoprotein). Both groups report highly effective removal and little to no non-target binding. These products were introduced during the conduct of the PPP pilot phase and were made available to investigators.

One way to maximize identifications is to analyze bound fractions as well as pass-through fractions, as done by He et al. [6] and by Labs 29 and 46 (Tables 1 and 2). He et al. [6] report large numbers of proteins in the top-6 immunoaffinity bound fraction when extensive LTQ-MS/MS is applied, utilizing the stringent PPP SEQUEST filters. They may not have used the full system optimized by the column manufacturer.

1.4.3
Comparing technology platforms

Li et al. [5] analyzed the PPP C1-serum specimen with five different proteomics technology combinations after immunoaffinity depletion of the top-6 proteins. In all, 560 unique proteins were identified, 165 with two or more peptides. Only 32 proteins were identified by all five approaches and 37 by 2-DE, 2-D HPLC, and shotgun approaches, primarily due to finding only 78 unique proteins among 1128 spots excised, digested, and analyzed with method 1, WAX-2-DE-MALDI-TOF-MS-MS. Protein 2-D-HPLC fractionation + RP-HPLC/microESI-MS-MS gave 179 proteins; an online SCX shotgun strategy ("bottom-up") gave 131, an offline SCX shotgun strategy gave 224, and an offline shotgun-nanospray strategy yielded 330 proteins. High and medium abundance proteins are found by all methods, while low abundance proteins are complementary, reflecting both different methods and inherent incompleteness of sampling and identifying peptide ions. Different technology combinations give different useful information; for example, the 2-DE method 1 provided more information about p*I*-altered isoforms and relative abundance of identified proteins. The offline strategies sharpen the peaks and improve separation of peptides, submit more fractions to the MS instrument, and allow the MS enough time to acquire the qualified spectra of more eluting peptides. Nanoflow accentuates the same advantages, permitting ultrahigh sensitivity. Overall, electrophoresis and chromatography, coupled respectively with MALDI-TOF/TOF-MS and ESI-MS/MS, identified complementary sets of serum proteins. Like Aebersold and Mann [2], they conclude that no single analytical approach will identify all the major proteins in any proteome. Others have recently used similar 2-D separation of peptides offline, intact protein fractionation prior to MS, or sensitive ESI-MS/MS analysis of fractionated peptides [36–39]. As far as cost-effectiveness, the 2-D HPLC approach required much more time and labor and was much less suited to automation than the other strategies; it has the advantage of being able to process large volumes of sample, when that is available and desired. Handling fractions also introduces more evidence of contamination; epidermal keratins are seldom found with the shotgun methods. Low abundance proteins are not only masked by medium abundance proteins on gels, but inefficient extraction of peptides from gels is a limitation for low abundance proteins.

Barnea et al. [30] expanded on their original submission as Lab 1 (Tables 1 and 2) with an analysis of several protein fractionation and several MS/MS methods on PPP reference specimen B2-serum. Albumin and IgG were depleted with are Bio-Rad mini-kit based on Affi-Gel Blue and Affi-Gel protein A, respectively. The aim was to increase the concentrations of individual proteins and then their tryptic peptides in each fraction submitted for MS/MS analysis, seeking to reach the threshold for detection. Combining pre-proteolysis fractionation with post-digestion fractionation was more effective than more extensive fractionation of the peptides. Each method has some advantages of avoiding loss of proteins with particular characteristics (pI, M_r, other). The base case was MudPIT analysis of unfractionated, digested proteins; then SDS-PAGE, SCX, and Rotofor fractionations were coupled with LC-MS/MS or with MudPIT. In each pair, MudPIT gave more protein IDs than LC-MS/MS. SCX gave the most IDs among the fractionation methods.

He et al. [6] analyzed ten pooled male and ten pooled female C1-sera, using top-6 depletion, tryptic digestion, then RP-HPLC, ESI-MS/MS shotgun analysis. They reported 944 non-redundant proteins under stringent PPP criteria based on [40], combining separate analyses of male (594) and female (622) sera; there were 206 with two or more peptides. Some lower abundance proteins were detected, including complement C5 and CA125. Instead of one analysis of serum, here there are eight analyses: male and female, bound and unbound, and a duplicate of each. The reproducibility of the duplicates is 40–50%; the overlap of bound and unbound is 16–18%, and of male and female 40–50% (*i.e.*, same as duplicates). They used four databases: IPI 2.20 (June 2003), IPI 2.32 (May 2004), Swiss-Prot 43 (March 2004), and NCBI (Dec 2003) and obtained quite similar protein groups for the first three and also for NCBI, though the pre-grouping numbers of proteins were 2.5 times larger for NCBI, demonstrating the known redundancy in the NCBI database.

1.4.4
Alternative search algorithms for peptide and protein identification

One of the important challenges for collaborative proteomic studies is the variety of search algorithms embedded in mass spectrometers. Some of these search algorithms are proprietary with key elements undescribed in the open literature or even for the user laboratory. Each investigator has many options in the choice of parameters for the software search to identify peptides from the mass spectra of ion fragments and then to deduce the best protein match from yet another broad array of gene and protein databases, including different versions of each evolving database. Expert curation of such collaborative datasets is required. In the PPP Jamboree Workshop of June 2004, the offer to generate cross-algorithm analyses with PPP data was strongly endorsed, and many months of effort were invested.

Kapp et al. [11] report a unique analysis of alternative search algorithms. They used one raw file from the Pacific Northwest National Laboratory LCQ-MS/MS data on serum depleted only of IgG published by Adkins et al. [41], which served as a basis for the later FT-ICR-MS analyses for the PPP (Lab 28). The same spectra were subjected to analyses with MASCOT, SEQUEST (with and without PeptideProphet), Sonar, Spectrum Mill, and X!Tandem by experts familiar with the use of each. Careful manual inspection was applied, as well, though it is always a challenge to understand what exactly were the criteria used in manual inspection. The paper provides a useful description and categorization of the features of each search engine into heuristic algorithms and probabilistic algorithms. The authors then present and compare their performance identifying peptides and proteins, benchmarking them based on a range of specified false-positive rates. In all, 600 peptides were identified, of which 355 were found with very high confidence (estimated error rate 1%) by all four of MASCOT, SEQUEST, Spectrum Mill, and X!Tandem. The authors concluded that no one of these algorithms outperforms the rest. Spectrum Mill and SEQUEST performed well in terms of sensitivity, but performed less well than MASCOT, X!Tandem, and Sonar in terms of specificity. Thus, they recommend using at least two search engines for consensus scoring, though the scheme for creating combined scores awaits further work. The probabilistic algorithm, MASCOT, correctly identified the most peptides, while the re-scoring algorithm, PeptideProphet, enhanced the overall performance of SEQUEST. This paper utilizes reversed-sequence searches, as well as probabilistic estimates of false-positive rates. Unfortunately, the spectra in this dataset were dominated by high abundance proteins, such that the 600 peptides were matched to only 40–60 proteins using a trypsin-constrained search.

1.4.5
Independent analyses of raw spectra or peaklists

After the original data submission protocol had been established, built upon peptide sequences and protein identifications, three groups emerged as having capability for centralized, independent analyses that would bypass the peculiarities of the search engine software embedded in particular MS instruments and the criteria applied by individual investigators in establishing thresholds for high and lower confidence identifications or applying manual inspection of the spectra.

Beer at IBM/Haifa developed PepMiner software [42], which processes very large numbers of raw spectra to generate clusters of spectra and then SEQUEST-like analysis and scoring for peptide and protein IDs. Beer et al. [43] applied this method to the spectra from laboratories 1, 2, 17, 22, 28, 29, 34, and 40. The data from laboratory 1 included those submitted for the Core Dataset(s) as well as those in the Barnea et al. [39] special project paper. They identified 14 296 peptides, which were assigned to 4985 proteins with one or more peptides, 2895 proteins with two or more peptides, and 1646 with three or more peptides. The 4985 IDs had 2245 in common with the 15 519 unintegrated and 1983 in common with the 9504 integrated PPP IDs. The 2895 based on two or more peptides compares with our 2868 based on two or more peptides for the same eight laboratories, with 865 in common with our Core Dataset.

Deutsch et al. [44], at the Institute for Systems Biology in Seattle, US, utilized SEQUEST with PeptideProphet/ProteinProphet software developed by the Eng group to estimate error rates and probability of correct assignment of spectra to peptide sequences and then to protein IDs [15, 45]. Analyzing the PPP datasets from laboratories 2, 12, 22, 28, 29, 34 (B1-heparin only), 37, and 40 with the PeptideAtlas process [46], they observed 6929 distinct peptides with a probability score ≥ 0.90, including 6342 which mapped to 1606 different EnsEMBL proteins and 1131 different EnsEMBL genes. Reduction of multiple mappings yielded 960 different proteins, of which 479 have matches in the PPP 3020.

Kapp et al. [11] at the Ludwig Institute in Melbourne are utilizing MASCOT and Digger software developed at Ludwig on submissions from 14 laboratories; incomplete analyses show more than 500 high-confidence, non-redundant proteins with trypsin-constrained searches.

In addition, Beavis at the Manitoba Centre for Proteomics created a dataset with 16 191 EnsEMBL proteins from the PPP raw spectra using X!Tandem [47], of which 9497 matched to IPI v2.21, 3903 to our unintegrated list, and 2828 to our 9504 proteins based on one or more peptides. Of 5816 IPI proteins with two or more peptides, 1259 matched to the 5102 unintegrated and 913 to the 3020 Core Dataset.

Martens et al. [48] noted the value of these independent analyses in overcoming numerous sources of variation from the search algorithm, the database, and the investigator. They recommend that m/z peaklists routinely be made publicly available, while deferring on the raw data, which currently lack standardized formats, let alone the required infrastructure for centralized storage and distribution. However, a plan to assure access to the raw spectra, as well as the peaklists, can facilitate wide dissemination and utilization of complex datasets, as we have demonstrated in this collaboration by both the participating laboratories and the independent analysts, the incorporation into PRIDE by EBI, into PeptideAtlas by ISB and ETH, and into the Global Proteome Machine DataBase by Beavis.

It is striking that these independent analyses not only differed in the proteins that they identified, but also in the peptides identified from the same MS/MS spectra that were the basis for the protein matches. Further improvements in software and analytical methods are needed, given the many sources of error in peptide identification [49]; automated *de novo* sequencing can help, and chemical synthesis of peptides to determine the spectra directly can be employed selectively.

1.4.6
Comparisons with published reports

Tab. 4 shows the numbers of proteins reported in human plasma or serum in the literature, the number of those proteins in the IPI database, and the congruence with our PPP 9504 and PPP 3020 protein lists. Our lists are integrated (see Section 3.1), while the others generally are not, and do not use the same methods. It is clear that the number and nature of proteins identified in serum and plasma depend greatly on the sample preparation and fractionation and on MS methods and analytical tools.

Tab. 4 Comparison of PPP integrated protein identification lists with published datasets for human plasma or serum

Published data	Total IDs	# IPI proteins	PPP_9504 dataset	PPP_3020 dataset
Anderson et al. [50]	1175	990	471	316
Shen et al. [38]	1682	1842	526	213
Chan et al. [54]	1444	1019	402	257
Zhou et al. [35]	210	107	68	51
Rose et al. [55]	405	287	159	142

Anderson et al. [50] published a compilation of 1175 non-redundant proteins reported in at least one of four sources (literature review plus three recent experimental datasets [51, 41, 52]); only 46 proteins were reported in all four sources, suggesting high false-positive rates from reliance on single-peptide hits [49]. The experimental papers used multidimensional chromatography, 2-DE, and MS; MudPIT analysis of a tryptic digest; or MudPIT of a tryptic digest of low-M_r plasma fractions. Of the 990 of these proteins which have IPI (version 2.21) identifiers, 316 are found in our 3020 protein Core Dataset. When we relaxed the integration requirement (5102 IPI IDs), as was the case for [50], this figure rose only to 356 matches. Using the full 9504 dataset, the corresponding matches were 471 with integration and 539 without integration (15 710 protein IPI IDs).

Shen et al. [38] used high-efficiency nanoscale RP LC and strong cation exchange LC in conjunction with ion-trap MS/MS and then applied conservative SEQUEST peptide identification criteria (with or without considering chymotryptic or elastic peptides) and peptide LC normalized elution time constraints. Between 800 and 1682 human proteins were identified, depending on the criteria used for identification, from a total of 365 µg of human plasma. With their cooperation, we re-ran their raw spectra using HUPO PPP SEQUEST parameters (high confidence: $X_{corr} \geq 1.9/2.2/3.75$ (for charges +1/+2/+3), deltaCn ≥ 0.1, and $R_{sp} \geq 4$; lower confidence: $X_{corr} \geq 1.5/2.0/2.5$ (for charges +1/+2/+3), deltaCn ≥ 0.1) and obtained 1842 IPI protein matches. Of these, 526 and 213 were found in the PPP 9504 and 3020 datasets, respectively.

Chan et al. [53] resolved trypsin-digested serum proteins into 20 fractions by ampholyte-free liquid phase IEF. These 20 peptide fractions were submitted to strong cation-exchange chromatography, then microcapillary RP-LC-MS/MS. They identified 1444 unique proteins in serum. When we mapped these proteins against the IPI v2.21 database, there were 1019 distinct proteins. From this set, 402 and 257 proteins matched with the 9504 and 3020 datasets, respectively.

Zhou et al. [35] identified an aggregate of 210 low M_r proteins or peptides after multiple immunoprecipitation steps with antibodies against albumin, IgA, IgG, IgM, transferrin, and apolipoprotein, followed by RP-LC-MS/MS. Only 107 proteins were mapped with IPI identifiers, of which 68 and 51 were found in the 9504 and 3020 PPP protein lists, respectively.

Finally, Rose et al. [54] reported fractionation in an industrial-scale approach, starting with 2.5 liters of plasma from healthy males, depleted of albumin and IgG, then smaller proteins and polypeptides separated into 12 960 fractions by chromatographic techniques. From thousands of peptide identifications, 502 different proteins and polypeptides were matched, 405 of which were included in the publication. Of the 287 which mapped to IPI identifiers, 159 and 142 are included in our 9504 and 3020 protein dataset, respectively.

Thus, across studies, as well as across the PPP participating laboratories, incomplete sampling of proteins is a dominant feature. A substantial depth of analysis is achieved with depletion of highly abundant proteins, fractionation of intact proteins followed by digestion and two or more MS/MS runs for each fraction. Standardized, statistically sound criteria for peptide identification and protein matching, and estimation of error rates are necessary features for comprehensive profiling studies.

1.4.7
Direct MS (SELDI) analyses

Ten laboratories requested PPP specimens for analyses with SELDI chip fractionation, MS analysis, and algorithm-based differentiation of m/z peaks across specimens. Rai et al. [56] report the cross-laboratory evaluation of eight submitted datasets, of which five were judged appropriate for comparison of plasma results and four for serum results. Intra-laboratory CV varied from 15 to 43%. Correlations across labs were 0.7 or higher for 37 of 42 spectra with signal/noise ratios >5. More detailed analyses were done to actually identify one protein, haptoglobin, and variation in the intensity/concentration of its subunits in the different PPP reference specimens. They recommend stringent standardization and pre-fractionation to increase the usefulness of this method.

1.4.8
Annotation of the HUPO PPP core dataset(s)

From the inception, HUPO has intended that the Plasma Proteome Project facilitate extensive and innovative annotation of the human plasma and serum proteome. A large element of the Jamboree Workshop was focused on collaborative annotation. Several papers in this issue report on those collaborations.

Ping et al. [56] emphasize use of peptide identification results from MS/MS to reveal cleavage of signal peptides, proteolysis within hydrophobic stretches in transmembrane protein sites, and PTMs. Using 2446 of the 3020 PPP from IPI that matched to EnsEMBL gene products, they highlight subproteomes comprised of glycoproteins, low M_r proteins and peptides, DNA binding proteins, and coagulation pathway, cardiovascular, liver, inflammation, and mononuclear phagocyte proteins. Surprises include 216 proteins matched by Gene Ontology to DNA binding and 350 to the nucleus, including histone proteins, suggesting detection of proteins released by apoptosis or other means of cell degradation. Using the Novartis Atlas of mRNA expression profiles for 79 human issues, liver dominated as the source of the majority of proteins, although many of these proteins are also produced in other tissues. Many classic protein markers of leukocytes were not detected, including markers of B-cell, T-cell, granulocyte, platelet, and macrophage lineages, presumably all at low abundance with little shedding. In contrast, some quite low abundance proteins were found repeatedly, such as VCAM-1 and especially IL-6.

Signal peptide cleavage sites are generally predicted based on presence of a hydrophobic stretch of amino acids flanked at one end by basic amino acids. Seeking experimental evidence for such cleavage sites, these authors focused on semi-tryptic peptides, presuming that the signal cleavage event does not involve trypsin *in vivo*. Such evidence may override database predictions, as, apparently, in the cited example of SERPINA3/alpha-1-antichymotrypsin. They also identified two previously unreported proteins that undergo regulated intramembrane proteolysis, one of which releases an extracellular immunoglobulin domain - a reason not to reject all immunoglobulin matches. The MS/MS spectra can be examined for evidence of unrecognized PTMs. Using the Osprey tool, they found an average of nearly six protein-protein interactions *per* protein for a subset of 652 proteins; if they are circulating as multi-protein complexes, they will be less likely to be cleared through the kidney glomeruli.

Berhane *et al.* [57] focused on 345 proteins of particular interest for cardiovascular research. They classified the proteins into eight categories, most of which have relevance to other organ systems, as well: markers of inflammation in cardiovascular disease, vasoactive and coagulation proteins, signal transduction pathways, growth and differentiation-associated, cytoskeletal, transcription, channels and receptors, and heart failure and remodeling-related proteins. Of particular interest were the detection for the first time in plasma of the ryanodine receptor, part of the intracellular calcium channel in cardiac (and skeletal) muscle, and smoothelin, a structural protein restricted to smooth muscle cells, co-localized with actin. They used a number of identified peptides as an indicator of abundance of the protein (as in Section 3.3, above); for the first two categories, about 50% of proteins were identified with less than ten peptides, whereas no proteins among transcription factors had more than ten peptides and 56% had the minimum of two peptides. No cardiac contractile proteins were identified, even though they are far more abundant than transcription factors or signaling proteins in the heart, suggesting that necrotic cell death and uncontrolled cell rupture had no part in the appearance of any of the detected proteins in the healthy donors studied.

Muthusamy *et al.* [58] utilized a Java 2 Platform literature search tool to facilitate manual curation of functional classes of proteins, starting with the PPP set of 3020 IPI proteins (2446 genes). They subjected protein and nucleotide sequences in NCBI to BLAST queries to identify splice isoforms; they report that 51% of the genes encoded more than one protein isoform (a total of 4932 products). A total of 11 381 single nucleotide polymorphisms involving protein-coding regions were mapped onto protein sequences.

The Core Dataset of 3020 proteins was annotated with use of Gene Ontology for subcellular localization, molecular processes, and biological functions, showing very broad representation of cellular proteins. Subcellular component classification of the 1276 IPI-3020 proteins included in GO showed a relatively high proportion of proteins from membrane compartments (26%), nuclei (19%), cytoskeleton (11%), and other cell sites (23%), compared with the expected predominance of secreted proteins ("traditional plasma proteins") (14%). GO analyses of molecular processes showed 39% binding, 28% catalytic, 7% signal transducer, 6% transporter, 4% transcription regulator, and 3% enzyme regulator. GO analyses of biological functions revealed 36% metabolism, 25% cell growth and maintenance, 5% immune response, 1% blood coagulation and 1% complement activation. Examination of specific Gene Ontology terms against a random sample of 3020 from the human genome (Supplementary Fig. 1) shows some proteins >3 SD from the expected line. Categories over-represented include extracellular, immune response, blood coagulation, lipid transport, complement activation, and regulation of blood pressure, as expected; on the other hand, surprisingly large numbers of cytoskeletal proteins, receptors and transporters also were identified.

An InterPro analysis similarly compared the 3020 protein dataset with the fine-grained protein families and domains described for the full IPI v2.21 56 530 human proteins dataset (Supplementary Fig. 2). Over-represented domains include EGF, intermediate filament protein, sushi, thrombospondin, complement C1q, and cysteine protease inhibitor, while underrepresented include Zinc finger (C2H2, B-box, RING), tyrosine protein phosphatase, tyrosine and serine/threonine protein kinases, helix-turn-helix motif, and IQ calmodulin binding region, compared with frequencies in the entire human genome.

Of the 1297 of the 3020 protein dataset that had identifiers in Swiss-Prot 44, 230 were annotated as transmembrane proteins. Another 25 have mitochondrial transit signals, and an N-terminal signal sequence occurred in 373 proteins. Putative PTMs were noted for 254, including 85 with phosphorylation and 45 with glycosylation sites. A separate analysis of nearly twice as many proteins based on EnsEMBL matches using the Human Protein Reference Database (www.hprd.org; Muthusamy et al., [58]) found 628 with a signal sequence, 405 with transmembrane domains, 153 with a total of 1169 phosphorylation events, and 112 with a total of 555 glycolysation events.

One of the aims of the HUPO initiatives, as noted in the Section, is to link organ-based proteomes (liver, brain) with detection of corresponding proteins in plasma, and with proteins that are mediators, or at least, biomarker candidates, of inherited or acquired diseases. Using the Online Mendelian Inheritance in Man (OMIM), we found 338 of our 3020 IPI proteins that match EnsEMBL genes in OMIM, including RAG 2 for severe combined immunodeficiency (SCID)/Omenn syndrome, polycystin 1 for polycystic kidney disease (PKD), and BRCA 1, BRCA 2, p53, and APC for inherited cancer syndromes.

In the final article of this special issue, Martens et al. [59] describe the development and usefulness of the EBI PRoteomics IDEntifications database (PRIDE). The HUPO PPP dataset was the first large dataset to populate this database. The aim is to make publicly available data publicly accessible, in contrast to voluminous lists in printed articles or, more often now, in journals' websites, with custom layouts not suited to computer-based re-analysis. PRIDE offers an Application Programming Interface. In contrast, tables in PDF are described as notoriously difficult to extract. As noted, the PPP established a short-term solution with a relational database using a Microsoft Structured Query Language (SQL) server, which centralized all data collection and served as the testbed for the centralized, project-independent database that is now PRIDE. In turn, PRIDE has been designed with several features intended to facilitate future collaborative studies.

1.4.9
Identification of novel peptides using whole genome ORF search

A fascinating annotation from the PPP database has been used by States to enhance the annotation of the human genome itself [60]. The mass spectra data obtained by PPP investigators represent a resource for identifying novel and cryptic genes that may have been missed in previous annotations of the human genome. A total of 583 proteins in the 3020 protein set, including 185 identifications supported by three or more peptides, is not associated with genes in EnsEMBL. These are confident to highly confident experimental observations. The fact that they are not associated with known genes demonstrates that the annotation of the human genome remains incomplete.

To test the feasibility of this approach, we searched all ORFs using peak list data from six PPP laboratories (17, 30, 37, 41, 52, 55). NCBI human genome sequence build 33 was translated in all three reading frames and both strands; all non-redundant ORFs were assembled into chromosome specific sequence collections. The open source tool X!Tandem [61] was used in these analyses, with requirements for multiple mass spectra and a threshold hyperscore of 30 to accept peptide matches and greatly reduce the likelihood of false positive matches to ORFs. In all, 118 novel peptides were identified as highly probable matches to ORFs in the human genome not previously known to have protein products. This kind of protein-to-DNA mapping of the human genome is a notable bonus of the Plasma Proteome Project.

1.4.10
Identification of microbial proteins in the circulation

Microbial organisms populate all orifices and surfaces of many organs in the body, and their proteins may enter the blood intact or after degradation, as well as through contamination during venepuncture. We separately matched our peak lists for six small datasets against microbial genomes in the NCBI Microbial (non-human) GenBank (June 2004 release), using X!Tandem for RefSeq protein sequence identification. In this preliminary analysis, we found matches to several *E. coli* proteins (including elongation factor EF-Tu, outer membrane protein 3a, and glutamate decarboxylase isozyme) and mycobacterial proteins (members of glycine-rich PE-PGRS family) based on at least three peptide matches. No peptides for these proteins were found in the IPI human database, so these sequences are independent of the human gene and protein collections.

1.5 Discussion

This Special Issue of PROTEOMICS presents papers integral to the collaborative analysis, plus many reports of supplementary work on various aspects of the PPP workplan. The Core Dataset of 3020 proteins based on two or more peptide matches provides an anchor for future studies and for meta-analyses of the growing literature. These PPP results advance our understanding of complexity, dynamic range, biomarker potential, variation, incomplete sampling, false-positive matches, and integration of diverse datasets for plasma proteins. These results lay a foundation for development and validation of circulating protein biomarkers in health and disease. For the present, we recommend use of EDTA-plasma or citrate-plasma as the specimen of choice. Few labs actually compared these two alternative methods for plasma (Tables 1 and 2).

There are many opportunities for the HUPO Plasma Proteome Project going forward. First, these papers document our present understanding and reveal several open questions which require more focused studies: (a) to generate guidelines and standardized operating procedures for specimen collection, handling, archiving, and post-archive processing, including the protease inhibitor issue; (b) to use high-resolution methods to optimize specific immunoaffinity depletion of abundant proteins with minimal non-target losses; (c) to combine separation platforms and MS capabilities with an aim to expand the portion of the plasma proteome that can be profiled with confidence; (d) to achieve quantitative comparisons across specimens, not just compositional analyses; (e) to achieve high concordance in repeat analyses of the same specimen with the same methods; and (f) to overcome the extremely low overlap between protein identification datasets within a large collaboration of this type and, of course, across the literature, especially addressing the discrepancies due to post-MS/MS spectral analysis and peptide and protein database matching.

Other challenges are not specific to the plasma proteome, so we should discuss them together with other HUPO initiatives: (a) the limitations of present sequence databases, which are incomplete, redundant, and constantly being updated with corrections and new splice variants and SNPs; (b) the need to improve the true-positive to false-positive ratio, which requires explicit optimization; (c) the lack of reference specimen materials, which should be prepared with specific objectives and user communities in mind; (d) the need for independent corroboration of initial findings; and (e) organized strategies to validate proteomic discoveries and lead to microarray analyses with well-characterized antibodies, so that many specimens from clinical trials and epidemiological studies can be assayed. A new generation of studies will be considered at the Munich 4th HUPO Congress on Proteomics.

Second, there is an opportunity for the HUPO PPP to play a leading role in the continuing development and analysis of datasets arising from all quarters, in collaboration with the HUPO Protein Standards Initiative led by EBI [62] and other leading bioinformaticians, many of whom have contributed to this pilot phase of the PPP [62]. An immediate role for PPP is the cross-initatives analysis of Human Liver Proteome and Human Brain Proteome datasets with the PPP datasets, explicitly including experimental analyses of plasma samples from the same people and animals whose liver and brain specimens are studied. Several of the challenges listed above which involve search engine performance and integration of peptide identifications and protein matches with different databases deserve systematic investigation. Furthermore, quantitative analyses of concentrations, interactions, and networks will be increasingly important and feasible [63].

Third, there is an opportunity for HUPO to facilitate, and possibly organize, major disease-related studies of candidate biomarkers for earlier diagnosis, better stratification of newly diagnosed patients, appropriate pathways-based monitoring of targeted therapies, and design of preventive interventions. There is great anticipation of the application of ever-improving proteomics technologies for disease studies [64, 65].

For the overriding strategic question of gaining much higher throughput, at least four options have emerged in preliminary discussions:

(a) LC-MS with highly accurate mass and elution time parameters for peptide identification. A combination of specific depletion of abundant proteins, slow (2 h) nano-flow LC for elution time standardization, and highly accurate mass determination (<1 ppm) may make it feasible to base identifications solely on enhanced mass fingerprints once a high-quality accurate mass x elution time database with adequate sequence coverage of proteins to differentiate variants due to splicing, SNPs, and protein processing is in place. Additions to the database would require prior MS/MS identification.

(b) High accuracy LC-MS/MS/MS for peptide identifications. At the HUPO 3rd World Congress on Proteomics in Beijing, Mann described remarkable mass precision and very good efficiency of analysis with MS3, comprising MS/FT-ICR/MS. Applications to intracellular localization and discovery-phase identification of PTMs have already been achieved. It is likely, as with other methods, that an MS/MS or MS/MS/MS-based discovery phase would be converted into a different methodology, such as protein capture microarrays for high-throughput analysis of large numbers of plasma (or serum) specimens once the biomarkers were validated.

(c) Protein affinity micro-arrays. Humphery-Smith [66] proposed that affinity ligands be designed and produced to recognize conserved regions in each Open Reading Frame for signal enrichment. The ligands could be antibodies, receptions, aptamers, or other capture agents. The conserved regions might be sequences uncomplicated by PTMs, not subject to cleavage, and exposed at the surface. Enhanced chemiluminescence, rolling circle amplification, isotopic labeling, light scattering, or other methods could serve as read-out technologies. This approach could improve protein identifications over a wide dynamic range.

(d) Isotope coded peptide standards for quantitative protein identification. Aebersold [67] proposed going from discovery using MS to "browsing" using unique chemically-synthesized peptides tagged with heavy isotope for each gene and even each protein isoform. This standard peptide mixture could be combined with specimen fractions on sample plates for MS. The double peaks would be examined with precise differential mass determination, using an ordered peptide array. This method would combine quantitation with identification, but the limits of dynamic range would persist.

In closing, the PPP Executive Committee expresses its appreciation to all the investigators and their associates, to the Technical Committee members, and to the government and corporate sponsors who have contributed greatly to the progress of the HUPO Plasma Proteome Project.

The HUPO Plasma Proteome Project received funding support under a trans-NIH grant supplement 84942 administered by the National Cancer Institute with participation from the National Institutes of Aging, Alcohol & Alcohol Abuse, Cancer (Prevention and Treatment Divisions), Diabetes, Digestive & Kidney Diseases, Neurological Diseases & Stroke, and Environmental Health Sciences. The Michigan Core had support from the Michigan Life Sciences Corridor grant MEDC-238. Corporate sponsors/partners provided funding, technology, specimens, datasets, and/or technical advice; we thank Johnson & Johnson, Pfizer, Abbott Laboratories, Novartis, Invitrogen, Procter & Gamble, BD Biosciences, Ciphergen, Agilent, Amersham, Bristol Myers Squibb, DadeBehring, Molecular Staging, Sigma-Aldrich, and BioVisioN.

1.6 References

[1] Omenn, G. S., *Proteomics* 2004, 4, 1235–1240.

[2] Mann, M., Aebersold, R., *Nature* 2003, 422, 198–207.

[3] Hanash, S., *Drug Discov. Today* 2003, 7, 797–801.

[4] Hanash, S. M., Celis, J. E., *Mol. Cell. Proteomics* 2002, 1, 413–414.

[5] Li, X., Gong, Y., Wang, Y., Wu, S. et al., *Proteomics* 2005, 5, DOI: 10.1002/pmic.200400425.

[6] He, P., He, H-Z., Dai, J., Wang, Y. et al., *Proteomics* 2005, 5, DOI: 10.1002/pmic.200400422.

[7] Adamski, M., Blackwell, T., Menon, R., Martens, L. et al., *Proteomics* 2005, 5, DOI: 10.1002/pmic.200500186.

[8] Adamski, M., States, D. J., Omenn, G. S., Data Standardization and Integration in Collaborative Proteomics Studies, In: Srivastava, S. (Ed) *Informatics in Proteomics*. New York, Marcel Dekker, 2004, Chapter 8, pp. 169–194.

[9] Carr, S., Aebersold, R., Baldwin, M., Burlingame, A. et al., *Mol. Cell. Proteomics* 2004, 3, 351–353.

[10] Kersey, P. J., Duarte, J., Williams, A., Karavidopoulou, Y. et al., *Proteomics* 2004, 4, 1985–1988; http://www.ebi.ac.uk/IPI/IPIhelp.html.

[11] Kapp, E. A., Schutz, F., Connolly, L. M., Chakel, J. A. et al., *Proteomics* 2005, 5, DOI: 10.1002/pmic.200500126.

[12] Adkins, J. N., Monroe, M. E., Auberry, K. J., Shen, Y. et al., *Proteomics* 2005, 5, DOI: 10.1002/pmic.200400633.

[13] Olsen, J. V., Mann, M., *Proc. Natl. Acad. Sci. USA* 2004, 101, 13417–13422.

[14] Tang, H-Y., Ali-Khan, N., Echan, L. A., Levenkova, N. et al., *Proteomics* 2005, 5, DOI: 10.1002/pmic.200401099.

[15] Nesvizhskii, A. I., Keller, A., Kolker, E., Aebersold, R., *Anal. Chem.* 2003, 75, 4646–4658.

[16] Sadygov, R. G., Liu, H., Yates, J. R., *Anal. Chem.* 2004, 76, 1664–1671.

[17] Peng, J., Elias, J. E., Thoreen, C. C., Licklider, L. J. et al., *J. Proteome Res.* 2003, 2, 43–50.

[18] Tabb, D. L., Saraf, A., Yates, J. R., III, *Anal. Chem.* 2003, 75, 6415–6421.

[19] Keller, A., Purvine, S., Nesvizhskii, A. I., Stolyar, S. et al., *OMICS* 2002, 6, 207–212.

[20] Sadygov, R. G., Yates, J. R., III, *Anal. Chem.* 2003, 75, 3792–3798.

[21] MacCoss, M. J., Wu, C. C., Yates, J. R., *Anal. Chem.* 2002, 74, 5593–5599.

[22] Haab, B. B., Geierstanger, B. H., Michailidis, G., Vitzthum, F. et al., *Proteomics* 2005, 5, DOI: 10.1002/pmic.200400470.

[23] Rai, A. J., Gelfand, C. A., Haywood, B. C., Warunek, D. et al., *Proteomics* 2005, *5*, DOI: 10.1002/pmic.200400537.

[24] Tammen, H., Schulte, I., Hess, R., Menzel, C. et al., *Proteomics* 2005, *5*, DOI: 10.1002/pmic.200400419.

[25] Misek, D. E., Kuick, R., Wang, H., Galchev, V. et al., *Proteomics* 2005, *5*, DOI: 10.1002/pmic.200500103.

[26] Echan, L. A., Tang, H.-Y., Ali-Khan, N., Lee, K., Speicher, D. W., *Proteomics* 2005, *5*, DOI: 10.1002/pmic.200400518.

[27] Zolotarjova, N., Martosella, J., Nicol, G., Bailey, J. et al., *Proteomics* 2005, *5*, DOI: 10.1002/pmic.200402021.

[28] Huang, L., Harvie, G., Feitelson, J. S., Herold, D. A. et al., *Proteomics* 2005, *5*, DOI: 10.1002/pmic.200400420.

[29] Yang, Z., Hancock, W. S., Richmond-Chew, T., Bonilla, L., *Proteomics* 2005, *5*, DOI: 10.1002/pmic.200400411.

[30] Barnea, E., Sorkin, R., Ziv, T., Beer, I., Admon, A., *Proteomics* 2005, *5*, DOI: 10.1002/pmic.200400412.

[31] Moritz, R. L., Clippingdale, A. B., Kapp, E. A., Eddes, J. S. et al., *Proteomics* 2005, *5*, DOI: 10.1002/pmic.200500096.

[32] Cho, S. Y., Lee, E.-Y., Chun, Y. W., Lee, J.-S. et al., *Proteomics* 2005, *5*, DOI: 10.1002/pmic.200400497.

[33] Kim, J. Y., Lee, J. H., Park, G. W., Cho, K. et al., *Proteomics* 2005, *5*, DOI: 10.1002/pmic.200400413.

[34] Zhang, H., Li, X. J., Martin, D. B., Aebersold, R., *Nat. Biotechnol.* 2003, *21*, 660–666.

[35] Zhou, M., Lucas, D. A., Chan, K. C., Issaq, H. J. et al., *Electrophoresis* 2004, *25*, 1289–1298.

[36] Vollmer, M., Horth, P., Nagele, E., *Anal. Chem.* 2004, *76*, 5180–5185.

[37] Marshall, J., Jankowski, A., Furesz, S., Kireeva, I. et al., *J. Proteome Res.* 2004, *3*, 364–382.

[38] Shen, Y., Jacobs, J. M., Camp, D. G., Fang, R. et al., *Anal. Chem.* 2004, *76*, 1134–1144.

[39] Qian, W. J., Liu, T., Monroe, M. E., Strittmatter, E. F. et al., *J. Proteome Res.* 2005, *4*, 53–62.

[40] Washburn, M. P., Wolters, D., Yates, J. R., *Nat. Biotechnol.* 2001, *19*, 242–248.

[41] Adkins, J. N., Varnum, S. M., Auberry, K. J., Moore, R. J. et al., *Mol. Cell. Proteomics* 2002, *1*, 947–952.

[42] Beer, I., Barnea, E., Ziv, T., Admon, A., *Proteomics* 2004, *4*, 950–960.

[43] Beer, I., Barnea, E., Admon, A., *Proteomics* 2005, *5*, DOI: 10.1002/pmic.200400457.

[44] Deutsch, E. W., Eng, J. K., Zhang, H., King, N. L. et al., *Proteomics* 2005, *5*, DOI: 10.1002/pmic.200500160.

[45] Keller, A., Nesvizhskii, A. I., Kolker, E., Aebersold, R., *Anal. Chem.* 2002, *74*, 5383–5392..

[46] Desiere, F., Deutsch, E. W., Nesvizhskii, A. I., Mallick, P. et al., *Genome Biol.* 2005, *6*, R9.

[47] Craig, R., Beavis, R. C., *Rapid Commun. Mass Spectrom.* 2003, *17*, 2310–2316.

[48] Martens, M., Nesvizhskii, A. I., Hermjakob, Adamski, M. et al., *Proteomics* 2005, *5*, DOI: 10.1002/pmic.200400376.

[49] Johnson, R. S., Davis, M. T., Taylor, J. A., Patterson, S. D., *Methods* 2005, *35*, 223–236.

[50] Anderson, N. L., Polanski, M., Pieper, R., Gatlin, T. et al., *Mol. Cell. Proteomics* 2004, *3*, 311–316.

[51] Pieper, R., Gatlin, C. L., Makusky, A. J., Russo, P. S. et al., *Proteomics* 2003, *3*, 1345–1364.

[52] Tirumalai, R. S., Chan, K. C., Prieto, D. A., Issaq, H. J. et al., *Mol. Cell. Proteomics* 2003, *2*, 1096–1103.

[53] Chan, K. C., Lucas, D. A., Hise, D., Schaefer, C. F. et al., *Clin. Proteomics* 2004, *1*, 101–225.

[54] Rose, K., Bougueleret, L., Baussant, T., Bohm, T. et al., *Proteomics* 2004, *4*, 2125–2150.

[55] Rai, A. J., Stemmer, P. M., Zhang, Z., Adam, B.-L. et al., *Proteomics* 2005, *5*, DOI: 10.1002/pmic.200400606.

[56] Ping, P., Vondriska, T. M., Creighton, C. J., Gandhi, T. K. B. et al., *Proteomics* 2005, *5*, DOI: 10.1002/pmic.200500140.

[57] Berhane, B., Zong, C., Liem, D. A., Huang, A. et al., *Proteomics* 2005, *5*, DOI: 10.1002/pmic.200401084.

[58] Muthusamy, B., Hanumanthu, G., Suresh, S., Rekha, B. et al., *Proteomics* 2005, *5*, DOI: 10.1002/pmic.200400588.

[59] Martens, M., Hermjakob, H., Jones, P., Adamski, M. et al., *Proteomics* 2005, *5*, DOI: 10.1002/pmic.200400647.

[60] Stein, L.D., *Nature* 2004, *431*, 915–916.

[61] Fenyö, D., Beavis, R.C., *Anal. Chem.* 2003, *75*, 768–774.

[62] Orchard, S., Hermjakob, H., Apweiler, R., *Mol. Cell. Proteomics* 2005, *4*, 435–440.

[63] Marko-Varga, G., Fehniger, T.E., *J. Proteome Res.* 2004, *3*, 167–178.

[64] Celis, J. E., Korc, M., *Mol. Cell. Proteomics* 2005, *4*, 345–593.

[65] de Hoog, C.L., Mann, M., *Annu. Rev. Genomics Hum. Genet.* 2004, *5*, 267–293.

[66] Humphery-Smith, I., *Proteomics* 2004, *4*, 2519–2521.

[67] Aebersold, R., *Nature* 2003, *422*, 115–116.

2
Data management and preliminary data analysis in the pilot phase of the HUPO Plasma Proteome Project*

Marcin Adamski, Thomas Blackwell, Rajasree Menon, Lennart Martens, Henning Hermjakob, Chris Taylor, Gilbert S. Omenn and David J. States

The pilot phase of the HUPO Plasma Proteome Project (PPP) is an international collaboration to catalog the protein composition of human blood plasma and serum by analyzing standardized aliquots of reference serum and plasma specimens using a variety of experimental techniques. Data management for this project included collection, integration, analysis, and dissemination of findings from participating organizations world-wide. Accomplishing this task required a communication and coordination infrastructure specific enough to support meaningful integration of results from all participants, but flexible enough to react to changing requirements and new insights gained during the course of the project and to allow participants with varying informatics capabilities to contribute. Challenges included integrating heterogeneous data, reducing redundant information to minimal identification sets, and data annotation. Our data integration workflow assembles a minimal and representative set of protein identifications, which account for the contributed data. It accommodates incomplete concordance of results from different laboratories, ambiguity and redundancy in contributed identifications, and redundancy in the protein sequence databases. Recommendations of the PPP for future large-scale proteomics endeavors are described.

2.1
Introduction

Data management was one of the key elements in the pilot phase of the HUPO Plasma Proteome Project (PPP). Data submission and collection approaches were defined collaboratively by the Bioinformatics and Technologies Committees, and were extensively discussed at the PPP Workshop in Bethesda, USA in July 2003 [1].

---

* Originally published in Proteomics 2005, 13, 3246–3261

Exploring the Human Plasma Proteome. Edited by Gilbert S. Omenn
Copyright © 2006 WILEY-VCH Verlag GmbH & Co. KGaA, Weinheim
ISBN: 3-527-31757-0

Ideally, experimental methods and the data generated by their execution would be fully described in a thoroughly decomposed manner, facilitating sophisticated searches and analyses. However, when dealing with the results from real experiments multiple compromises must be made. The first concerns the level of detail that can be requested: while it is, in principle, desirable to have all methodological steps, parameters, data, and analyses described in full detail, many laboratories lack automated laboratory information management systems and manual record keeping is laborious, limiting the granularity of information that can be captured. The second compromise concerns the degree to which experimental reports will be decomposed and structured by the submitter: from a long run of free text as in a journal paper to a fully annotated list of all the relevant items of information, arranged in an elaborate and well-specified hierarchy that captures the interrelationships of those items. It is notoriously difficult to automatically extract even the simplest information from free text [2, 3]. However, thoroughly classifying information for submission is burdensome. Indeed, developing standards, data definitions, forms or submission tools, and the associated documentation and training material is a substantial task. Third, the pilot phase of the PPP was designed to encourage individual laboratories to push the limits of their technologies to detect and identify low-abundance proteins; the Technology Committee was not able to define in advance all the parameters that emerged as desirable inputs for analysis in this broad, largely voluntary collaboration. The fourth compromise concerns the design and implementation of the data systems used for storage of the data at the central repository. It is desirable to retain as close a link as possible to the original submissions from the participating laboratories in the central repository, but this implies that the details of which data sets superseded earlier submissions, exceptions encountered in the data loading, and other detailed information on submission processing need to be encoded in subsequent queries, complicating the task of writing and debugging software to analyze the data.

Finally, a compromise at the level of the overall project relates to the choice of sequence database used for analysis and whether to "freeze" on a particular release of the sequence database. The results of protein identification by search of mass spectra against a database are necessarily dependent on the database being searched. Freezing on a particular protein sequence database release not only facilitates comparison of identification data sets but also prevents corrections and revisions to the protein sequence collection from being incorporated into the identification process. Further, freezing on a particular protein sequence database release complicates the task of linking the findings of the current study to evolving knowledge of the human genome and its annotation, because many of the entries in the protein sequence database available at the initiation of the project have been revised, replaced, or withdrawn over the course of this project, and continue to be revised.

The major aim of the pilot phase of the HUPO PPP was the comparison of protein identifications made from multiple reference specimens by all participating laboratories. An additional important aim was the development of an efficient method of data acquisition, storage, and analysis in such a big collaborative proteomics experiment [3]. Here we describe the data management system developed during the pilot phase of the HUPO PPP.

2.2
Materials and methods

2.2.1
Development of the data model

To encourage participation by laboratories, the data model focused on identifications of whole proteins as a high-level, concise description of experimental results, requiring a minimum of data input, transmission, and potential reformatting. The guidance specified the collection of the protein accession numbers and names, binary descriptions of the confidence of the protein identifications (high or lower), lists of identified peptides, and free text descriptions of experimental protocols. Analysis of the preliminary results brought to the fore a major problem with a data integration and validation process based exclusively on protein accession numbers. Participating laboratories used not only different search databases but also different algorithms to assemble protein identifications from their database search results. Additionally, the estimation of confidence of the identification, based on search scores and laboratory binary judgment, was inconsistent. To address these problems, the original data model was enhanced to include the peak lists used to obtain protein identifications, and raw spectra in the instrument native format.

The expanded data model is generally in concert with recently proposed guidelines for publication of protein and peptide identification data [4]. Since our studies were started before publication of these guidelines, our data collecting decisions do not reflect all of the requirements proposed by Carr *et al.* [4] Tab. 1 compares the guidance proposed in [4] with the information collected in the present study. The HUPO PPP data model consists of the following main objects:

2.2.1.1 Laboratory
Information about the participating laboratories, such as principal investigator, contact person, postal and email addresses, identifiers, descriptions, *etc.*

2.2.1.2 Experimental protocol
Free text descriptions sufficiently detailed to allow the work to be reproduced. The level of experimental detail was specified to be sufficient for the protocol to be considered for publication in Proteomics or the Journal of Biological Chemistry.

2.2.1.3 Protein identification data set
The identified protein accession numbers, names, search database and version, sequences of the identified peptides, and an estimate of confidence for each protein identification, plus any supporting information about PTMs (from experimental measurements, or other sources), and estimates of relative protein abundance in the specimen. Identification data sets were stored as peptide lists, reflecting the fact

Tab. 1 Comparison of the HUPO PPP data model with guidance for publishing peptide and protein identification data by Carr et al. [4]

Guideline proposed by Carr et al.[4]	HUPO PPP data model
1. Supporting information	
The method and/or program used to create the "peak list" from raw data and the parameters used in the creation of this peak list.	Data were collected as a part of free text description of performed experiments. Recommendation to use PEDRO tool was moot, since tool was not ready for use.
The name and version of the program(s) used for database searching and specific parameters used for its (their) operation.	Name of the search program collected, but not version or operation parameters.
Scores used to interpret MS/MS data and thresholds and values specific to judging certainty of identification, whether any statistical analysis was applied to validate the results, and a description of how it was applied.	Scores and thresholds were collected.
The name and version of sequence database used; the count of number of protein entries in it at the time searched.	Both name and version of the sequence database were collected. The sequence database itself was also recorded.
2. Information regarding the observed sequence coverage	
Table that lists for each protein the sequences of all identified peptides.	Peptides (sequences) identified for each protein were collected.
To calculate the sequence coverage different forms of the same peptide are to be counted as only a single peptide.	All forms of identified peptides were collected, but as long as they have the same amino acid sequence they were counted only once.
The total number of MS/MS-interpreted spectra assigned to peptides corresponding to each protein.	Raw spectra were collected.
3. Protein assignments based on single-peptide assignments	
The sequence of the peptide used to make each such assignment, together with the amino acids N- and C-terminals to that peptide's sequence.	Sequence of the peptide was collected but not the terminal information.
The precursor mass and charge.	The precursor charge state was collected as a part of the peptide data. The mass was requested as part of the peak list information.
The scores for this peptide.	Scores were collected.
4. Biological conclusions based on observation of a single peptide matching to a protein	
Such conclusions must be supported by inclusion of the corresponding MS/MS spectrum.	Raw spectra were requested for all the MS/MS identifications (including single peptide).
5. Peptide mass fingerprint data	
In addition to listing the number of masses matched to the identified protein, authors should also state the number of masses not matched in the spectrum and the sequence coverage observed.	Only peptides matched to the identified protein were collected. Sequence coverage was calculated.

Tab. 1 Continued

Guideline proposed by Carr et al.[4]	HUPO PPP data model
Parameters and thresholds used to analyze the data.	Data collected only as a part of free text description of performed experiments. No particular information was requested.
6. Ambiguous protein identifications	
The same protein appears in many cases under different names and accession numbers in the database. When matching peptides to members of such a family, it is the authors' responsibility to demonstrate that they are aware of the problem and have taken reasonable measures to eliminate redundancy. In cases where a single-protein member of a multiprotein family has been singled out, the authors should explain how the other members of the group were ruled out.	A data integration workflow was specially designed to address this problem. It is described in the following sections.
7. Submission of MS/MS spectra	
Submission of all MS/MS spectra mentioned in the paper as supplemental material. The dta, pkl, and mgf files are accepted.	Raw spectra in the instrument native format were collected and are available on request. They may be converted to the other formats with use of special software.

that some laboratories applied significant protein fractionation prior to tryptic digest and mass spectral analysis. In a pure "bottom up" strategy, any protein can contribute any peptide and no information is gained by retaining group structure for peptides. However, when protein fractionation is used, knowledge that a group of peptides were all derived from the same protein fraction can enhance the power of identification.

2.2.1.4 Peak list
Lists of mass over charge peaks used by search engines for protein identifications. The peak lists were accompanied by amino acid modifications catalogs, lists of all modified residues, including the symbol for and mass of the modified residue, and the type of modification.

2.2.1.5 Summary of technologies and resources
This included estimates of the time, capital, and operating costs of the analyses.

2.2.1.6 MS/MS spectra
The unprocessed data from spectrometers.

2.2.1.7 SELDI peak list
Peak lists from direct MS/SELDI experiments (registered for a separate analysis; see Rai *et al.*, this issue).

2.2.2
Data submission process

The data submission strategy was designed to make the submission process simple for the participants and at the same time error-proof and relatively easy to process for the data collection and integration center. As stated above, the consensus data model of the PPP pilot phase included only a limited representation of methods and results, to minimize the time commitment for participating experimentalists. Two methods for submitting were offered: (a) a combination of Microsoft Excel™, Microsoft Word™, and text forms, or (b) an XML (http://www.w3.org/XML) schema-based file format (PEDRO [5, 6]). Those who chose the form-based submission were asked to fill out a set of preformatted Excel/Word/text document templates, and submit them online using a web-based submission server at the University of Michigan. Those who chose the XML format were asked to email their submissions to the European Bioinformatics Institute, after generating one or more XML documents using the provided XML schema. The schema of the XML document allowed for the collection of all the information in one, hierarchically organized file. To generate the XML documents the participants were encouraged to use the PEDRO data entry tool [6], or to export XML directly from their existing LIMS system. The XML documents were checked for compliance with the schema and forwarded to the University of Michigan for further processing.

During the course of the project, we decided to request the raw MS/MS spectra in the form of instrument files in spectrometer native format. The size of these files, sometimes in excess of several gigabytes, did not allow for their collection by the standard data submission route; instead, CD or DVD disks were submitted to the University of Michigan Core and distributed to three groups for special cross data set analyses (see Omenn *et al.*, Kapp *et al.*, and Beer *et al.*, this issue).

At the beginning of the project each participating laboratory received two distinct identifiers: the first, a numeric public identifier used for interactions with the submission centers and other laboratories, and the second, a three-character private code known only to the laboratory and the central data analysis group. These private identifiers were used to create data surveys without disclosing the identity of submitters.

2.2.3
Design of the data repository

The project data repository was built with a Structured Query Language (SQL) relational database server. The data structure was divided into two main parts: (1) an intermediate structure presenting an exact copy of the data from documents submitted by the project participants, to make the data available for further processing, and for checking correctness of the submitted documents; and (2) the main data structure designed to hold the integrated project data.

The structure can be divided into four main sections: (1) experiment description, (2) protein identifications made by data producers from peptide sequences, (3) MS/MS peak lists, and (4) protein identifications from database searches made by groups other than the data producers.

In the database design (Fig. 1), experiments performed by the project participants are stored in the entity Experiment. This entity is referenced directly by the entity Laboratory and by a set of look-up entities: Specimen, Depletion, SeparationProtein, ReductionAlkylation, SeparationPeptide, and MassSpec. Experiment also has a many-to-many relationship with a free text protocol description (entities Protocol and ExperimentProtocol). At the experiment level the database structure branches into two sections. The first section started by the entity IdentificationSet stores protein identifications submitted by the participants. The second section started by the entity MsRun stores MS peak lists and the results from their analysis. The two-branched database structure reflects the changes in the project data collection model, from identification-oriented at the beginning to a more fine-grained description utilized later.

The database can capture three sets of protein identifiers from the same experiment. The first set stores protein identifications made by data producers in the entity Identification. The second set stores the results of peptide list searches done by the data integration center, in the entity ProteinByPeptides. This set captures peptide group information. The third set of identifiers (multiple subsets of these identifiers are possible) is derived from the same experimental results, but this time by an analytical group other than the data producer, through the MsRun branch of the database (entities MsRun, MzPeak, and ProteinByMsSearch).

The main project database does not store SELDI peak lists or MS/MS raw spectra. These data are available as downloadable files.

2.2.4
Receipt of the data

The data documents were uploaded using a web-based submission site established at the University of Michigan. During submission each document received a unique ID number used subsequently by the document tracking and transforming mechanism. The XML documents submitted by email were processed separately. Data from the received documents were transferred to an intermediate database. The transfer was done automatically for each web-submitted document and separately for the emailed XML submissions. The data in the intermediate structure represent an exact copy of the data from the original documents, without any transformation or integration. The intermediate database allows checking the correctness of the structure of the submitted documents and makes the data available for the integration procedures. Verified data were then rewritten using a consistent format for protein accession numbers, database names, peptide sequences, peak lists, and experimental categories.

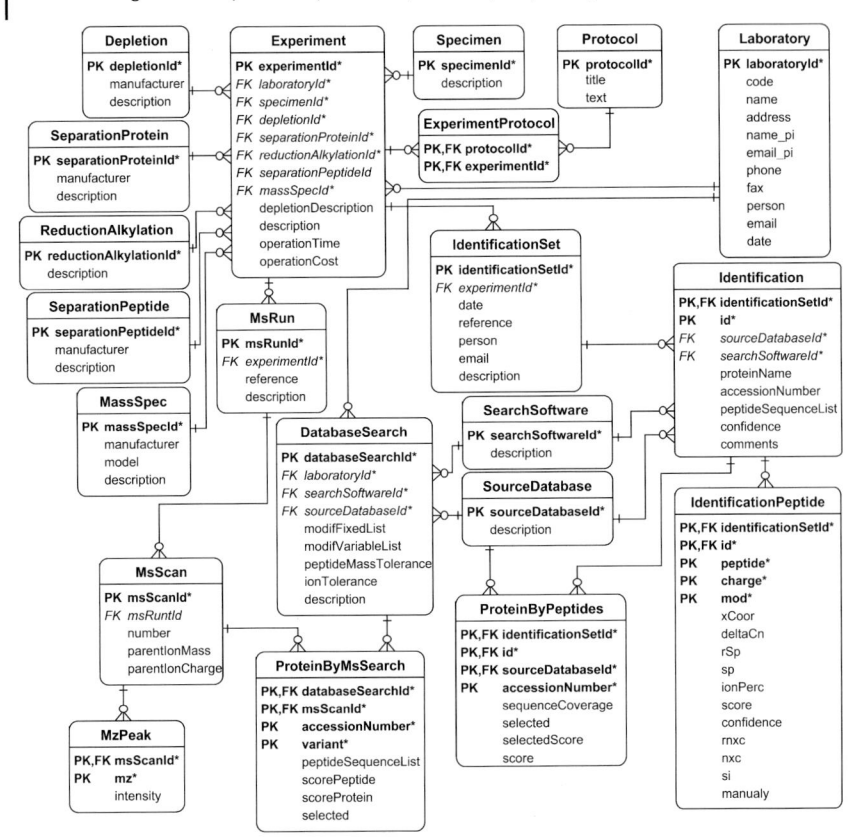

Fig. 1 Entity-relationship diagram of the HUPO PPP data repository. Boxes symbolize entities or tables; connecting lines represent relations between the entities.

2.3
Inference from peptide level to protein level

In the pilot phase of the HUPO PPP, proteins were identified by MS experiments, followed by searches of protein databases to find peptide sequences matching observed spectra. Often, such a search returns a cluster of proteins, all of which contain the same set of matching peptides. Problems with ambiguity of protein identifications obtained from searches of tandem mass spectra and methods for managing them have been widely discussed, *e.g.*, by Nesvizhskii *et al.* [7] and Sadygov *et al.* [8]. In these earlier works, protein identifications were inferred from lists of assigned peptides accompanied by probabilities that those assignments are correct. In the present report, however, we integrated lists of peptides obtained using several different search algorithms and different search databases, which frequently lacked identification probabilities. Although during the course of the project, participants were asked to additionally submit peptide and protein identification probabilities or scores, as well as peak lists and raw MS spectra, the main part of integrating the results was based solely on the sequences of the submitted peptides. The raw spectra and peak lists were subject to separate analysis and will be described elsewhere.

The integration workflow we describe here benefits from the collaborative character of the studies and is based on a heuristic approach that assumes that the proteins most likely to be truly present in the sample are those supported by the largest number of maximally independent experiments. The workflow additionally takes into account the "level of annotation" of the protein, thus preferentially selecting the proteins with the most extensive description available.

The workflow algorithm includes several consecutive steps:

(1) Assemble peptide sequence lists: Protein identifications submitted by the participating laboratories were accompanied by lists of sequences of matched peptides. All the lists were collected to form a set of distinct peptide sequence lists. Each list in that set preserves all references to its origin, *e.g.*, if a particular list is reported from more than one experiment, it has more than one reference.

(2) Search the peptide lists: Each peptide sequence list obtained in the previous step was subsequently searched against the IPI version 2.21 (July 2003) database [9]. This was selected as the standard database of the project. Each match requires 100% identity between sequences and disregards flanking residues.

(3) Select one representative protein from each cluster of equivalent protein hits: Often, more than one entry in the reference protein database matches all of the components of a peptide sequence list. We call this set of matching entries a "cluster of equivalent protein hits" for that peptide sequence list. The clusters for different lists may overlap. When they do, we wish to choose one protein entry from the intersection of several clusters to represent all proteins in each of the overlapping clusters, that is, the proteins identified by each of the associated peptide sequence lists. The selection is done as follows.

Each protein entry in the reference database receives three integer scores:

(a) The number of different laboratories reporting a peptide sequence list whose cluster includes this protein.

(b) The number of distinct experiments (laboratories × specimens × protocols) reporting a peptide sequence list whose cluster includes this protein.

(c) The number of identifications (laboratories × specimens × protocols × clusters) for clusters including this protein. For each peptide sequence list, the cluster member with the largest value of score (a) is chosen as the representative protein entry. Scores (b) and (c), followed by criteria (d–g) listed below, are applied in succession to break numeric ties at higher levels.

(d) Well-described protein – product of a well-described gene. The EnsEMBL gene model was used for the annotation. The "well-described" proteins and genes are those with a nonempty description line, and without words like "fragment", "similar to", "hypothetical", "putative", *etc.* in their description.

(e) Well-described protein-product of any gene.

(f) Well-described protein not assigned to any gene.

(g) Protein not assigned to any gene and described as a fragment, by its similarity to another protein, or with no IPI description line at all. Any remaining ties are broken by selecting the protein having the lower IPI number.

As a result, one protein will generally be chosen as the representative entry from several overlapping clusters of equivalent protein identifications. This simplifies later comparisons between laboratories and experiments. This particular choice for a representative protein is motivated by the idea that the protein whose identification is supported by the largest number of independent experiments is the protein most likely to be actually present in the specimen. Score (a) counts each laboratory only once, no matter from how many specimens or with how many different peptide sequence lists the laboratory identified this protein. Next in importance, score (b) counts the number of independent experiments in which the protein was identified. Score (c) counts all reported peptide sequence lists, even if several results are from the same experiment. Criteria (d–g) indicate the level of annotation for each database entry. They facilitate selection of the best-described proteins.

2.4
Summary of contributed data

Laboratories participating in the project submitted a total of 12 667 distinct protein accession numbers. This number includes 11 253 accession numbers from MS/MS – both MALDI and LC-ESI, and an additional 1414 IDs from FT-ICR-MS. FT-ICR-MS identified 2230 proteins, but 816 were also identified by the MS/MS technologies. In addition, participating laboratories contributed 653 identifications from MALDI-MS peptide mass fingerprints. These data were analyzed separately and will be reported elsewhere.

The majority of reported protein identifications from the MS/MS and FT-ICR-MS experiments (11 960 of 12 667 – 94%) were obtained by searching the tandem mass spectra against the IPI database. The remaining 6% were generated using either the Swiss-Prot or NCBInr databases (Tab. 2). Almost all of the submitted peptide sequence lists (12 388 of 12 667 – 98%) were matched in the standard database for the project, *i.e.*, IPI version 2.21. The 2% of peptide sequence lists for which no exact match was found in this database most likely represent up to 5% mismatch between database entries, which is permitted when constructing the IPI database (see [9]). We believe that the submitting laboratory searched one of the source databases for IPI, rather than IPI itself, and matched the spectrum to a source entry which is included in IPI as a secondary rather than a master entry.

The 12 388 reported identifications with peptides matching the IPI 2.21 database correspond to 18 098 distinct peptide sequence lists. Searching these lists against IPI 2.21 results in 15 710 matching entries. For each of 12 303 of these lists (68%), exactly one of 6601 IPI entries was matched. These were reported with 7000 different protein accession numbers, including Swiss-Prot and NCBI identifiers. The 6% reduction from 7000 to 6601 distinct identifiers comes from converting Swiss-

Tab. 2 Usage of the search databases

Category	Search database			
	IPI	Swiss-Prot	NCBInr	All three
Submitted protein identifications	11 960	199	508	12 667
Submitted identifications with peptide sequence lists found in IPI database	11 741 98%	196 98%	451 89%	12 388 98%
Entries in IPI database matching submitted peptide sequence lists	15 463	488	552	15 710
Average number of IPI entries *per* submitted protein identification	1.3	2.5	1.2	1.3

Tab. 3 Effectiveness of the integration process

Category	Number of IPI entries matching single-peptide sequence list		
	One (distinct IDs)	More than one (indistinct IDs)	One or more (all IDs)
Submitted peptide sequence lists	12 303	5795	18 098
Submitted protein accession numbers	7000	5388	12 388
Matching entries in IPI database	6601	9668	15 710
Matching entries in IPI database after the integration	6601	3273	9506
Reduction level of submitted accession numbers to IPI entries	6%	39%	23%

Prot and NCBI identifiers to IPI identifiers. As these identifications are already unique, the integration workflow did not additionally reduce these 6601 accession numbers.

In the remaining 5795 (32%) cases, each peptide sequence list matches more than one IPI protein sequence, resulting in an ambiguous identification or a cluster of equivalent hits (Tab. 3). In this group of ambiguous identifications, searches of the 5795 peptide sequence lists return 9668 distinct IPI protein accession numbers. The integration workflow reduces this group to a set of 3273 distinct proteins, which explain the presence of all reported peptides. In the next step, the 6601 accession numbers from the group of uniquely identified proteins are combined with the 3273 accession numbers from the group of ambiguous identifications. Of the resulting 9874 identifications, 9506 represent distinct accession numbers.

Details of the integration process for the 5795 clusters of ambiguous hits are presented in Tab. 4. Scores (a–c) evaluate the level of confirmation of each protein identification by the number of completely independent experiments.

Tab. 4 Number of clusters qualified on different levels of the integration

Integration level		Number of clusters	
A	Number of laboratories	1680	2288
B	Number of experiments	419	
C	Number of reports	189	
D	Well-described EnsEMBL gene	2429	3507
E	Any EnsEMBL gene reference	99	
F	No EnEMBL reference	286	
G	Poorly described protein	693	
	Total number of potentially ambiguous peptide sequence lists processed		5795

Tab. 5 Distribution of numbers of entries from the HUPO PPP and complete IPI databases in the integration categories

Integration category	Complete IPI database		HUPO PPP database		
	No. of entries	Fraction of all entries	No. of proteins	Fraction of all identifications	Fraction of IPI entries
D	13 588	24%	3900	41%	29%
E	855	2%	220	2%	26%
F	6633	12%	716	8%	11%
G	35 454	63%	4670	49%	13%
All	56 530		9506		17%

In 2044 (35%) of the cases, the decision of protein selection was done on the basis of the score (a): selecting a protein detected by the largest number of laboratories. In 1680 (82%) of those cases it was a single protein, and no additional selection step was required. In the remaining 18% of the cases, selection by score (a) returned more than one protein. The tie was then broken using additional scoring categories (d–g). In 2966 (51%) of the cases, all proteins in the cluster were indistinguishable using scores (a–c) and the decisions were made exclusive using categories (d–g).

The categories (d–g) classify IPI database entries by the amount of detail in their description. It is then reasonable to compare such a classification of proteins in the project database with the same classification of proteins in the complete IPI database. Details of this comparison are given in Tab. 5. This shows that 41% of entries from the HUPO PPP database and 24% of the entries from the IPI database belong to the highest category (d) – the best-described proteins. The intermediate categories (e) and (f) include relatively few proteins while category (g) – the least described proteins – contains the majority of the entries, 49 and 63% for the HUPO PPP and IPI databases, respectively. For the HUPO PPP database, the ratio between the percentage of entries from categories (d) and (g) is 41/49% = 0.84. This ratio for the IPI database is 24/63% = 0.67. Thus, the laboratories were more likely to identify better-described proteins. This result can be interpreted as confirming the presence of proteins that were previously studied in detail, possibly because of their relative abundance or ease of identification. Alternatively, the integration workflow itself preferred the best-described proteins wherever possible, pushing the ratio toward category (d).

To further compare results from the HUPO PPP with all the proteins from IPI, we compared the distributions of peptide sequence length (number of amino acid residues *per* peptide) in both data sets (Fig. 2). The distribution of peptide length from the HUPO PPP database is noticeably shifted toward longer peptides – median equal to 12.9 residues – in comparison to the distribution of the lengths of tryptic peptides in IPI-median equal to 10.5 residues. We hypothesize that the under-representation of short peptides may be explained by the nature of the tandem mass spectrum search algorithms which require the spectra from short peptides to be of much better quality than spectra from longer peptides, to result in a significant match. Many laboratories did not report any peptides shorter than five residues. The fraction of nontryptic peptides in each peptide length bin is very small. These peptides were identified in a few nonenzyme-specific database searches and, as they passed quality control in the participating laboratories, they were included in our analysis. The origin of these peptides is not analyzed in this paper, but we speculate that they may be products of other endogenous proteases present in the tissue of origin or in human plasma [10].

Based on the nonuniform reporting of short peptides from participating laboratories, the limited spectral data available for short peptides, and the limited power for protein identification using a peptide present in multiple protein sequences, we decided to eliminate peptides shorter than six residues from further analysis. In doing so, we disregarded two protein identifications, each based on a single peptide of five amino acids. This reduces the number of accepted protein identifications from 9506 to 9504 accession numbers.

2.4.1
Cross-laboratory comparison, confidence of the identifications

The distribution of the number of protein identifications among participating laboratories is shown in Fig. 3. Individual laboratories are encoded using their numeric identifiers. The 18 laboratories identified a total of 9504 distinct IPI proteins. The number identified by individual laboratories varied from 52 to 4569. The laboratories were asked to mark as "high confidence" those identifications that passed more stringent criteria, chosen by each laboratory individually, although the PPP did issue guidance after the June 2004 Jamboree Workshop for SEQUEST searches to use $X_{corr} \geq 1.9$, 2.2, 3.75 for 1+, 2+, and 3+ ions, respectively, plus $\Delta C_n \geq 0.1$ and $R_{Sp} \leq 4$ for tryptic peptides. The number of these lab-reported high-confidence identifications ranged from 21 to 789. To further assess the confidence of protein identifications from individual laboratories, we counted the number of proteins, which were also reported by a second laboratory. We considered such

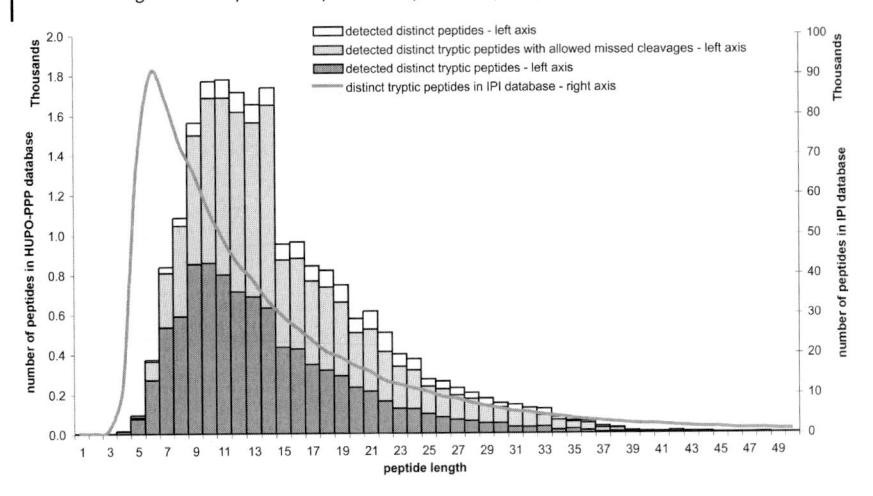

Fig. 2 Comparison of distributions of length of tryptic peptides (dark gray bars), tryptic peptides with missed cleavages allowed (light gray bars), and all peptides, including nontryptic peptides (white bars) detected in the course of the project using MS/MS (both MALDI and LC) and FT-ICR-MS methods, to the distribution of the length of tryptic peptides from the complete IPI database (gray line).

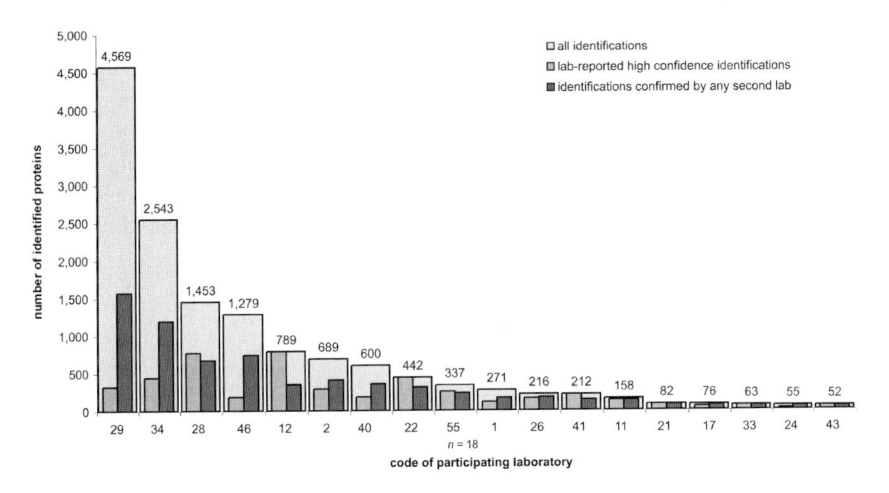

Fig. 3 Distribution of MS/MS and FT-ICR-MS protein identifications among 18 participating laboratories, encoded using their numeric identifiers.

identifications to be confirmed. The fraction of confirmed identifications is higher for laboratories, which submitted lower numbers of proteins. This may be caused by several factors including the followings. (1) Different stringencies for acceptance of the identifications – smaller sets may mean that more stringent criteria have been used and the resulting proteins are more likely to be true identifications. (2) Differences in experimental techniques – smaller sets of proteins may be obtained by shallower sampling, picking up only the more abundant, *i.e.*, more frequently identified proteins. (3) The intrinsic nature of the confirmation process – the more sensitive the procedures used by a particular laboratory are, the more likely it is that it will be the only laboratory reporting a particular identification. Thus, the requirement for confirmation penalizes the laboratories that submitted the largest data sets.

The level of cross-laboratory confirmation of the identifications, as a function of the number of peptides detected across experiments and laboratories, is shown in Fig. 4. The first category – all identifications – has a confirmation level equal to 25%. The second category, resulting from elimination of single-peptide identifications, dramatically reduces the number of proteins from the original 9504 to 3020, and at the same time raises the confirmation level to 75%. The absolute number of confirmed identifications in these two categories is virtually the same, meaning that of 6484 single-peptide protein identifications almost none was confirmed. Limiting the identifications to those which are supported by an even larger number of peptides causes a further reduction in the number of proteins and a rise in the confirmation level.

The analysis described above led us to categorize protein identifications into four classes, based on the level of the identification confidence. The four categories are organized in a diamond-shaped parallelogram (Fig. 5). Identifications from the least stringent category – "all identifications" (9504 proteins) – are divided into two more stringent, parallel categories: "high-confidence identifications" (2857 proteins), including proteins reported at least once as high-confidence, and "multi-peptide identifications" (3020 proteins), including proteins for which two or more distinct peptides were reported project wide, following data integration. The most stringent category "high-confidence multipeptide identifications" (1555 proteins) includes proteins from the intersection of the preceding categories. Proteins in this category are identified with two or more distinct peptides, requiring at least one to have been reported as part of a high-confidence protein identification.

2.5
False-positive identifications

False-positive peptide identifications exist and are widely acknowledged to be a problem [7, 8, 11–15]. One arises whenever the top-scoring database match for a particular spectrum has a score which passes all reporting thresholds, yet the matched database sequence is not the same as that of the biological specimen in the instrument. This will occur for a variety of reasons. The spectrum may represent a mixture of different peptides with almost equal parent masses and elution times. The biological specimen may be a contaminant or an allelic variant not recorded in the database being searched. Even if the database contains the correct amino acid sequence, this sequence may fall outside the scope of the search, due to PTMs or requirements for proteolytic cleavage. In each of these cases, the top-scoring match

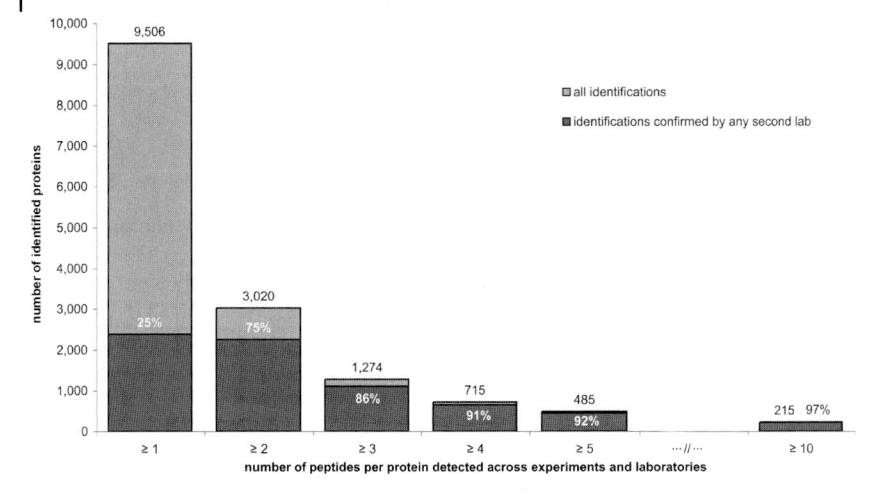

Fig. 4 Distribution of MS/MS and FT-ICR-MS protein identifications as a function of the number of peptides detected *per* protein.

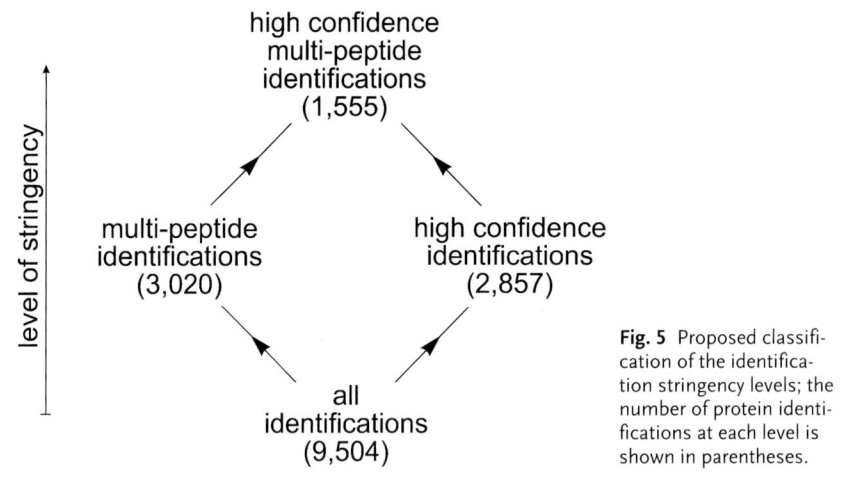

Fig. 5 Proposed classification of the identification stringency levels; the number of protein identifications at each level is shown in parentheses.

within scope and within the database is returned by the search software. If its score passes reporting thresholds, the (mis)match will be accepted and reported as a peptide identification.

False-positive and true-positive peptide identifications show opposite behavior when we accumulate large numbers of peptide identifications, as in this project [7, 11]. One expects false-positive peptide identifications to accumulate roughly proportional to total peptide identifications. However, the chance that two or more false-positive peptide identifications coincide on the same database entry should be no better than random. On the contrary, a protein which is present at a detectable concentration in the specimen will produce many tryptic peptides in nearly stoichiometric quantities. Increased sampling should increase the number of distinct peptides, which are reported, and all of these will map to the same (correct) database entry. This means that, as we accumulate more and more peptide identifications, the class of protein identifications based on a single peptide reported project wide is simultaneously depleted of correct peptide identifications (as these are promoted to multiple-peptide protein identifications) and refilled with false-positive protein identifications. Below, we consider a range of values for the fraction of such false-positive identifications. One major participating HUPO laboratory, after manually reviewing several hundred of their protein identifications, concluded that a single peptide constituted sufficient evidence in perhaps 20% of the cases where only one peptide from a protein had been seen. The acceptance rate after manual review was much larger for proteins identified using two or three peptides, precisely because of the selection described above. Manual review of all the spectra was not feasible, and all of their identifications were submitted to the database.

To assess the confidence of protein identifications, we use a Poisson model for the distribution of false-positive peptide matches. Two parameters are needed to specify the model: the total number of database proteins and the number of peptide level matches that are incorrect.

The IPI version 2.21 database contains 56 530 sequences, with some redundancy and overlap between entries. To model the database integration procedure, the two largest tryptic peptides from each database entry were calculated, and all entries containing exact matches to these two peptides were collapsed into a sequence group. This process resulted in 49 924 sequence groups. This is used as the number of bins in the random model.

Lower and upper bounds for the number of false peptide level matches are estimated by assuming either that all of the lower confidence single-peptide identifications are erroneous or that all single-peptide identifications, regardless of confidence, are erroneous. Of the 6484 identifications based on a single peptide project wide, 1956 were assigned with high confidence by at least one participating laboratory and 4528 are lower confidence identifications. The Poisson distribution parameter λ is chosen so that the random model predicts the assumed number of false single-peptide identifications. The range for λ lies between 0.146 and 0.211. The estimate of 80% false-positive rate cited above gives $\lambda = 0.168$, within this range. Values for λ larger than 0.211 would predict more protein-level identifications due to false positives alone than the 9504 total identifications reported, and are inconsistent with the random model.

For each $k = 0, 1, 2, 3, \ldots$ the expected number of database entries (out of 49 924) supported by exactly k false-positive peptide matches is calculated from a Poisson distribution. These are allocated in proportion among the reported protein identifications with $s \geq k$ supporting peptides. Only the predictions for which $s = k$ result in false-positive identifications at the protein level. The principle here is that a protein identification is considered correct if at least one of its supporting peptide identifications is correct. The allocation is illustrated in Tab. 6, and protein-level confidence is summarized in Fig. 6 and Tab. 7.

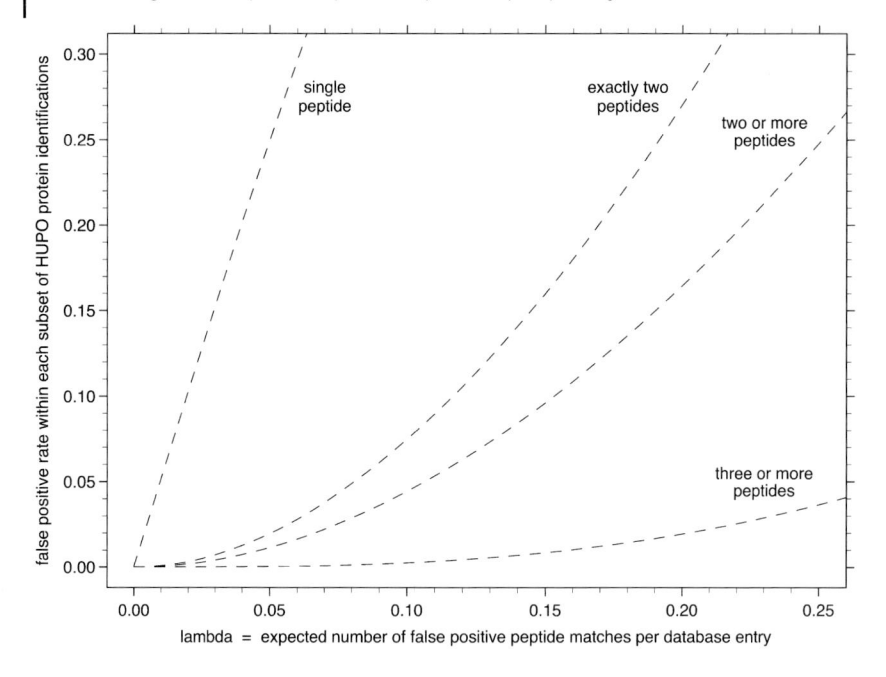

Fig. 6 At protein level, false-positive identifications are strongly concentrated among the protein identifications based on a single peptide project wide. This figure shows predicted error rates (1-confidence, vertical axis) from the Poisson model as a function of λ (horizontal axis, expressed as the expected number of false-positive peptide reports *per* IPI database entry). Four curves represent the classes of protein identifications based on exactly one, exactly two, two or more, and three or more distinct peptides reported project wide.

At the lower bound, the random model predicts 268 false-positive identifications at protein level among 1746 proteins with exactly two distinct peptides reported project wide, and 10 false positives among 1274 proteins with three or more distinct peptides project wide. The confidence within each class is the observed number of identifications minus predicted false positives, divided by the observed number of identifications. A lower bound on error becomes an upper bound on confidence. These upper bounds are a confidence of 85% for identifications based on exactly two peptides and 99% for those based on three or more peptides. Corresponding worst-case estimates are 70 and 97% for exactly two and for three or more peptides, respectively.

We acknowledge uncertainty in the exact value for λ. However, qualitative interpretations of the data are not sensitive to λ. For the quantity of data accumulated in this study, and throughout the range of choices for λ, the confidence in protein identifications based on four or more peptides easily exceeds 0.99 and for identifications based on exactly three peptides project wide, it varies from 0.95 ($\lambda = 0.211$) to 0.99 ($\lambda = 0.146$). Both classes achieve the traditional 95% confidence threshold for accepting an assertion as true, regardless of λ. The confidence for identifications based on exactly two peptides project wide varies from 0.7 ($\lambda = 0.211$) to 0.85

Tab. 6 Allocating predicted false positives among observed identifications for $\lambda = 0.146$. Predicted total number of proteins with exactly k false-positive supporting peptides (right-hand column) is allocated proportionally among the observed identifications with $s \geq k$ supporting peptides (preceding columns). Each column total is the number of observed identifications with exactly s supporting peptides, and each row total is the number of identifications predicted to have exactly k false-positive supporting peptides. Only the cases where $s = k$ (main diagonal, bold type) produce false-positive identifications at the protein level

	s	0	1	2	3	4	≥ 5	Total number of proteins with k false-positive peptides predicted from Poisson model
k								
0		40 420	1 956	445.87	140.24	57.64	121.53	43 141.28
1			4 528	1 032.16	324.65	133.42	281.33	6 299.56
2				267.97	84.29	34.64	73.04	459.94
3					9.83	4.04	8.52	22.39
4						0.26	0.55	0.82
≥ 5							0.02	0.02
Number of observed protein identifications		40 420	6 484	1 746	559	230	485	49 924

s, number of distinct peptides project wide; k, number of distinct false-positive peptides.

Tab. 7 Confidence in protein identifications as predicted by the Poisson model

Number of peptides s	Reported identifications	Predicted false positives		Confidence	
		$\lambda = 0.146$	$\lambda = 0.211$	$\lambda = 0.211$	$\lambda = 0.146$
1	6484	4528	6484	0	0.302
2	1746	268	533	0.695	0.847
3	559	10	28	0.950	0.982
4	230	0.26	1.08	0.995	0.999
≥ 2	3020	278	562	0.814	0.908
≥ 3	1274	10	29	0.977	0.992
≥ 4	715	0.27	1.12	0.9984	0.9996
≥ 5	485	0.01	0.04	0.9999	0.9999

($\lambda = 0.146$). Again, regardless of λ, these identifications would be described in lay language as "probably correct, but by no means sure". The majority of single-peptide identifications are false under any reasonable values for λ.

We have chosen to concentrate further analysis on the 3020 identifications made with two or more peptides project wide for two reasons. Excluding identifications based on exactly two peptides would exclude a large number of identifications that we believe are probably correct. Second, it would introduce a strong bias toward highly abundant proteins. Since the goal of the PPP is to identify a representative set of blood proteins, we chose to base subsequent analyses on the 3020 core data set, realizing that we are including a number of false-positives, but yielding a more representative view of the human plasma proteome.

The wide range of concentrations for proteins in blood plasma and serum presents an additional complication. Clinical ELISA assays, where available, report a measurable concentration for many proteins that were never reported by MS. Almost every protein in the body is potentially present at some concentration in blood plasma or serum, whether as an intact protein or as degradation products. There is no set of proteins we can exclude as known negatives; a large number of potential positives are present at unknown but low concentrations. A similar situation is found in *Saccharomyces cerevisiae*. A recent tagging experiment [16] measured protein concentrations spanning four orders of magnitude for 4251 proteins, roughly 80% of all proteins expressed in log-phase yeast. Two separate MS/MS surveys conducted earlier [11, 17] show low concordance in protein identifications. They reported roughly 1500 proteins each, with 57% of proteins in common and 43 or 41% reported in one survey but not in the other. In yeast, as well as in this project, the reporting of low-abundance proteins is highly variable.

2.6
Data dissemination

The project participants accessed the database through a web-based SQL interface developed specifically for project needs. During the data submission process, before the official in-project data release, each laboratory could retrieve only its own data submitted to date. After the in-project data release, laboratories could freely access data from all the participants. The database access was limited to the project laboratories by a user and password mechanism. Each laboratory could use a set of predefined SQL queries to perform standard data requests as well as define its own, private queries for more specific tasks and save these for future use.

For the dissemination of the data gathered by the HUPO-PPP, the *ab initio* construction of a novel data structure was decided upon. Indeed, the PPP, as the first HUPO project to complete the pilot milestone, is uniquely positioned for fulfilling the pioneering role in establishing such a data (infra) structure. The finalized data are publicly available in the proteomics identifications (PRIDE) database (http://www.ebi.ac.uk/pride) (see Martens *et al.*, this issue). The results of a PRIDE web query can be visualized either as an HTML page or in the PRIDE XML format. The complete PRIDE database is also available for download in XML format. The PRIDE project site offers an Application Programmers Interface (API), which provides the tools necessary to efficiently access the PRIDE XML format and reference database implementation programmatically.

2.7
Discussion

The PPP integration workflow is based on a heuristic approach that the protein identifications most likely to be true are those which are supported by the largest number of independent experiments. The strength of the "independent experiment" term is gradually loosened in consecutive steps of the algorithm to select a single protein, which represents a whole cluster of equivalent identifications.

Such an optimization approach, by its nature, may not always lead to the smallest set of proteins possible. For example, let us consider a simplified problem where there are only six protein identifications in the database – A, B, C, D, E, and F. All of them are products of independent experiments. Furthermore, they are single-peptide identifications associated with distinct peptides a, b, c, d, e, and f, respectively. Searching for these peptide sequences in the protein database shows that the peptides can be found in three different proteins with overlapping sequences – p1, p2, and p3.

Fig. 7 depicts the problem: rows represent the three proteins, columns the six peptide identifications. If a particular peptide can be found in a specified protein, it appears in the appropriate row.

Scoring the proteins using the algorithm results in: p1 = 4 (four different identifications), p2 = 3 (three different identifications), and p3 = 2 (two different identifications). This leads to the following assignment of the protein accession numbers to the identifications: ID A → p1, ID B → p1, ID C → p1, ID D → p1, ID E → p2, ID F → p3. Although it complies with the algorithm, the selection of protein p2 for identification E is not optimal from a mathematical point of view. If protein p3 were assigned instead of p2, the size of the set of proteins would reach its minimum. In a real experiment, the coincidence of such a particular overlap of the protein sequences and specific scoring conditions necessary to cause the algorithm to fail is very rare. Processing a subset of the HUPO PPP MS/MS and FT-ICR-MS data resulting in 9504 distinct protein identifications caused the algorithm to fail (*i.e.*, not to reach the minimum) in only ten cases.

Maximizing the number of independent supporting experiments also biases the selection of representative proteins towards those with the longest sequence, as illustrated in Fig. 8. The algorithms used to construct the IPI database also systematically select longer precursor sequences in preference to shorter forms [9].

A more sophisticated approach might incorporate additional sources of biological information in choosing a representative protein for each group. Sources of such information include protein annotation databases like GO [18] or HPRD [19]. We chose not to pursue this option because current annotation databases have limited coverage and might introduce historical biases into the protein identification process.

The integration algorithm seeks to assign the minimum number of proteins necessary to account for the observed peptide sequence lists. With no *a priori* knowledge of which proteins are present in the blood, an alternative, and equally valid, approach would be to list all proteins from which each peptide might have been derived. Fig. 9 compares the results of this latter approach with the integration algorithm presented above. Note that many proteins not selected by the integration algorithm may, nevertheless, have been the source of a large number of observed peptides.

```
    |    id A           id B           id C          id D          id E          id F
    |   <pept a>       <pept b>       <pept c>      <pept d>      <pept e>      <pept f>
 ___|_____
    |
p1  |   <pept a>---<pept b>---<pept c>---<pept d>
    |
p2  |                                  <pept c>---<pept d>---<pept e>
    |
p3  |                                                                <pept e>---<pept f>---
```

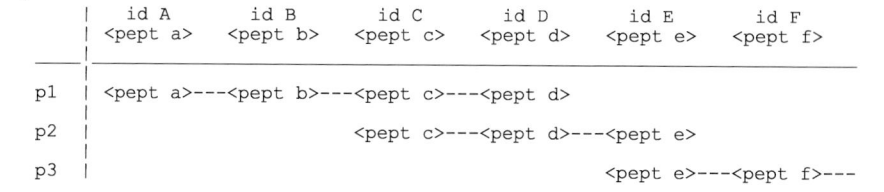

Fig. 7 Theoretical example presenting a situation where the integration workflow may not produce the minimal possible set of proteins.

```
    |    id A           id B           id C          id D          id E          id F
    |   <pept a>       <pept b>       <pept c>      <pept d>      <pept e>      <pept f>
 ___|_____
    |
p1  |   <pept a>---<pept b>---<pept c>---<pept d>---<pept e>---<pept f>---
    |
p2  |   <pept a>---<pept b>---<pept c>---<pept d>
    |
p3  |                                                                <pept e>---<pept f>---
```

Fig. 8 Length bias in representative protein selection. Shown in the figure are a precursor, p1, and two proteolytically cleaved products, p2 and p3. Precursor contains all the identifying peptides contained in the products. As a result, the integration algorithm will select the precursor independent of other knowledge about which form might be present in the sample.

2.8
Concluding remarks

The pilot phase for the HUPO PPP is the first large-scale collaborative proteome project ever undertaken, and our experience highlights the challenges in data integration that are likely to be encountered in future high-throughput and collaborative proteomics studies. Several issues are identified.

A key decision was to define one recommended protein database and release, IPI 2.21 of July 2003, for all subsequent work in the project. Although this was not universally adhered to by all project participants, it simplified early data comparisons and later merging of results. However, the decision to standardize on IPI release 2.21 also complicated the annotation process. By the time the data-gathering phase of the project had concluded, this release was necessarily out of date. The process of mapping version 2.21 identifiers to version 3.01 identifiers proved to be challenging because of the large number and complex nature of the changes that have taken place in the underlying sequence collection.

We overestimated the laboratories' ability to use XML data formats. Although tools and support for XML were offered, the vast majority of laboratories chose to submit data in Word/Excel formats.

We underestimated the importance of collecting peak lists and raw spectra. The decision to collect data at the level of protein identifications rather than individual peptide identifications meant that information defined at the peptide level, such as peak lists and SEQUEST scores, were not collected.

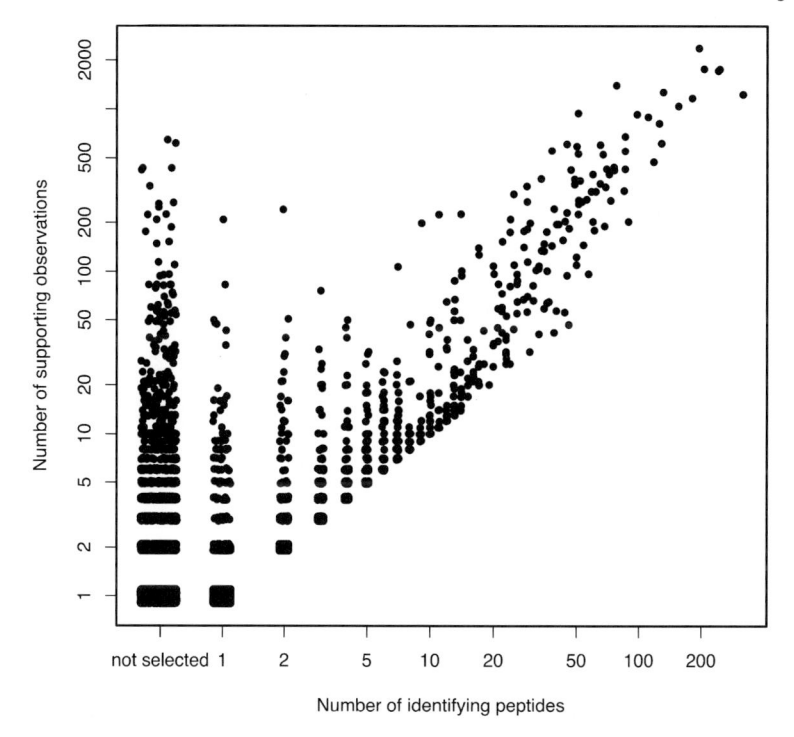

Fig. 9 Number of identifying and supporting observations. This figure shows a scatterplot for all the 15 695 proteins in IPI version 2.21, which contain at least one peptide observed in the project. X axis is the number of distinct peptides assigned to a protein by the integration algorithm. Y axis is the number of distinct (laboratory × experiment × specimen) observations of a peptide which could have been derived from the protein. Note that for some proteins not selected or assigned only one peptide by the integration algorithm, a large number of supporting observations are present in the data set.

In order to use tools like PeptideProphet and ProteinProphet [15, 7] to assess the reliability of protein identifications, search results or complete sets of peak lists are required, including those which match with extremely low scores. At the inception of the project, the decision was to perform all data analysis at the participating laboratories and to submit only protein identifications to the central repository. The initial submission forms specified only a minimal set of supporting data. As the project progressed and the data repository group assumed more responsibility for quality assurance, we requested more supporting data from the contributing laboratories including mass spectrum peak lists and full binary data files.

The decision to request a pilot round of data submissions proved invaluable in allowing the data repository group to assess the data and identify the problems described above. As a result of this pilot round of data submissions, significant changes were introduced during the project's operation. As a consequence, the data collection/integration center had to deal with the data formatted according to both the old and the new protocols, but the final product of the project was greatly enhanced.

A revised database schema for future projects has been developed; this more extensive, finer-grained schema will better serve the future needs of the PPP, and will also serve as the core for schemata tailored to meet the requirements of other HUPO tissue projects (*e.g.*, liver, brain). In this revised protocol, all entries, whether they contain new data or reanalysis of existing data, are assigned an accession number as a point of reference for use in the publications. The schema is straightforwardly extensible to accommodate additional technologies. For example, we are coordinating with project participants that generate quantitative data. Reliable quantitations, both relative and absolute, can come from a variety of methods such as differential gel electrophoresis, isotope tagging or chemical modification for MS, and protein array technologies [20].

There is also a need to "point outwards" to different resources, often done by creating a field to capture a Uniform Resource Indicator or URI (a generalized version of the familiar URL web address). Such resources include annotation resources such as UniProt (http://www.uniprot.org), EnsEMBL (http://www.ensembl.org), HPRD (http://www.hprd.org), and PeptideAtlas (http://www.peptideatlas.org) [21]. Importantly, URIs can also link to "raw" mass spectrum data repositories (the original output of a mass spectrometer scan as opposed to the heavily processed peak list); these data are increasingly in demand for in-depth analyses [22], but require special handling separate from the main project database, due to their size (see also Martens *et al.*, this issue).

In addition to its main goals of beginning the map of the human plasma proteome and assessing the power of different techniques to resolve proteins, the HUPO-PPP pilot phase has generated an extensive "real world" collection of data that will be invaluable in developing and testing enhanced software tools for proteomics. Both the structure of the revised schema and the experience gained in the pilot phase of the PPP will contribute to other HUPO proteome initiatives, in particular the Liver and Brain Proteome Projects, and the HUPO Proteomics Standards Initiative [23], which seeks to provide general standards for proteomics, both for the level of detail required when reporting work (the Minimum Information About a Proteomics Experiment, MIAPE) and the file format in which such information should be captured.

2.9
Computer technologies applied

The main project data repository was established with use of the Microsoft SQL server 2000™ working on a Dell Power Edge™ server running operating system Microsoft Windows 2003™. Templates of documents for the data transfer were produced with use of Microsoft Word and Microsoft Excel packages. The data submission site was established on Dell Power Edge server running Microsoft Windows 2003 and Internet Information Services. The online data submission and data access sites were created using Microsoft Visual Studio 2000™ and written in language C#. Data integration procedures were written either in C# or as stored procedures in the MS SQL server native language.

We thank all the participating laboratories for their dedication to this project, Yin Xu of the University of Michigan for data analytical support, and Rolf Apweiler of the European Bioinformatics Institute for overall guidance. We acknowledge support from the National Institutes of Health and from the multiple corporate partners of the Human Proteome Organization Plasma Proteome Project. L.M. would like to thank Kris Gevaert and Joel VandeKerckhove for their useful comments and support. L.M. is a research assistant of the Fund for Scientific Research – Flanders (Belgium), F.W.O. – Vlaanderen.

2.10
References

[1] Omenn, G. S., *Proteomics* 2004, *4*, 1235–1240.
[2] Blaschke, C., Hirschman, L., Valencia, A., *Brief Bioinform.* 2002, *3*, 154–165.
[3] Shatkay, H., Feldman, R., *J. Comput. Biol.* 2003, *10*, 821–855.
[4] Carr, S., Aebersold, R., Baldwin, M., Burlingame, A. *et al.*, *Mol. Cell. Proteomics* 2004, *3*, 531–533.
[5] Taylor, C. F., Paton, N. W., Garwood, K. L., Kirby, P. D., Stead, D. A. *et al.*, *Nat. Biotechnol.* 2003, *21*, 247–254.
[6] Garwood, K. L., Taylor, C. F., Runte, K. J., Brass, A. *et al.*, *Bioinformatics* 2004, *20*, 2463–2465.
[7] Nesvizhskii, A. I., Keller, A., Kolker, E., Aebersold, R., *Anal. Chem.* 2003, *75*, 4646–4658.
[8] Sadygov, R. G., Liu, H., Yates, J. R., *Anal. Chem.* 2004, *76*, 1664–1671.
[9] Kersey, P. J., Duarte, J., Williams, A., Karavidopoulou, Y. *et al.*, *Proteomics* 2004, *4*, 1985–1988.
[10] Qian, W. J., Liu, T., Monroe, M. E., Strittmatter, E. F. *et al.*, *J. Proteome Res.* 2005, *4*, 53–62.
[11] Peng, J., Elias, J. E., Thoreen, C. C., Licklider, L. J., Gygi, S. P., *J. Proteome Res.* 2003, *2*, 43–50.
[12] Tabb, D. L., Saraf, A., Yates, J. R., 3rd, *Anal. Chem.* 2003, *75*, 6415–6421.
[13] Keller, A., Purvine, S., Nesvizhskii, A. I., Stolyar, S. *et al.*, *OMICS* 2002, *6*, 207–212.
[14] Sadygov, R. G., Yates, J. R., 3rd, *Anal. Chem.* 2003, *75*, 3792–3798.
[15] Keller, A., Nesvizhskii, A. I., Kolker, E., Aebersold, R., *Anal. Chem.* 2002, *74*, 5383–5392.
[16] Ghaemmaghami, S., Huh, W. K., Bower, K., Howson, R. W. *et al.*, *Nature* 2003, *425*, 737–741.
[17] Washburn, M. P., Wolters, D., Yates, J. R., 3rd, *Nat. Biotechnol.* 2001, *19*, 242–247.
[18] Ashburner, M., Ball, C. A., Blake, J. A., Botstein, D. *et al.*, *Nat. Genet.* 2000, *25*, 25–29.
[19] Peri, S., Navarro, J. D., Kristiansen, T. Z., Amanchy, R. *et al.*, *Nucleic Acids Res.* 2004, *32 (Database issue)*, D497–D501.
[20] MacBeath, G., *Nat. Genet.* 2002, *32 Suppl*, 526–532.
[21] Desiere, F., Deutsch, E. W., Nesvizhskii, A. I., Mallick, P. *et al.*, *Genome Biol.* 2005, *6*, Epub R9.
[22] Beer, I., Barnea, E., Ziv, T., Admon, A., *Proteomics* 2004, *4*, 950–960.
[23] Orchard, S., Hermjakob, H., Apweiler, R., *Proteomics* 2003, *3*, 1374–1376.

3
HUPO Plasma Proteome Project specimen collection and handling: Towards the standardization of parameters for plasma proteome samples*

Alex J. Rai, Craig A. Gelfand, Bruce C. Haywood, David J. Warunek, Jizu Yi, Mark D. Schuchard, Richard J. Mehigh, Steven L. Cockrill, Graham B. I. Scott, Harald Tammen, Peter Schulz-Knappe, David W. Speicher, Frank Vitzthum, Brian B. Haab, Gerard Siest and Daniel W. Chan

There is a substantial list of pre-analytical variables that can alter the analysis of blood-derived samples. We have undertaken studies on some of these issues including choice of sample type, stability during storage, use of protease inhibitors, and clinical standardization. As there is a wide range of sample variables and a broad spectrum of analytical techniques in the HUPO PPP effort, it is not possible to define a single list of pre-analytical standards for samples or their processing. We present here a compendium of observations, drawing on actual results and sound clinical theories and practices. Based on our data, we find that (1) platelet-depleted plasma is preferable to serum for certain peptidomic studies; (2) samples should be aliquoted and stored preferably in liquid nitrogen; (3) the addition of protease inhibitors is recommended, but should be incorporated early and used judiciously, as some form non specific protein adducts and others interfere with peptide studies. Further, (4) the diligent tracking of pre-analytical variables and (5) the use of reference materials for quality control and quality assurance, are recommended. These findings help provide guidance on sample handling issues, with the overall suggestion being to be conscious of all possible pre-analytical variables as a prerequisite of any proteomic study.

3.1
Introduction

The major goal of the HUPO Plasma Proteome Project (HUPO PPP) is comprehensive characterization of the plasma proteome. The first phase of this project serves to

---

* Originally published in Proteomics 2005, 13, 3262–3277

Exploring the Human Plasma Proteome. Edited by Gilbert S. Omenn
Copyright © 2006 WILEY-VCH Verlag GmbH & Co. KGaA, Weinheim
ISBN: 3-527-31757-0

establish the advantages and disadvantages of the various technologies, and allows for an initial assessment of the different sample types that can be used. The technologies for analysis include MS-based methods, (*e.g.*, MALDI-TOF, SELDI-TOF, MS/MS, and FT-ICR-MS), 2-DE, antibody microarrays, and other methodologies.

The Specimen Collection and Handling Committee (SCHC) was created in order to evaluate a number of pre-analytical variables that can potentially impact the outcome of results, but are not related to inherent sample (*e.g.*, patient or donor) differences. These include the choices of sample type and collection system after collection, and handling issues which include stabilization, processing, storage, and potential effects of additives. We focused on these issues in the PPP pilot phase. Other issues such as patient status, venipuncture, phlebotomy technique and collection devices were not addressed, as a centralized collection procedure and strictly controlled conditions were necessary for such a large scale collection of a limited number of sample types.

The HUPO PPP elected to use limited pooled specimens in order to reduce the number of variables that could potentially confound the analysis and comparison of methods. While inter-individual variation is also important, this finer discrimination between donors will be reserved for later phase studies. After careful consideration of the objectives, a rigorous, standardized specimen collection procedure was established and the collection of HUPO reference specimens was initiated [1, 2, and Section 2]. The decision was made to collect and pool sufficient quantities from two individuals, derived from each ethnic group studied.

Another important question was that of comparing serum and plasma specimens with regard to the human proteome. Historically, serum samples dominate archives, however they were amassed based on requirements of conventional assays and not necessarily because they represent the most appropriate sample for protein analytics. For serum samples, clotting only by glass/silica-based activation was used, to eliminate any variability from the addition of other unwanted *ex vivo* effects, such as protease activation. On the other hand, the acquisition of plasma requires the use of anticoagulants. Plasma specimens were derived using the three most common anticoagulants, namely potassium-EDTA, lithium-heparin, and sodium-citrate. Depending on analytical objectives and/or target peptides or proteins, use of either plasma or serum may impact both method and results.

The HUPO PPP specimens were collected from three ethnic groups: Caucasian American, Asian American and African American. In addition, the National Institute of Biological Standards and Control (NIBSC, UK) provided a lyophilized plasma specimen, which was compared with the frozen specimens. The use of protease inhibitors was also discussed. The committee developed protocols that were comprehensive in nature, applicable to most methods or techniques and practical for use in a clinical laboratory setting, where the criteria of specificity, sensitivity, quality, reproducibility, and consistency are of critical importance.

After specimen collection, the impact of processing time and storage temperature on sample integrity was investigated. The effect of freeze-thaw cycles was also deemed important and was discussed. We sought to better define many of these issues and conditions related to blood collection and handling. What follows are the collective observations from PPP participants, related to each of these parameters.

These results and suggestions come from this first comprehensive effort designed to address the many aspects of sample selection and multiple issues that may result in pre-analytical variation. Clearly, these observations indicate that a great deal more work must be done to define a true "best practices" approach to plasma or serum samples. Through this manuscript, we hope to emphasize that a better appreciation of these issues will increase awareness and understanding of their consequences and sensitize researchers to the confounding artifacts that can result from non-ideal conditions. Sound experimental design, an educated choice of specimen type, and careful sample collection and handling procedures can have a profound impact on results, especially in inter-laboratory or multi-facility studies. The experimental observations and the resulting recommendations are described herein.

3.2 Materials and methods

3.2.1 HUPO reference sample collection protocol

HUPO reference samples were collected at, and using materials from BD Diagnostics. For each ethnic group (Caucasian, Asian American, African American), blood was acquired by venipuncture from one male donor and one female donor. Donors were tested and determined negative for HIV-1 and HIV-2 antibody, HIV-1 antigen (HIV-1), Hepatitis B surface antigen (HBsAg), Hepatitis B core antigen (anti-HBc), Hepatitis C virus (anti-HCV), HTLV-I/II antibody (anti-HTLV-I/II), and syphilis. Blood intended for plasma preparation was collected into the following tubes: BD Plus Plastic K_2EDTA, 10 mL, reorder# 367525; BD Glass Sodium Citrate, 0.105 M, 10 mL, reorder# 366007; BD Plus Plastic Lithium Heparin, 10 mL, 16 × 100 mm, reorder# 367880; BD Glass Serum with silica clot activator, 10 mL, 16 × 100 mm, reorder# 367820. The collection procedure was as follows. Under the direction of a qualified and licensed physician, trained phlebotomists collected blood from each donor into evacuated blood collection tubes. From each individual consenting donor (n = 6 donors), approximately 400 mL of blood was collected into 40 tubes by two venipunctures, 20 tubes *per* venipuncture. Discard tubes (serum) were drawn first, with each venipuncture, to prime the tubing. With the first venipuncture, the order of draw was to fill 10 citrate tubes followed by 10 serum tubes. With the second venipuncture, the order of draw was 10 heparin tubes first followed by 10 K_2EDTA tubes. Tube handling conditions are detailed in Tab. 1.

The specimens were centrifuged appropriately (see Tab. 1) under refrigerated conditions (2–6°C). The resultant serum and plasma from the 10 spun tubes of each type from each donor were pooled into one secondary 50 mL conical bottom BD (TM) Falcon tube for each tube type. The secondary tube was centrifuged at 2400 RCF for 15 min to remove potentially remaining cellular material from the serum and to prepare platelet poor plasma from the EDTA, heparin and citrate secondary tubes. Equal volumes of serum or plasma were pooled (across gender,

Tab. 1 Tube handling conditions

Tube type	No. of inversions	Clotting time	Centrifugation speed	Centrifugation time
K_2EDTA	10	None	$\leq 1300 \times g$	10 min
Lithium heparin	10	None	$\leq 1300 \times g$	10 min
Serum	None	30 min	$\leq 1300 \times g$	10 min
Sodium citrate	4	None	$1500 \times g$	15 min

but within ethnicities) from each secondary tube into media bottles, leaving approximately 10% at the bottom of the secondary tube to ensure no cellular material is collected. Serum/plasma was mixed gently and kept on ice while being distributed into 250 µL aliquots in cryovials. All aliquoting was completed within 75 to 90 min of specimen collection.

3.2.2
Differential peptide display

After separation on a RP-HPLC column employing an ACN-gradient (4 to 40%) and collection of 96 fractions, a 15 µL equivalent of plasma was subjected to MALDI-TOF MS (for details see [5]) (Voyager DE STR, Applied Biosystems). The MALDI matrix consisted of alpha-cyano-4-hydroxycinnamic acid (matrix) and 6-desoxy-l-galactose (co-matrix) in ACN containing 0.1% TFA. Data were analyzed, including peak recognition and visualization using the software package Spectromania®. Quantification of mass spectrometric signals was performed after baseline correction by integrating absolute signal intensities in 1 Da bins. Using the software, the mass spectrum of each fraction was transformed to a virtual lane. The molecular weight of each peptide is indicated by its position within the virtual lane, whereas the MALDI-signal intensity for each peptide is depicted by the color intensity of the corresponding bar. The converted mass spectra of all 96 fractions were combined resulting in a 2-D display of peptide masses termed peptide map display. Detection of differentially expressed peptides was achieved by calculation of subtractive peptide maps displays or correlation analysis (differential peptide display) by referring to mass spectrometric data [3–5].

3.2.3
Stability studies and SELDI analysis

Aliquots of the samples were prepared, sealed, and stored under the designated condition for the appropriate length of time. When the incubation period was complete, samples were denatured in a solution containing 9 M urea, 2% CHAPS, in PBS, and stored at −80°C. After all the samples were collected, they were simultaneously thawed and processed in duplicate for analysis using the carboxymethyl surface employing optimized SELDI-TOF protocols that are detailed in Rai et al., in this issue [6]. This simultaneous processing effectively removed any bias related to sample processing over multiple runs.

3.2.4
SDS-PAGE analysis for stability studies

To assess the stability of serum samples stored at 23°C, 4°C, −20°C, and −80°C and liquid nitrogen (LN$_2$), samples were simultaneously thawed and separated using precast NuPAGE™ (Invitrogen) SDS gels. For low molecular weight protein separation, 4 μg serum was separated using 10% Bis-Tris NuPAGE™ gels, electrophoresed using MOPS running buffer, and stained with SYPRO Ruby fluorescent gel stain (Molecular Probes). For analysis of large proteins, 1 μg of serum was run on 3–8% Tris-Acetate NuPage™ gradient gels with Tris-Acetate running buffer, and silver stained (Silver Quest silver staining kit; Invitrogen). SYPRO Ruby stained gels were scanned using the ProXpress Proteomic Imaging System (Perkin Elmer Life and Analytical Sciences).

3.2.5
2-DE for stability studies

IPG strips were purchased from GE Healthcare and proteins were isoelectrofocused using the PROTEAN IEF Cell™ (Bio-Rad Laboratories) IEF system, essentially as described by Görg et al. [7]. Frozen samples were briefly thawed and diluted into IEF sample buffer containing 9 M urea, 2 M thiourea, 4% CHAPS, 0.1 M DTT, 1.6% pH 3–10 linear carrier ampholyte buffer to yield 100 μg of protein in 400 μL sample buffer. IPG strips (18 cm) were rehydrated with sample buffer containing serum proteins at 50 V for 12 h and then the applied voltage was increased in a linear fashion to a maximum of 10 000 V until a total of 60 000 Vh was reached. The IPG strips were equilibrated and applied to 18 × 19 cm, 1.0 mm-thick second dimension 10% Tris-tricine polyacrylamide gels [8], cast without stacking gels and with sodium thiosulfate added to reduce silver stain background, as described by Hochstrasser et al. [9]. Gels were run at 100 mA constant current with external cooling (∼4 h) until the tracking dye migrated to within 1 cm of the bottom of the gel. Gels were then stained with SYPRO Ruby and imaged using the ProXpress Proteomic Imaging System.

3.2.6
SELDI-TOF analysis for protease inhibitor studies

Plasma samples from 20 individuals were collected into identical EDTA-containing tubes with a patented mechanical separator, either with or without the presence of a protease inhibitor cocktail. After collection, samples were processed as quickly as possible (within 15 min) into plasma, hence "time zero" analyses were from aliquots that were frozen at this earliest time point. Analysis was performed using H50 chips, as *per* manufacturer's instructions. Briefly, the chips are preincubated with 50% ACN, twice for 5 min each, followed by activation with binding buffer (10% ACN, 0.1 M NaCl, and 0.1% TFA) for 5 min and removed. Additional binding buffer (90 μL) is added, followed by 10 μL of plasma. After a 30 min incubation period, the chip is removed, and then washed twice by two 50 µL washes of binding buffer and another two washes with water. Surfaces are allowed to air dry, and matrix (1.5 µL of CHCA in 50% ACN, 0.1% TFA) is added, and allowed to air dry again prior to reading on a PBS2 instrument (Ciphergen Biosystems).

3.2.7
2-DE for plasma protease inhibition studies

All materials mentioned in this section were sourced from Sigma-Aldrich, except where noted below. Plasma samples (HUPO PPP No. BDAA01-Cit) were thawed and either water or protease inhibitor cocktail (P 8340) added at a 1% level (1:100 dilution). The protease inhibitor stock solution consists of 104 mM 4-(2-aminoethyl) benzenesulfonyl fluoride (AEBSF), aprotinin (80 mM), leupeptin (2 mM), bestatin (4 mM), pepstatin A (1.5 mM), and E-64 (1.4 mM). The final concentrations in the plasma samples were 1/100th of these values. Plasma was also treated with each of the individual protease inhibitors, as detailed above. These inhibitors were dissolved in DMSO (D 8418) at the concentrations indicated above for the cocktail. The treated plasma samples (50 µL each) were depleted of albumin and IgG using an antibody based depletion resin, (Proteo-Prep Immunoaffinity Albumin and IgG Depletion Kit (PROT-IA)), then equilibrated with 1% protease inhibitors or water in equilibration buffer. Depleted plasma was precipitated in TCA/deoxycholate (PROT-PR) and the pellets dissolved in 7 M urea, 2 M thiourea, 1% w/v C7BzO, and 40 mM Trizma base (C 0356). Protein quantitation of each depleted sample was determined using a Bradford protein assay (B 6916), with a 1 mg/mL BSA (P 0914) standard. Each depleted plasma sample was reduced and alkylated with a final concentration of 5 mM tributylphosphine and 15 mM iodoacetamide (PROT-RA) respectively for 30 min each, at room temperature. For first dimension IEF, bromophenol blue (B 0126) was added as a tracking dye to each sample at a final concentration of 0.001% w/v. The IPG strips, 11 cm, pH 4–7, (GE Healthcare) were passively rehydrated for 4 h at room temperature, until the entire sample was absorbed into the strips. The strips were then focused at 6000 V for a total of 85 000 Vh at 20°C, using a PROTEAN IEF cell (Bio-Rad). For second-dimension SDS-PAGE, the IPG strips were equilibrated with an SDS equilibration buffer (I 7281) for 15 min, then loaded onto 4–20% SDS-PAGE gels (Bio-Rad) and electrophoresed at 170 V for 80 min in Tris-glycine-SDS running buffer (T 7777). SigmaMarker™ Wide Range molecular weight markers (M 4038) were added to the lane on the extreme right of the gels. The gels were then Coomassie stained with EZBlue Gel Staining Reagent (G 1041), and subsequently destained in water. Stained gels were imaged and then evaluated using the Phoretix Expression 2-DE imaging software (Nonlinear Dynamics). Spots of interest were manually excised from the gel and the proteins tryptically digested overnight at 37°C using the Trypsin Profile IGD Kit (PP0100). The extracted peptide digests were dried at 30°C using a vacuum centrifuge.

3.2.8
Tryptic digestion and protein identification for protease inhibition studies

Tryptic peptides derived from each gel spot were submitted for identification by PMF using MALDI-TOF MS. Dried samples were reconstituted in 0.1% TFA (typically 10–20 µL), and mixed 1:1 with matrix solution (α-cyano-4-hydroxycinnamic acid, 5 or 10 mg/mL in 70% ACN, with 0.03% TFA). Aliquots (1 µL) were spotted on the MALDI target and allowed to dry at room temperature before introduction into the mass spectrometer. Mass spectra were acquired in positive ion reflectron mode after close-external calibration using bradykinin 1–7 and insulin oxidized B-chain as standards. Spectra were typically summed over 100 shots. A Shimadzu-Biotech Axima CFR^{TM}*plus* instrument was employed for this work. Protein identification was performed using MASCOT database searching at http://www.matrixscience.com. Search parameters were restricted to *Homo sapiens* taxonomy for use with the NCBInr database. Enzyme selection was trypsin, with up to one missed cleavage permitted. Carbamidomethylation of cysteines was selected as a fixed modification. Protein mass was unrestricted, and peptide mass tolerance typically set at 250 ppm. Mass values were entered as monoisotopic MH+. Mass lists were generated automatically using the MASCOT Wizard software application, with a S/N threshold of 100.

3.2.9
Antibody microarray analysis using two-color rolling circle amplification

All assays were performed as described in Zhou *et al.* [10]. Briefly, the labeled samples (different serum types) were incubated on antibody microarrays containing 48 unique antibodies that were printed on NC-coated glass microscope slides (FAST™ slides; Schleicher & Schuell Bioscience, Keene, NH USA). Labeling was performed by mixing the serum samples 1:15 with 50 mM carbonate buffer, pH 8.3, and 1/20th volume 6.7 mM *N*-hydroxysuccinimide (NHS) ester linked biotin or digoxigenin (Molecular Probes-Invitrogen) in DMSO. The reactions were quenched and unreacted dye was removed by size exclusion spin chromatography (Bio-Spin-P6, Bio-Rad), molecular weight cutoff = 6000 Da). The arrays were detected using the TC-RCA procedure, as detailed in [10].

3.3
Results

The SCHC within the HUPO PPP comprises a diverse group of researchers, each investigating various parameters relating to a specific topic. Given the complexity of the blood proteome, collection and handling present a very broad range of specific technical challenges, far too many to be completely investigated in the PPP pilot phase. A list of some of these issues is presented in Tab. 2. The data that follow represent available data contributed from various SCHC member laboratories. The

Tab. 2 List of pre-analytical variables. There are a number of factors that can alter the composition of samples prior to proteomic analysis. Some of these are listed here; these issues can be divided into those relating to patient history, sample collection and handling, and specimen storage.

- Patient information
  - Gender
  - Age
  - Diet
  - Genetics
  - Medical background
  - Health background
  - Special conditions: pregnancy, post/pre- menopausal, medications
  - Social history: alcohol intake, smoking status
  - Nature of sample: Fasting, post-prandial, or random
- Venipuncture
  - Needle gauge
  - Details of blood collection set
- Phlebotomy
  - Tourniquet technique
  - Patient position: seated/standing/lying
  - Tube order- first *versus* last
  - Blood source: venipuncture or from existing line
- Collection device
  - Tube or bag
  - If tube, glass or plastic
  - Gel or non-gel separator
  - Other tube additives
  - Manufacturer & device information
- Blood derivative and processing
  - Sample type- plasma *versus* serum
  - If plasma, nature of anticoagulant: EDTA, citrate, heparin
  - If serum, clotting procedure used, and/or type of clot activator
  - Details of processing: time, protocol, *etc.*
    - Separation of blood from cells
    - Centrifugation, speed & duration
    - Aliquotting before analysis, and handling/storage of those aliquots, *e.g.*, time and temperature conditions
    - Length of time before analysis
- Storage
  - Frozen before analysis: snap-frozen, *e.g.*, dry ice-ethanol, or slowly cooled, *e.g.*, storage at $-20°C/-80°C$?
  - Elapsed time and temperature, prior to freezing
  - Short or long-term storage
  - Storage temperature
  - Expiration dating
  - Storage materials: Glass *versus* plastic
  - If plastic, type of plastic
  - Number of freeze/thaw cycles

results are based on very defined and specific experiments covering several parameters. They are presented to suggest the scope and tone of the important task of refining collection and handling into a concrete set of recommendations. These experiments and results fall into three broad categories: comparison of sample types, studies of basic sample handling and storage issues, and evaluation of the use of protease inhibitors.

3.3.1
Comparisons of specimen types

In the following section, special emphasis is given to the peptide content of human blood samples [11] using differential peptide display [3–5]. Additional details are provided in a separate manuscript in this issue, Tammen *et al.* [12].

3.3.1.1 Analysis of serum
A significant number of peptides in the mass range of 1–15 kD were found to be different between serum and all plasma specimens of the HUPO-PPP standard sample set (Fig. 1). Approximately 40% of all detected signals are only present in serum and not detectable in plasma. Sequence analysis of these differences revealed intracellular, coagulation dependent and enzymatic activity-derived peptides.

3.3.1.2 Analysis of plasma
The HUPO plasma samples show significant differences compared to reference plasma collected according to the BioVisioN sample protocols (detailed in [12]). In our study, different protocols for EDTA/citrate specimen collection were investigated. A major finding is the detection of platelet-derived peptides in HUPO specimens. We searched the Swiss-Prot (www.expasy.org) database using the sequence retrieval system (entry tissue = platelet). Out of 104 retrieved precursor proteins, we found 35 to be present in the HUPO database and some have been identified by several groups (Supplementary Tab. 1).

3.3.2
Evaluation of storage and handling conditions

The quality of data obtained from any study will be limited by the quality of the specimens used and thus, the integrity of samples is an important consideration. To delineate and identify the changes that can occur to specimens as a result of differences in storage conditions, a study was designed to systematically test the effects of storage under various time and temperature conditions (Tab. 3).

SELDI-TOF analysis was performed on these timed samples and results are displayed in Tab. 4. Although there were no major differences observed between storage at $-20°C$, $-80°C$, and liquid nitrogen, over the 2 month time period as detected by SELDI-TOF under our conditions, there were multiple peak differences noted that were present or absent at both room temperature and refrigerated storage

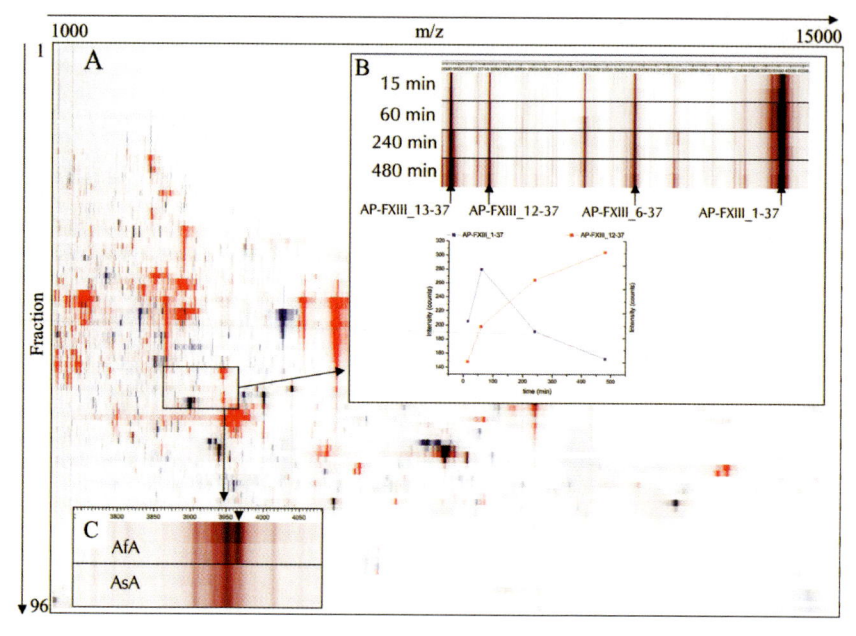

Fig. 1 Differential peptide display of plasma and serum specimens from African American donors. (A) This heat map indicates differences between plasma (blue) and serum (red). (B) This panel depicts peptides, which are generated by thrombin and plasma kallikrein during coagulation of serum samples in a time dependant manner (15 min to 8 h). The graph illustrates the kinetics of generation and degradation of the activation peptide of coagulation factor XIII (AP-FXIII). Time kinetics have been carried out in separate experiments at BioVisioN. (C) This panel depicts a zoom view into mass spectra representing a distinct chromatographic area and m/z region. Four lanes from peptide displays (African American (AfA) and Asian American (AsA)) serum specimens are shown. The AP-FXIII is solely displayed in serum samples. The additional peak in AfA samples (arrow) represents the Leu34 polymorphism of AP-FXIII.

(Tab. 4). A subset of these samples, *i.e.*, those that had been incubated for 30 days at each tested temperature, were selected for more in depth investigation using SDS-PAGE (Fig. 2) and 2-DE (Fig. 3). There were a number of protein bands on 1-D SDS gels that decreased or appeared, presumably relating to proteolytic fragments and/or modifications, after long term storage at higher temperatures. Similarly, specific protein spots decreased or appeared on 2-D gels (Fig. 3) under the various time and temperature conditions.

To facilitate correlation of these results to more standard clinical testing, the levels of fourteen protein analytes in serum were also measured (Supplementary Tab. 2). A number of alterations in protein levels and enzyme activities under the various conditions tested were noted. The results demonstrate that there are changes in enzyme activity and protein stability for several proteins: lactate dehydrogenase (LDH), aspartate-amino transferase (AST), and lipase show decreases at room temperature over the 2-month period. Amylase and alanine-amino transferase (ALT) show changes at both room temperature and $-20°C$ storage, during the same time period.

Tab. 3 Time and temperature conditions used for stability testing. Serum samples were prepared and stored under the appropriate condition for the requisite time period. Samples were then denatured and stored until testing was initiated (see text for details)

R/T (23°C)	Refrigerated (4°C)	Frozen at −20°C	Frozen at −80°C	Liquid nitrogen
0h				
2h				
4h	4h			
8h	8h			
1d	1d	1d		
2d	2d	2d		
7d	7d	7d		
30d	30d	30d	30d	30d
		60d	60d	60d

Interestingly, alkaline phosphatase activity decreased by storage at −20°C, but was not altered by storage at room temperature. It was also noted that two of the cholesterol components, LDL and HDL, showed increases by extended storage at room temperature.

3.3.3
Evaluations of the use of protease inhibitors

3.3.3.1 Analysis with SELDI-TOF MS of "time zero" effects of protease inhibitors in plasma

We used SELDI-TOF analysis to compare twenty samples in two groups, distinguished only by the presence or absence of protease inhibitors (PIs) at the time of collection. A number of peptide and protein mass/charge (m/z) peaks show statistically significant differences between the two groups, with $m/z \sim 6803$ as representative data shown in Fig. 4. The particular m/z peaks show consistently higher signals in the presence of protease inhibitors, while far lower signal is observed for the untreated plasma samples.

3.3.3.2 **Analysis by 2-DE**

Samples of albumin and IgG depleted citrate-plasma were subjected to the addition of protease inhibitor cocktail (Fig. 5, Panel A), water (Panel B) or individual protease inhibitor components (Panels C–H), and separated by 2-DE. The presence of the PI cocktail appears to perturb the native pI profiles of several proteins, particularly higher abundance proteins including apolipoprotein A1, apolipoprotein A2, haptoglobin (alpha and beta subunits), transthyretin and fibrinogen – as identified by in-gel tryptic digestion and MALDI-TOF MS (data not shown). Furthermore, the PI cocktail presence induces general smearing of

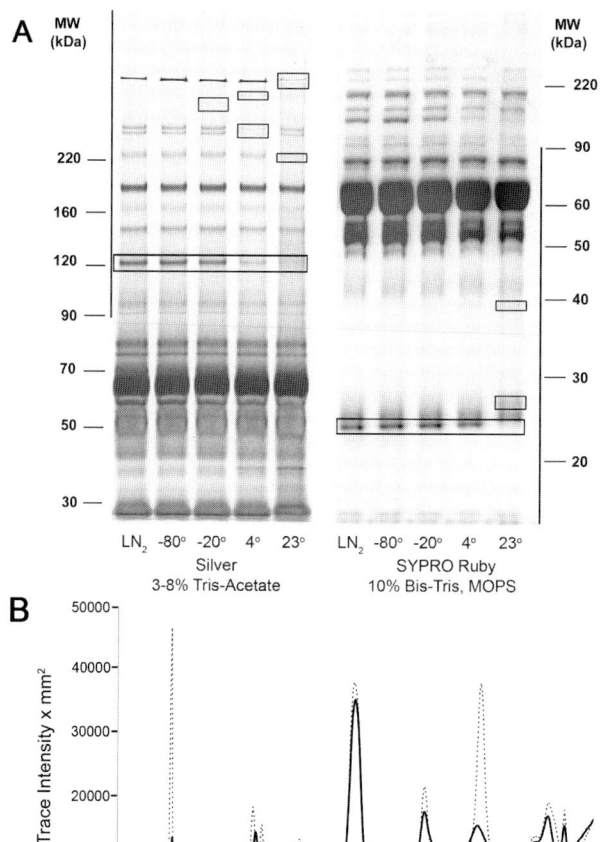

Fig. 2 The assessment of sample integrity under various storage conditions as assessed by SDS-PAGE, using (A, left panel) 3–8% Tris-Acetate and (A, right panel) 10% Bis-Tris/MOPS. (B) A trace diagram is included showing alterations in selected protein bands with two samples stored at room temperature and liquid nitrogen.

the entire 2-DE protein profile, and 13% fewer resolved spots, although it is possible that addition or alteration of sample preparation steps could eliminate at least the smearing.

The changes in the isoform profiles are exemplified by apolipoprotein A1 (Fig. 5, boxes in panels) in which the intensity of the two high pI isoforms are substantially increased with the inclusion of the PI cocktail. The relative intensities of the protein spots from panels A and B were compared using the 2-DE gel analysis software. The intensities of the lower pI isoform spots decrease with PI added, and there is a concomitant increase in the intensities of the higher pI isoforms (data not shown).

Tab. 4 Changes in SELDI-TOF profiles of serum samples under different storage conditions. Samples were analyzed in duplicate and alterations in peak intensities were noted after the generated profiles were analyzed using Ciphergen's Biomarker Wizard®. The presence of a peak is denoted by a check mark, whereas the absence is denoted by an X. The control sample corresponds to a sample that was prepared, aliquoted and denatured at $t = 0$ (see Section 2 for details) and serves as a basis for comparison for the other timed samples

m/z	Control (t = 0)	23°C 8h	23°C 7d	4°C 2d	4°C 7d	−20°C 30d	−20°C 60d	−80°C 30d	−80°C 60d	Liquid nitrogen 30d	Liquid nitrogen 60d	Notes
3900	✓	✓	✓									Specific to 23°C
4100	✓	✓	✓	✓								Specific to 23°C and 4°C
7200	✓	✓	✓									Specific to 23°C
7600	✓			✓	✓	✓	✓	✓	✓	✓	✓	Undetected at 23°C
13600	✓	✓	✓	✓	✓							Specific to 23°C and 4°C
14500	✓	✓	✓									Specific to 23°C
17300	✓			✓	✓	✓	✓	✓	✓	✓	✓	Undetected at 23°C
28000	✓			✓	✓	✓	✓	✓	✓	✓	✓	Undetected at 23°C
90000		✓	✓									Specific to 23°C
95000	✓			✓	✓	✓	✓	✓	✓	✓	✓	Undetected at 23°C

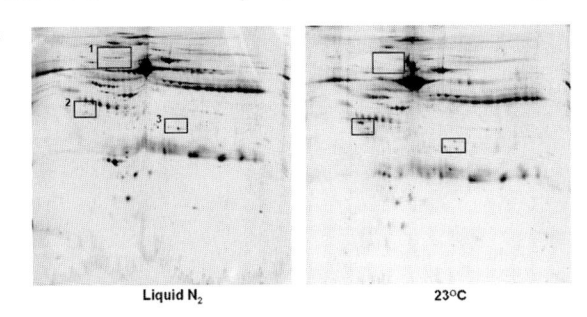

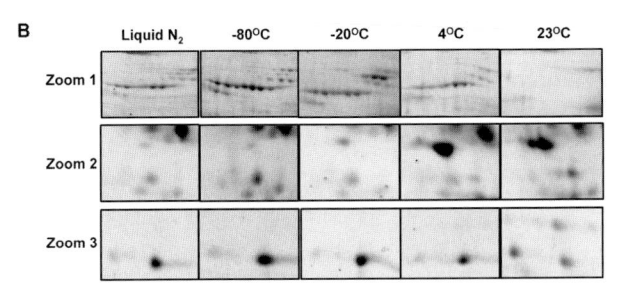

Fig. 3 2-D analysis of sample storage conditions. (A) Human serum stored at extreme low (LN$_2$, left) and high (23°C, right) temperature conditions. 100 µg serum was separated on linear pH 3–10 IPG strip (first dimension) 2-D gels and stained with SYPRO Ruby. Several representative areas where protein spots appear to be missing, or changing in intensity have been highlighted. (B) Enlargements of areas highlighted in panel A are shown for all temperature conditions. These enlargements illustrate the disappearance of some protein spots at higher temperatures and the appearance of new spots indicating proteolytic fragments.

In order to determine which PI component(s) were responsible for this observed effect, citrate-plasma samples were subjected to each individual protease inhibitor at the same concentration as is present in the cocktail, using otherwise identical methods as above. Gels were compared for each of the six protease inhibitor treated samples. The gel in which 4-(2-aminoethyl) benzenesulfonyl fluoride (AEBSF) was added shows similar perturbations of native patterns to those seen using the complete protease inhibitor cocktail, (Fig. 5, Panels A and C). Samples run in the presence of the other protease inhibitors show sharply resolved spots and the same isoform profiles as seen without the inhibitor cocktail.

3.3.3.3 Analysis with antibody arrays

Protease inhibitor (PI) cocktail effects on antibody microarray experiments using two-color rolling circle amplification (TC-RCA) detection [10] were investigated. Three different serum samples were used, and each sample had four preparations, namely: no PI added, PI added prior to sample labeling, PI added after the sample labeling, and PI added both before and after the sample labeling. Representative images from each of the four conditions are shown in Fig. 6. The background level

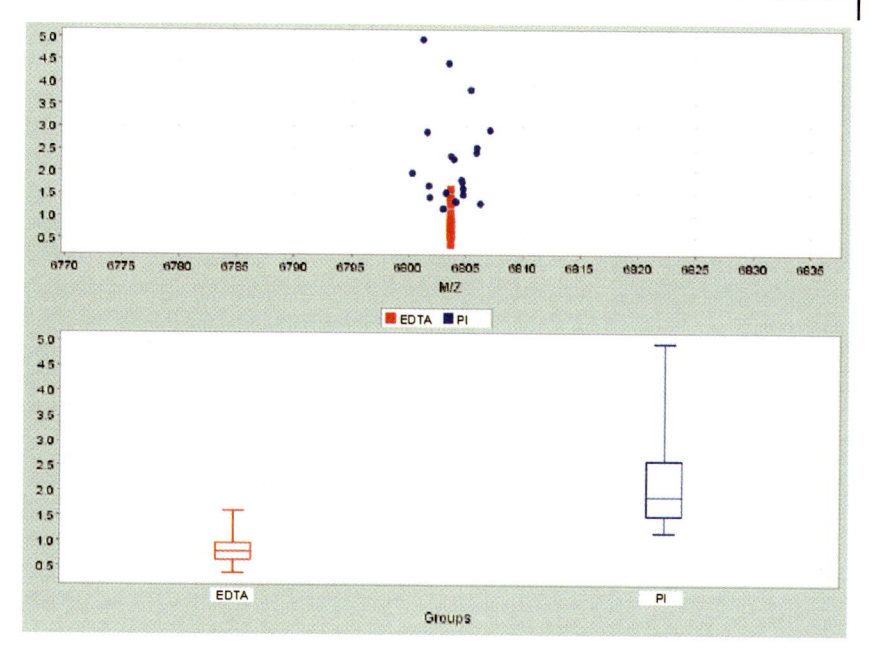

Fig. 4 Comparison of twenty blood samples collected into tubes having either PI cocktail and EDTA (blue circles) or EDTA alone (red squares). Samples represent "time zero" timepoints, processed from whole blood into plasma, aliquoted, and frozen, within 15 min from blood draw. Upper graph: 20 data points for each tube type, showing m/z ratio of approximately 6803 and signal intensity, after normalization of each of the forty SELDI mass spectra represented. Lower graph: whisker plot of the same data. Note that the samples with PI display both higher overall signal intensities and a wider intensity distribution than the untreated plasma samples. These SELDI experiments do not provide information on the identity of the discriminatory peaks.

of nonspecific binding to the NC matrix was highest in the sample without PI, and the background appeared lowest when PI was used both before and after the labeling. Plots of the average backgrounds for each serum sample in each condition show the decrease in background with each addition of PI. This effect was observed more in the 543 nm (green) channel (Fig. 7A) than in the 633 nm (red) channel (Fig. 7B). The average signal-to-background (S/B) ratios increase with the addition of PI, both in the 543 nm channel (Fig. 7C) and the 633 nm channel (Fig. 7D), showing that the signals do not decrease at the same rate as the backgrounds.

3.4 Discussion

The goal of the SCHC is to develop a standard operating procedure (SOP) for blood collection and handling for proteomic studies. This is a daunting task considering the large number of variables that must be thoroughly studied and the complica-

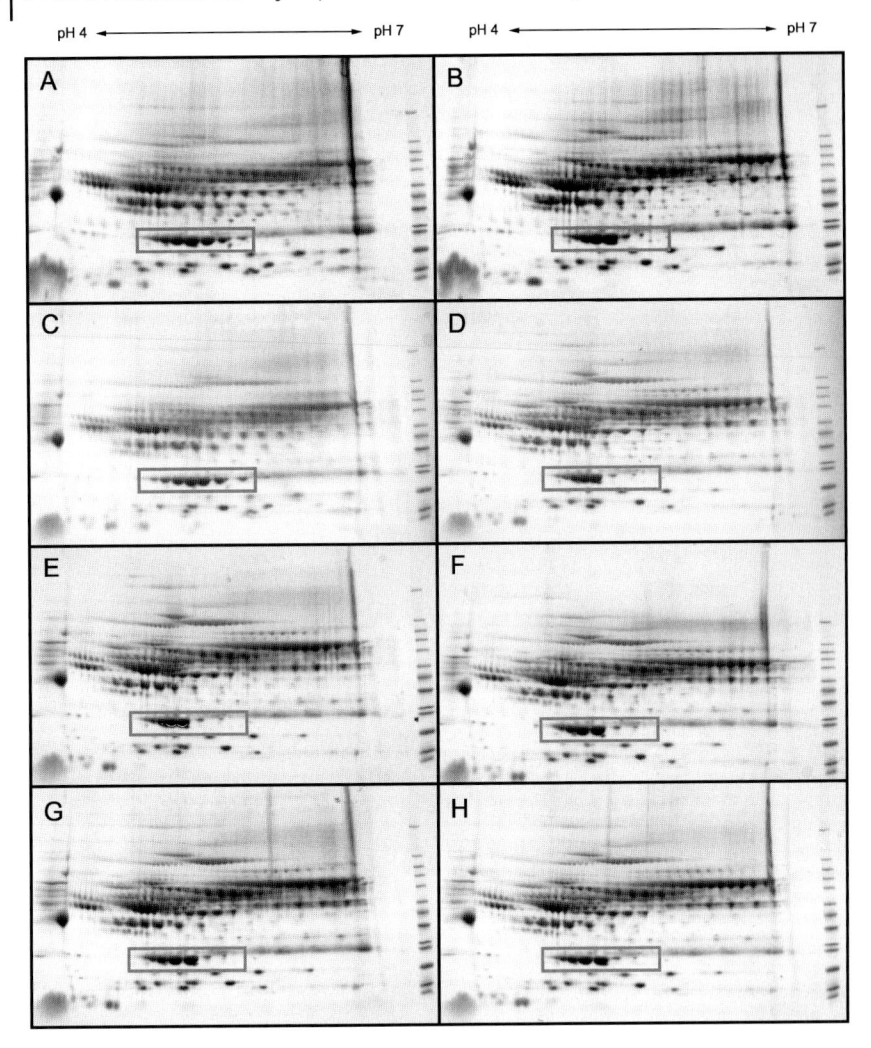

Fig. 5 Examination of the effect of a protease inhibitor cocktail and individual protease inhibitors on the 2-DE profile of human plasma. Eight samples (50 µL) of heparin plasma (HUPO PPP BDAA01-Hep) were subjected to the addition of the following reagents: Protease inhibitor cocktail (Panel A), water (Panel B), AEBSF (Panel C), Aprotinin (Panel D), Leupeptin (Panel E), Bestatin (Panel F), Pepstatin A (Panel G), and E-64 (Panel H). All samples were depleted of albumin and IgG. A portion of each of the depleted samples (Panels A and B – 500 µg, Panels C–H – 200 µg) was separated on 2-DE gels as described in Section 2.

tion and difficulties in standardization, as evidenced within the HUPO PPP efforts to date. A microcosm of this complexity is captured in the data presented above: a wide variety of analytical techniques addressing specific but varied parameters, from which to draw concrete and broadly applicable conclusions. As such, we do not put forward

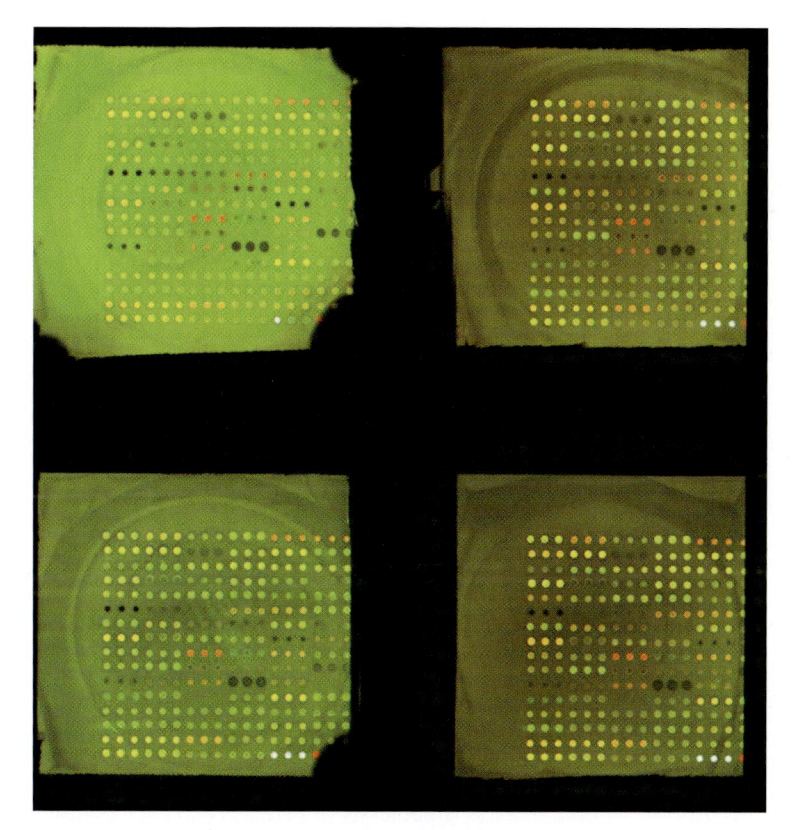

Fig. 6 Representative antibody microarray images from four experimental conditions: (A) No PI added; (B) PI added after the labeling; (C) PI added before the labeling; (D) PI added both before and after the labeling. The composite images from the 543 nm (green) and the 633 nm (red) channels are shown.

broad, sweeping recommendations based on the individual data sets. However, we believe that our data represent important lessons for the field of plasma proteomics, so we present here general recommendations and cautions that can help guide researchers toward a more thoughtful experimental design, hopefully leading to even more robust plasma proteome studies in the future. As the collective knowledge of the field grows, eventually we may achieve a definition for a handling and collection SOP.

The data as presented, although derived from independent experiments, provides insight into pre-analytical variables that could prove detrimental to proteomics experimental design and outcome. The following discussion of the data illustrates particular pre-analytical variables, and how the proteomic analyses in question were altered. Topics touched by this data include platelet contamination, storage, and the use of protease inhibitors as a protective mechanism. This is the first step towards developing a sample acquisition and handling SOP for plasma proteomics.

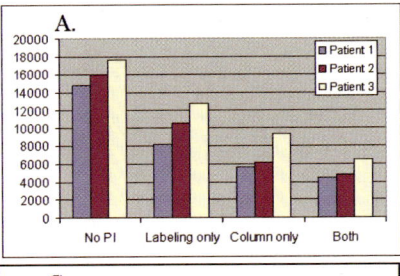

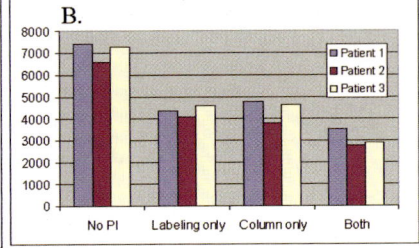

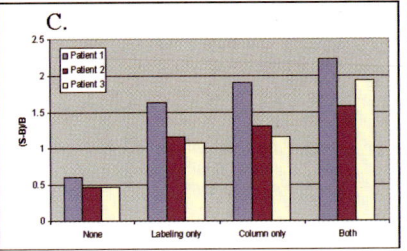

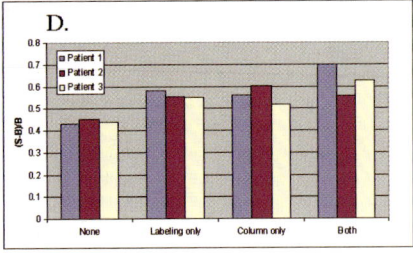

Fig. 7 Average backgrounds and signal-to-background ratios for the four conditions. (A & B) For each of the three serum samples, the local backgrounds from all the antibodies on the array were averaged and plotted with respect to experimental condition for both the 543 nm (A) and the 633 nm (B) channels. (C & D) The signal-to-background ratios from all the antibodies on the array were averaged and plotted with respect to experimental condition for both the 543 nm (C) and the 633 nm (D) channels. The signal-to-background ratio was calculated as (S-B)/B, where S = fluorescence in the spot and B = local background.

The topic of platelet contamination and also that of choice of sample type, *i.e.*, serum *versus* plasma, are addressed by an analysis of peptide contents of various samples. During processing of venous blood into serum, various *ex vivo* processes occur which lead to neo-generation of many peptides. Since serum generation relies on a biochemical process, it is reasonable to expect that various parameters like temperature after sample collection, time for sample processing/clot formation or medication of patients, can alter the peptide content of serum. These issues are difficult to standardize in routine clinical practice, in cohorts, and among different centers. Furthermore, certain proteins may bind to the clot in an uncontrolled fashion (causing a concominant decrease in free protein concentration during clot formation). Finally, serum shows many highly concentrated and intense peptide signals, which impede the detection of endogenous peptides. Therefore the use of serum samples for peptidomic mono/oligo biomarker discovery should be avoided with these caveats in mind. Different research groups [13, 14] use serum peptide patterns for prediction of early stage cancers and the controversial debate [15, 16] about this approach is still ongoing. At this time it is not clear whether the proteomic patterns reflect directly disease related peptides or peptides which are generated due to secondary effects during *ex vivo* coagulation and may only be loosely connected to the disease. Nevertheless, serum is the most commonly archived sample and may be helpful to validate results from biomarker studies, so care must be exercised when using archived samples, given the potential pitfalls described.

With regard to selection of the optimal sample type, we suggest that this will be dependent on the downstream analysis that is performed. Each of the individual sample types: serum, EDTA-plasma, heparin-plasma, and citrate-plasma, all exhibit shortcomings and should not be used under particular circumstances. We have discussed above the issues relating to serum ([12] and Fig. 1). EDTA is an aminopolycarboxylic acid and is negatively charged. It forms soluble complexes with metal ions and prevents them from further reaction [17]. If the endpoint measurement of interest involves assays wherein divalent cations, such as Mg^{2+} or Ca^{2+}, are necessary, EDTA-plasma is not an ideal sample choice. This can occur in cases where the assay is used to measure the free ion, the metal ion is required as an enzyme cofactor, or when the metal ion is an intermediate in the assay reaction. Heparin can result in interference in some affinity processes, such as SELDI-TOF analysis [6]. Heparin is a sulfated glycosaminoglycan that prolongs the clotting time of blood [17]. It is a highly charged molecule, and its presence in solution can compete for or prevent binding of molecules to charged surfaces. This step is important for the surface adsorption of proteins to protein chips; in the SELDI process, these bound proteins are laser-desorbed, ionized, and then detected after traversing the length of the flight tube. Finally, citrate can bind calcium [17] and has been shown to cause falsely lowered readings of immunoassay measurements for multiple analytes [18]. The citrate anticoagulant is present in collection tubes in liquid form, and thus, exerts a dilutional effect when blood is added during sample collection. Thus, an educated choice for the selection of optimal sample type requires one to be cognizant of each of these caveats and to take the necessary precautions in choosing the most appropriate specimen.

The HUPO plasma samples show significant differences compared to reference plasma collected according to the BioVisioN sample protocols. In our study, different protocols for EDTA/citrate specimen collection were investigated. A major finding is the detection of platelet-derived peptides in HUPO specimens. After correlation analysis (detailed in [12]), we estimate that at least 14% of peptides are platelet related. This was confirmed by sequencing a subset of 20 peptides. The appearance of platelet related peptides is either because residual platelets were present in HUPO samples, or platelets were activated prior to their removal. This may have led to release of platelet derived peptides or enzymes causing further proteolysis of proteins. The elimination of platelets prior to freezing is recommended; in addition, activation of platelets prior to their removal has to be kept to a minimum. We established protocols to remove platelets (total residual platelets <10/nL, see [12]) to obtain a platelet-poor plasma by combining centrifugation with a filtration step. Alternatively, sequential centrifugation at room temperature may be useful. We suggest the use of platelet-poor plasma as the preferred specimen for peptidomic analysis. Further, we suggest that protein sequence identifications corresponding to intracellular proteins should be carefully characterized to distinguish between possible biomarker candidates from distant organ systems and platelet derived, contaminating proteins.

Regarding storage conditions, the purpose of our studies was to determine what changes occur to a reference serum sample upon storage under different conditions of temperature and length of time (Tab. 3). Interestingly, in contrast to degradation in cell lysates where incubations of minutes to hours is often sufficient to completely destroy most protein bands and convert gel images to smears, it was observed that even very long term degradation at room temperature in serum only marginally or moderately degrades most high abundance proteins. Thus, the general pattern displayed by serum proteins upon analysis by SDS-PAGE or 2-D gels remains relatively similar, because such read-out is dominated by the abundant proteins. However, as the above data shows, while serum is more stable to proteolysis than cell lysates, substantial proteolysis and other modifications to a subset of proteins is clearly occurring, even at 4°C or −20°C. In addition, the changes in enzyme activity are similar to those described for other analytes in the literature [19], and are not unexpected. Although different enzymes are susceptible to varying degrees, a survey such as this, which comprises a panel of proteins, is likely to reveal changes to a subset of the analyzed proteins.

Further, the use of glycerol, or other similar additives, to stabilize protein structure and/or function during storage, crystallization, lyophilization, and other harsh conditions, has been previously described [20–25]. Results of these studies demonstrate that such additives can help mitigate damage to protein structure and/or stability, demonstrating the sensitivity that certain proteins can have towards their physical environment.

Extending beyond this generic use of alcohol additives, several approaches have been used to measure the specific benefits of protease inhibitors to protect plasma proteins. Interestingly, on 2-D gels, there was no direct evidence that PIs protected against a decrease in molecular weight. Instead, findings suggest that inclusion of a particular protease inhibitor cocktail causes isoforms of certain human plasma proteins (depleted of albumin and IgG) to be shifted to higher pI isoforms. These phenomena are fully discussed at a high level of detail in a separate manuscript [26]. Systematic analysis of the PI cocktail leads to the conclusion that AEBSF was the cause of this significant distortion of the 2-DE gel profile using citrated human plasma. Similar results were observed using heparinized and potassium-EDTA plasma (results not shown). AEBSF (also called Pefabloc) is a serine protease inhibitor, which inhibits proteases such as trypsin, chymotrypsin, plasmin, kallikrein and thrombin [27]. Inhibition of these proteases takes place by covalent (irreversible) modification of serine residues in the active site through the formation of sulfonate esters. AEBSF has also been shown to derivatize other proteins by serine modification [28]. Such a covalent modification would add an amine functional group, thereby shifting the proteins to higher pI forms.

In contrast, the SELDI study presented suggests that a PI cocktail has immediate and detectable benefits to stabilize proteins at the time of phlebotomy. For processed blood samples, either serum or plasma, the concept of "time zero" is such that at least several minutes pass during processing before any newly acquired blood sample is available for further analysis. Even under optimal collection conditions, blood processing requires up to 10 min post-phlebotomy for plasma, and at least 30 min or more for serum. During this processing time, the intrinsic biochemical processes of blood continue to operate, including enzymatic proteolysis that could possibly affect proteomic analyses [29]. The potential beneficial effects of protease inhibitors have been analyzed. The presence of protease inhibitors at the moment of blood draw, could reduce proteolytic damage, and thus prevent any time dependence of sample processing to overall proteomic and diagnostic outcomes. A representative result (Fig. 4) is included here, and a more thorough description of the experiment and results will be described in a subsequent manuscript (Yi and Gelfand, in preparation).

Interestingly, these observed differences are present at "time zero," suggesting the requirement for and benefit of the immediate mixing of PIs with blood during phlebotomy. These data are consistent with PI-mediated preservation of the species corresponding to this peak of interest. There are a number of other peaks (Yi and Gelfand, unpublished observations) that appear "preserved" by PI presence, and also a number of peaks, generally in the peptide range, that are more intense in the EDTA samples, so the figure presented does not represent an isolated incident. All data thus far are consistent with the benefits of blocking protease activity, and, perhaps more importantly, of blocking this activity immediately, during sample acquisition.

Another observed benefit of PI use has been seen with an antibody microarray approach. The decrease in background signal with the addition of PI could be the result of cleavage products being more "sticky," or likely to bind nonspecifically, as compared to their whole-protein counterparts. Thus, when the production of cleavage products is suppressed, nonspecific binding to the nitrocellulose is also suppressed. The cumulative effect from addition of PI both before and after labeling likely reflects the amount of time the samples were incubated with or without inhibitors. Further research will focus on the effects of PI on data quality when using other surfaces and on the quantification of specific proteins.

In summary, with regard to the use of PIs, we conclude that their use is recommended for top-down analyses. However, the researcher must be careful to avoid some of the potential drawbacks that might arise with use of PIs, with particular regard to the analytical method and targets being evaluated. Care should be exercised in the selection of the specific inhibitor. Those which act through a non-covalent mechanism should not cause any unexpected modifications. The observations of alterations in isoelectric forms in the presence of AEBSF provide an example of artifactual modification of proteins that are possible when using covalently modifying PIs, potentially adding analytical complexity.

Beyond potential artifactual modifications, there are process considerations as well. Obviously, inclusion of PIs immediately prior to a complete proteolytic sample digestion (e.g., for MS/MS analysis in a "bottom-up" approach) is counterproductive. Further, the use of low molecular mass PIs, or other chemical additives, can mask the presence of species at similar mass ranges, such as for peptide analyses.

3.4.1
Other pre-analytical variables and control considerations

Proteomic investigations involving blood products (serum/plasma) are especially complicated for a variety of reasons. Not only is the plasma proteome very complex, there are a wide range of other variables that can affect the source, acquisition and treatment of the blood sample itself, and many of these variables could lead to alteration of the detected proteome. These effects may range in severity from subtle or not detectable, to those having a great impact. The variables fall into distinct categories, as shown in Tab. 2. An appreciation of these conditions and their effects, and the diligent tracking of these variables, may prove to be essential for reproducibility of results.

For the HUPO PPP effort, BD Diagnostics prepared reference specimens under strictly controlled conditions and these were subsequently made available to all laboratories [1, 2, and Section 2], thus eliminating many of the potential variations. Additional studies will be necessary to evaluate all of these variables in Tab. 2, and the severity of their impact upon the plasma proteome. Such studies can be conducted in the future with national or regional proteomics organizations. Until a more conclusive body of data concerning these variables can be generated, pre-analytical effects on individual studies may only be discovered in retrospect. Thus, meticulous tracking of pre-analytical variables is critical for analysis, evaluation and ensuring reproducibility of results.

With regards to the storage and handling of samples, there are a number of issues that are worthy of mention. It is critical to accurately document the temperature storage history of each individual archived serum or plasma sample, in addition to establishing realistic and well defined sample processing times. Conventional wisdom, in the absence of other data, suggests that samples should be aliquoted and stored in liquid nitrogen as soon as possible after collection. However, this approach will probably be impractical for most sample collection efforts. A common practical compromise is to process and aliquot serum and plasma samples within 1 to 2 h of collection followed by flash freezing and storage at $-80°C$ in a freezer where temperature fluctuations are minimized. However, potentially degrading enzymatic processes may be damaging, even during this 1–2 h timeframe (as demonstrated by the SELDI data presented above). Further, the number of freeze-thaw cycles should be minimized, ideally using sample without refreezing, but certainly to no more than once, with an absolute limit of two. Such procedures, and accompanying rigorous documentation of temperatures and times, will help to maintain the integrity of samples, which is paramount in any proteomic analysis. Researchers will need to be aware of developing data on such variables, to improve the quality of plasma proteome research in general.

3.4.2
Reference materials

Reference materials (RMs) are applied in a variety of analytical procedures to perform harmonized quantifications of analytes. Methods can be standardized by the use of common RMs [30]. The use of international RMs ensures world wide uniformity in most analytical measurements and provides users with reliable and comparable quantitative information based on operationally defined "conventional true values" in the sense of a "gold standard" of the certified analytes. The agreement of a measured average value with a given "conventional true value" displays the systemic error component or the trueness of the measurement expressed as a bias, *i.e.*, a deviation from a "conventional true value" [31]. Thus, RMs are an important tool for quality control and quality assurance, and are crucial for the determination of the accuracy of a measurement determined by both its trueness and precision. In conjunction with reference procedures, RMs are the best guarantee for correct calibration and thus trueness of results [32].

Besides these benefits, the application of appropriate RMs is the basis for the cross validation of technologies and methods. The use of RMs enables the diligent evaluation of pre-analytical influences, *e.g.*, sampling, storage and processing, and analytical influences. These aspects are of definite importance for the objectives of the HUPO PPP [1, 2]. An example of their use is provided by Favaloro *et al.* [33] who showed that refrigerated storage of blood samples can have an effect on the measurement of Factor VIII and von Willebrand Factor proteins. These values were shown to be lowered, when compared to reference sample results, as a result of storage under refrigerated conditions and such measurements could potentially result in the misdiagnosis of von Willebrand disorder or hemophilia A.

RMs can also aid in the determination of reference limits [30]. This can be helpful in the discovery of new diagnostic markers, and can also affect their validation, which is a crucial step towards their clinical application [34]. In this respect, the use of RMs or related standards may also be leveraged in proteomics research, as one major goal of proteomics' activities is the discovery of new markers for diagnostic applications. This will necessarily apply to HUPO PPP efforts in the initial phase and also for its long term goal.

RMs are standards developed by international and national organizations that contain certified quantities of substances. The working group on Plasma Protein Standardization of the International Federation of Clinical Chemistry (IFCC) established a Certified Reference Material (CRM) for plasma proteins called CRM 470 [35–37 and http://www.irmm.jrc.be/mrm.html]. CRM 470 contains 15 proteins with certified concentrations; these are listed in Tab. 5. The CRM 470 preparation has been obtained by a very well described protocol encompassing sample collection, pooling, processing, stability testing, accelerated degradation, and finally, value assignment. Molecular characteristics of all the proteins in CRM 470 were established by electrophoretic and immunochemical techniques. The values of the 15 constituents are within the reference intervals for those of healthy individuals, and are detailed in Tab. 5.

Based on primary and secondary international RMs like CRM 470, industry standards, calibrators, and controls are produced that serve laboratories as references. For example, among other calibrators from Dade Behring used in the immunoassay analysis of the HUPO PPP reference specimens, Dade Behring's N Protein Standard SL was applied [18]. The N Protein Standard SL is based on CRM 470, WHO 67/97, WHO 80/578, IRP Code 75/502, and highly purified proteins in case international RMs are not available. N Protein Standard SL contains 25 proteins with certified concentrations, listed in Tab. 5.

Undoubtedly, the use of common RMs involves compromises and falls short of an ideal solution, in particular for proteomics solutions as only a limited number of proteins have certified concentrations, and serum or plasma is processed to gen-

Tab. 5 List of proteins with certified concentrations for CRM 470 and Dade Behring's N Protein Standard SL (Lot #083629)

Protein	Concentrations [g/L]	
	CRM 470[a]	Dade Behring N Protein Standard SL Lot No. 083629[b]
Total protein		72.8[c]
Albumin	39.7	47.80[d]
IgG	9.68	9.33[d]
IgG 1		6.06[e]
Transferrin	2.45	2.95[d]
IgG 2		2.75[e]
Ig/L chain type kappa		2.24[b]
IgA	1.96	1.85[d]
Alpha-2-macroglobulin	1.64	1.60[d]
Alpha-1-antitrypsin (alpha-1-proteinase-inhibitor)	1.206	1.57[d]
Haptoglobin	0.893	1.41[d]
Complement 3c	1.091	1.40[d]
Ig/L chain type lambda		1.23[b]
Hemopexin		0.943[f]
Alpha-1-acid glycoprotein (orosomucoid)	0.656	0.889[d]
IgM	0.797	0.735[d]
IgG 4		0.553[e]
IgG 3		0.368[e]
Prealbumin (transthyretin)	0.243	0.314[d]
Ceruloplasmin	0.205	0.31[d]
Alpha-1-antichymotrypsin	0.245	
Complement 4	0.151	0.241[d]
Retinol-binding protein		0.045[f]
C-reactive protein	0.0392	
Beta-2-microglobulin		0.0013[f]
Soluble transferrin receptor		0.0012[f]
Ferritin		0.000348[g]
IgE		568 IU/mL[h]

a) Whicher, J. T. [36]
b) Dade Behring N Protein Standard SL, product data sheet
c) Measurement by TCA method-Dade Behring BN
d) IFCC/BCR/CAP (International Federation of Clinical Chemistry/Community Bureau of Reference/College of American Pathology)- CRM 470 = RPPHS (Reference Preparations for Proteins in Human Serum) Lot 5
e) WHO (World Health Organization) 67/97
f) Based on highly purified protein hemopexin (Lot No. 070792/IA), retinol binding protein (Lot No. 100990/IA), β2-microglobulin (Lot No. 111093/IA), soluble transferrin receptor (Lot No. 97.2)
g) SRP WHO 80/578
h) IRP (International Reference Preparation) Code 75/502

erate RMs. The latter aspect indicates that RMs do not reflect a native serum or plasma composition. Nevertheless, quality control and quality assurance by RMs provide more than just information on the trueness of measurements of certified proteins. They also help in the assessment of the quality and comparison of data regarding proteins that are not certified. Thus, RMs and derived standards will allow for the systematic inter-laboratory and inter-technology comparisons that will be part of the continuing HUPO-PPP initiative. Clinical chemists, clinicians, and scientists working in proteomics should follow the reference values strategy [38], as it will improve standardization in preparation for the subsequent phases of HUPO PPP efforts of larger, population-based studies.

3.5
Concluding remarks

Plasma proteomic analysis is a complicated undertaking due to factors ranging from the dynamic range of protein analytes to the sheer variation in physical properties of the individual proteins comprising the human proteome. The examples in this article all illustrate further complications, including aspects inherent to the sample itself, in addition to sample handling, transport, storage, and processing. These variables are all pre-analytical processes, and influence the quality or reproducibility of the data, in addition to the myriad of analytical process variables themselves.

The examples cited in this article include many aspects of pre-analytical variation, from the phlebotomy event itself, to the addition and selection of protease inhibitors. These studies were conducted independently to address the many different issues. It is difficult to draw detailed sample handling recommendations that can be applied in broad situations. However, several important outcomes can be derived from this assembly of data. First, it is clear that particular analytical techniques may have certain sample and/or pre-analytical requirements that are transparent to other methods. Thus, it is advisable for researchers to consider all aspects of the sample and its handling, in relation to the analysis that is performed. Although it may be an obvious statement, certain variables may only be discovered to be important in retrospect, or may easily be overlooked entirely by those not familiar with the details of clinical sample acquisition and handling (Tab. 2). The data presented herein suggest that it is important to match the sample type, *e.g.*, selection of plasma *versus* serum, to the analytical method, whatever that may be. For example, rigorous peptidomic analyses may be incompatible with the use of serum samples. In addition, sample aging during storage can alter analytical data, possibly in unanticipated ways.

Second, and perhaps more important, is the fact that, in the nascent field of proteomics, there are a large number of factors, maybe even currently unidentified, that could alter data or its reproducibility. Our most important collective suggestion is to carefully track every possible variable, and to eliminate as many as possible, during all steps between phlebotomy and the final proteomic data generation. Thus, full annotation of the donor's history information, the sample, and the numerous processes in its handling are recommended. Use of reference materials for calibration, control, traceability, and commutability can help improve reproducibility and strengthen comparisons between experiments and laboratories. Proper specimen storage, including minimizing the exposure of samples to elevated temperatures and for extended periods of time, is critical to maintaining specimen integrity. Protease inhibitors may be protecting plasma proteins, as early as during the phlebotomy process itself, and protease inhibitor use seems likely to provide a more reproducible sample. However, some inhibitors, e.g., AEBSF, have the potential to alter proteins and thus careful consideration of the desired analytical outcome is important. Protease inhibitor presence has also been demonstrated to have a beneficial effect on one analytical method outcome itself (Fig. 6), and thus protection against proteolysis can have benefits beyond merely preserving the original sample proteins. Careful consideration of all aspects of sample handling and the pre-analytical process will strengthen studies, improve reproducibility, and accelerate the growth in the utility of proteomics for further characterization of the plasma proteome.

We would like to thank Lynn Echan at the Wistar Institute for performing the 1-D and 2-D gel analyses associated with the stability studies, and Randall Orchekowski at the Van Andel Institute for performing the antibody microarray experiments. A portion of this work was supported by a grant from the HUPO to A.J.R.

3.6
References

[1] Omenn, G. S., *Proteomics* 2004, *4*, 1235–1240.

[2] Omenn, G. S., *Clin. Chem. Lab. Med.* 2004, *42*, 1–2.

[3] Schrader, M., Schulz-Knappe, P., *Trends Biotechnol.* 2001, *19*, 55–60.

[4] Schulz-Knappe, P., Zucht, H. D., Heine, G., Jurgens, M. et al., *Comb. Chem. High Throughput Screen.* 2001, *4*, 207–217.

[5] Tammen, H., Mohring, T., Kellmann, M., Pich, A. et al., *Clin. Chem.* 2004, *50*, 545–551.

[6] Rai, A. J., Stemmer, P., Zhang, Z., Adam, B. L. et al., *Proteomics* 2005, *5*. DOI 10.1002/pmic.200400606

[7] Görg, A., Obermaier, C., Boguth, G., Weiss, W., *Electrophoresis* 1999, *20*, 712–717.

[8] Schagger, H., von Jagow, G., *Anal. Biochem.* 1987, *166*, 368–379.

[9] Hochstrasser, D. F., Patchornik, A., Merril, C. R., *Anal. Biochem.* 1988, *173*, 412–423.

[10] Zhou, H., Bouwman, K., Schotanus, M., Verweij, C. et al., *Genome Biol.* 2004;5, R28. Epub.

[11] Richter, R., Schulz-Knappe, P., Schrader, M., Standker, L. et al., *J. Chromatogr. B Biomed. Sci. Appl.* 1999, *726*, 25–35.

[12] Tammen H., Schulte, I., Hess, R., Menzel, C. et al., *Proteomics* 2005, *5*. DOI 10.1002/pmic.200400419

[13] Villanueva, J., Philip, J., Entenberg, D., Chaparro, C. A. et al., *Anal. Chem.* 2004, *76*, 1560–1570.

[14] Petricoin, E. F., Ardekani, A. M., Hitt, B. A., Levine, P. J. et al., *Lancet* 2002, *359*, 572–577.

[15] Check, E., *Nature* 2004, *429*, 496–497.

[16] Diamandis, E. P., *Mol. Cell. Proteomics* 2004, *3*, 367–378.

[17] *Tubes and Additives for Venous Blood Specimen Collection; Approved Standard- Fifth Edition*, NCCLS, Wayne PA 2003, H1-A5, 23(33).

[18] Haab, B. B., Geierstanger, B. H., Michailidis, G., Vitzthum, F. et al., *Proteomics* 2005, *5*. DOI 10.1002/pmic.200400470

[19] Procedures for the Handling and Processing of Blood Specimens; Approved Guideline – Second Edition, NCCLS, Wayne PA 1999, 19(21).

[20] Radha, C., Muralidhara, B. K., Kumar, P. R., Tasneem, R. et al., *Indian J. Biochem. Biophys.* 1998, *35*, 76–85.

[21] Gupta, C. K., Leszczynski, J., Gupta, R. K., Siber, G. R., *Vaccine* 1996, *14*, 1417–1420.

[22] Gekko, K., Morikawa, T., *J. Biochem.* 1981, *90*, 39–50.

[23] Sousa, R., *Acta. Crystallogr. D. Biol. Crystallogr.* 1995, *51*, 271–277.

[24] Ruan, K., Xu, C., Li, T., Li, J. et al., *Eur. J. Biochem.* 2003, *270(8)*, 1654–1661.

[25] Liao, Y. H., Brown, M. B., Quader, A., Martin, G. P., *Pharm. Res.* 2002, *19*, 1854–1861.

[26] Schuchard, M. D., Mehigh, R. J., Cockrill, S., Lipscomb, G. T. et al., *Biotechniques* 2005, in press.

[27] Mintz, G. R., *BioPharm* 1993, *6*, 34.

[28] Sweeney, B., Proudfoot, K., Parton, A. H., King, L. M. et al., *Anal. Biochem.* 1997, *245*, 107–109.

[29] Marshall, J., Kupchak, P., Zhu W., Yantha, J. et al., *J. Proteome Res.* 2003, *2*, 361–372.

[30] Steinmetz, J., Tarallo, P., Fournier, B., Caces, E. et al., *Eur. J. Clin. Chem. Clin. Biochem.* 1995, *33*, 337–342.

[31] www.iso.org, International Standards Organization (ISO) ISO/DIS 5725-1, Geneva: ISO, 1990.

[32] Büttner, J., *Eur. J. Clin. Chem. Clin. Biochem.* 1995, *33*, 981–988.

[33] Favaloro, E. J., Soltani, S., McDonald, J. et al., *Am. J. Clin. Pathol.* 2004, *122*, 686–692.

[34] Granger, C. B., Van Eyk, J. E., Mockrin, S. C., Anderson, N. L., *Circulation* 2004, *109*, 1697–1703.

[35] Baudner, S., Haupt, H., Hubner, R., *J. Clin. Lab. Anal.* 1994, *8*, 177–190.

[36] Whicher, J. T., *Clin Biochem* 1998, *31*, 459–465.

[37] Johnson, A. M., Whicher, J. T., *Clin. Chem. Lab. Med.* 2001, *39(11)*, 1123–1138.

[38] Petersen, P. H., Henny, J., *Clin. Chem. Lab. Med.* (special issue) Reference Values 2004, *42*, 7.

4
Immunoassay and antibody microarray analysis of the HUPO Plasma Proteome Project reference specimens: Systematic variation between sample types and calibration of mass spectrometry data*

Brian B. Haab, Bernhard H. Geierstanger, George Michailidis, Frank Vitzthum, Sara Forrester, Ryan Okon, Petri Saviranta, Achim Brinker, Martin Sorette, Lorah Perlee, Shubha Suresh, Garry Drwal, Joshua N. Adkins and Gilbert S. Omenn

Four different immunoassay and antibody microarray methods performed at four different sites were used to measure the levels of a broad range of proteins ($N = 323$ assays; 39, 88, 168, and 28 assays at the respective sites; 237 unique analytes) in the human serum and plasma reference specimens distributed by the Plasma Proteome Project (PPP) of the HUPO. The methods provided a means to (1) assess the level of systematic variation in protein abundances associated with blood preparation methods (serum, citrate-anticoagulated-plasma, EDTA-anticoagulated-plasma, or heparin-anticoagulated-plasma) and (2) evaluate the dependence on concentration of MS-based protein identifications from data sets using the HUPO specimens. Some proteins, particularly cytokines, had highly variable concentrations between the different sample preparations, suggesting specific effects of certain anticoagulants on the stability or availability of these proteins. The linkage of antibody-based measurements from 66 different analytes with the combined MS/MS data from 18 different laboratories showed that protein detection and the quality of MS data increased with analyte concentration. The conclusions from these initial analyses are that the optimal blood preparation method is variable between analytes and that the discovery of blood proteins by MS can be extended to concentrations below the ng/mL range under certain circumstances. Continued developments in antibody-based methods will further advance the scientific goals of the PPP.

* Originally published in Proteomics 2005, 13, 3278–3291.

Exploring the Human Plasma Proteome. Edited by Gilbert S. Omenn
Copyright © 2006 WILEY-VCH Verlag GmbH & Co. KGaA, Weinheim
ISBN: 3-527-31757-0

4.1
Introduction

Antibody-based analytical methods can provide quantitative, reproducible, and sensitive measurements of specific analytes. These capabilities are valuable both for routine clinical analysis and for the high-throughput exploration of hypotheses regarding specific proteins. The multiplexing of antibody-based assays through the use of planar microarrays [1–6] and bead arrays has opened up new research opportunities. The various formats of antibody microarrays and the applications and relative merits of each are reviewed elsewhere [7]. Several goals of the Plasma Proteome Project (PPP) of the HUPO can be advanced through the use of antibody-based methods.

One of the major goals of the pilot phase of the PPP was to determine the effects of the blood preparation method on the quality of proteomic data. Blood may be prepared as serum (the soluble portion of clotted blood) or as plasma (the soluble portion of anticoagulated blood), and various anticoagulants may be used to make plasma. Before attempting a large-scale study of the human plasma proteome, it is necessary to determine if the preparation method introduces systematic alterations to the levels of all proteins or specific proteins, or whether certain preparation methods are desirable or not for certain applications. Antibody-based methods are well suited to study that question, since the levels of multiple proteins may be precisely and accurately measured in multiple samples. An additional valuable use of antibody-based methods for the PPP is to provide complementary information to the broad-based discovery capabilities of separations and MS methods.

The exploration of these topics was facilitated by the assembly of human serum and plasma reference specimens by BD Diagnostics (Franklin Lakes, NJ), the National Institute for Biological Standards and Control (NIBSC, UK), and the Chinese Academy of Medical Sciences (CAMS, Beijing) [8]. Blood samples, each pooled from a male and female donor, were prepared in four ways: as serum, as plasma anticoagulated with sodium citrate, as plasma anticoagulated with K-EDTA, and as plasma anticoagulated with lithium heparin. Four different laboratories used antibody-based methods to analyze the reference specimens, with each laboratory using a distinct method. The combined data sets were used to investigate the level of systematic variation in protein levels introduced by the preparation methods and to gain further insight into the suitability of the various methods for proteomic analyses. We evaluated the following: evidence for bias in the concentrations of all the proteins in general; evidence for protein-specific alterations in concentration as a function of preparation method; and the relationship between the detectability of proteins by MS and their concentrations in plasma or serum.

4.2
Materials and methods

4.2.1
Reference specimens

Reference specimens were prepared by BD Diagnostics, NIBSC, and CAMS [8]. BD prepared three different specimens (designated BDAA, BDAF, and BDCA), each a pool from a male and female donor, in four different ways – as serum, as plasma anticoagulated with sodium citrate, as plasma anticoagulated with K-EDTA, and as plasma anticoagulated with lithium heparin – resulting in 12 different samples. CAMS prepared one specimen, pooled from a male and a female donor, with the four methods. The NIBSC made available its Thrombosis and Hemostasis standard, a lyophilized citrate-anticoagulated-plasma [8]. The samples were shipped frozen on dry ice to the four sites. The receiving sites were not blinded to the sample types. Dade Behring (DB) received all the specimens, Van Andel Research Institute (VARI) received the three BD specimen sets, and the later participants Genomics Institute of the Novartis Foundation (GNF) and Molecular Staging (MSI) received the BDAA, BDAF, CAMS, and NIBSC and the BDAA, BDAF, and NIBSC specimen sets, respectively.

4.2.2
DB immunoassays

DB immunoassays (see Supplemental Tab. 1, http://www.vai.org/vari/labs/haab. asp) were performed on a Behring Nephelometer (BN) II (2.2/D, serial no. 330135) and on a Dimension (DIM) RxL (serial no. 970933-AX) from DB (Deerfield, IL) with the HUPO PPP specimens [8]. Most tests performed are approved by the Food and Drug Administration (FDA) only for serum samples, as outlined in the manufacturer data sheets. Tests for ferritin (FERR), soluble transferrin receptor (sTfR), cardiac troponin I (cTNI), and myoglobin (MYO) on the Dimension system are also approved for heparinized plasma. Tests for C-reactive protein (CRP), IgE, β2-microglobulin, and MYO on the BN system are also approved for EDTA and heparinized plasma. The creatine kinase MB (mass assay, MMB), human chorionic gonadotropin (HCG), and thyroid stimulating hormone (TSH) assays are FDA-approved for use in serum, EDTA-plasma, and heparin-plasma samples. The fibrinogen, plasminogen, antithrombin III, and fibronectin tests are approved only for EDTA- and citrate-plasma samples, not for serum. In case test formats were not compatible with a sample type (e.g., fibrinogen in serum), data were not considered.

Two assay systems were used at DB: the BN and the Dimension methods. Both are rapid, specific, precise, and accurate [9–11]. For each analysis, appropriate Dade Behring standards, calibrators, and controls were utilized, along with a PSA control from Bio-Rad Laboratories (Hercules, CA). These standards are based on highly purified proteins and/or common international reference materials (IRMs) [12–14]. BN systems are dedicated protein analyzers that apply either antiserum or particle-enhanced immunonephelometric quantitation of analytes [10, 15]. Pro-

Tab. 1 Number of individual assays with consistent maxima or minima in each preparation type for each data set. Each column gives the number of proteins for a given preparation method that showed a maximum (top) or minimum (bottom) value in that preparation method for every sample and every replicate. Total number of assays in each data set is given in the right column

	Data set	Citrate	EDTA	Heparin	Serum	Total	Total assays
Maxima	DB	0	0	2	10	12	33
	GNF	1	13	3	4	21	88
	MSI	0	10	0	9	19	168
	VARI	0	1	4	1	3	28
	Total	1	24	9	24	55	317
Minima	DB	24	0	0	1	25	33
	GNF	4	1	0	3	8	88
	MSI	0	1	2	1	4	168
	VARI	3	0	0	0	3	28
	Total	31	2	2	5	40	317

teins in the human sample form immune complexes with specific particle-bound or antiserum antibodies. These complexes scatter a beam of light, with intensity proportional to the relevant protein concentration. Dimension methods on routine clinical analyzers are enzyme immunoassays based on the "sandwich" principle. A sample incubated with chromium dioxide particles coated with an mAb and a conjugate reagent labeled mAb specific for the protein to be analyzed forms a particle/protein/conjugate sandwich. Unbound conjugate is removed by magnetic separation and washing. The sandwich bound conjugate enzyme triggers an amplification cascade, which produces a colored product [9].

4.2.3
Antibody arrays at GNF

4.2.3.1 Antibodies, reagents, microarray printing, and platform
Antibodies and antigens (Supplemental Tab. 1) were purchased from various vendors. Resonance light scattering particles (RLS) refer to colloid gold particles coated with an antibiotin antibody [16, 17] purchased from Genicon Sciences, now Invitrogen (Carlsbad, CA). A total of 88 sandwich immunoassays were assembled and optimized in two antibody array panels, panels A and B (Brinker *et al.*, in preparation).

4.2.3.2 Microarray layout and processing
Forty-eight identical antibody microarrays with up to 48 different capture antibodies were printed onto single glass microscope slides. Four such slides were mounted onto a slide holder effectively generating a microtiter plate with 384 spacing and an antibody microarray at the bottom of each well. On each slide, eight of the wells were incubated with standard mixtures of purified antigens in diluent, resulting in an

eight-point standard titration curve that was used to quantify the analyte concentrations in each sample. The 40 remaining wells *per* slide were incubated with four dilutions (2-, 20-, 200-, and 200 000-fold) of ten samples. The diluent used throughout contained Roche "Complete" protease inhibitor cocktail at one tablet *per* 50 mL. After incubation for 1 h, all arrays were washed; a mixture of biotinylated detection antibodies was applied for 1 h, followed by washing. In a final 1 h incubation, RLS gold particles were applied to the arrays. Excess material was removed by washing. Slides were dipped twice into 50 mL deionized water and spun dry before coating with "RLS archiving" solution. For further details see Saviranta *et al.*, [18].

4.2.3.3 Array imaging and data analysis

Microarray slides stained with RLS particles were imaged at a resolution of 10 µm with a 16 bit CCD camera-based scanner (Invitrogen) and images analyzed with ArrayVision, version 8.0 (Imaging Research, St. Catharine's, Canada). Median-trimmed mean signal values for each spot on a slide were imported into EXCEL. For each slide, standard curves for each analyte were generated by four-parameter logistic fitting. Unknown sample concentrations were calculated using the corresponding signal values, the curve fitting parameters, and the dilution factors. An average concentration (derived from the three replicate spots) was calculated for each of the dilutions. To obtain a single concentration value, the program automatically chose the lowest dilution that gave a signal in the dynamic range of the assay. We performed one four-slide experiment for each of the two antibody array panels. For each of the four HUPO reference specimen preparations, three aliquots of the Asian-American, African-American, and Chinese samples and one aliquot of the NIBSC citrate-plasma reference sample were incubated on the same slide and measured against the same set of standard curves.

4.2.4 Antibody microarrays at MSI

4.2.4.1 Chip manufacture

A Teflon mask was applied to each slide creating 16 individual sample wells with 0.65 cm diameter. Prior to printing, glass slides were cleaned and derivatized with 3-cyanopropyltriethoxysilane. Panels of 25–37 capture antibodies were spotted in quadruplicate into each sample well using a Perkin-Elmer SpotArray Enterprise noncontact arrayer equipped with piezoelectric tips, delivering ~350 pL for each 120 µm antibody spot. Antibodies were applied at a concentration of 0.5 mg/mL at defined positions within each of the six production chips. Each well of a slide was printed with a single array type, containing panels of 26, 27, 26, 37, 25, and 28 antibodies, respectively, for chips 1–6. (See Supplemental Tab. 1 for a complete listing of antibodies surveyed.)

4.2.4.2 Rolling circle amplification (RCA) immunoassay

The manual RCA microarray immunoassay reported previously [4] was modified to optimize performance on an automated platform (Protedyne BioCube). Incubation times were increased from 30 to 45 min for two of the assay steps (RCA signal amplification and detector incubation), and the number and volume of washes between steps increased from 2 to 4–5 and from 20 to 30 µL, respectively. Slides were scanned using an LS200 scanner (TECAN). Scanned images were analyzed using proprietary software. The fluorescence intensity was analyzed for each sample analyte with the resulting mean intensity measurements converted into concentration values.

4.2.4.3 Conversion of mean fluorescent intensity to concentration

Preparations of standardized multiplex analyte titration series were manufactured using recombinant analytes diluted in buffer covering the range from 12 pg/mL to 81 ng/mL at 14 discrete points plus zero analyte buffer blanks. These titration points were distributed among the 16 available wells on three control slides. The standard titrations, designed to overlap the linear range of detection for each individual analyte, were used to generate standard curves from which sample analyte concentrations were determined. A four-point standard titration was run on every slide for normalization and quality control purposes. The four wells, designated "anchor point" controls, were derived from the standard 14-point titration series run on separate control slides to generate standard curves for each analyte. Anchor point controls contained a cocktail of all cytokines corresponding to the printed capture antibodies for a given array. The anchor points were prepared at four concentrations that fell within the linear range of detection for each analyte. Individual sample values were normalized using linear regression of the anchor points to reduce assay imprecision observed among replicates. Fluorescence intensities of the four spot replicates for each analyte within an anchor point well were averaged on a logarithmic (base two) scale to generate within-slide titration curves. Linear regression coefficients (slope and intercept) were calculated between individual titration curves from each slide to generate an "average" titration curve. Calculated slope and intercept were used to transform averaged analyte values for each sample well. Data normalization was performed on the data set after removal of outliers.

4.2.5
Antibody microarrays at VARI

4.2.5.1 Fabrication of antibody microarrays

Antibodies (Supplemental Tab. 1) prepared at ~500 µg/mL in 1 × PBS were printed in microarrays on the surfaces of NC-coated microscope slides (FAST™ slides, Schleicher & Schuell) using a custom-built contact arrayer.

4.2.5.2 Serum labeling
The 12 PPP reference specimens were received from BD. An aliquot from each of 12 serum samples was divided into a portion to be labeled with NHS-digoxigenin and a portion to be labeled with NHS-biotin (Molecular Probes). The digoxigenin-labeled samples were pooled, and equal amounts of the pool were transferred to each of the biotin-labeled samples. Each labeled protein solution was supplemented with nonfat milk to a final concentration of 3%, Tween-20 to a final concentration of 0.1%, and $1 \times$ PBS to yield a final serum dilution of 1:100.

4.2.5.3 Processing of antibody microarrays
One hundred microliters of each labeled serum sample mix was incubated on a microarray with gentle rocking at room temperature for 2 h. After washing, the arrays were detected by two-color RCA (TC-RCA) (see [19] for experimental details). The biotin-labeled proteins were detected with green fluorescence; the digoxigenin-labeled proteins were detected with red fluorescence.

4.2.5.4 Analysis
The microarrays were scanned (ScanArray; Perkin-Elmer Life Sciences) for fluorescence using laser excitation at 543 and 633 nm; GenePix 5.0 (Axon Laboratories) was used to quantify the images. For spots with fluorescence signal surpassing an intensity threshold in both color channels [3], the ratio of background-subtracted median sample-specific fluorescence to background-subtracted median reference-specific channel fluorescence was calculated, and ratios from replicate antibody measurements within the same array were averaged.

4.2.6
Retrieval and matching of IPI numbers for the analytes

International protein index (IPI) accession numbers were obtained for each analyte in the quantitative assays of this study using two search methods. In the first search, the analyte names were subjected to an internet search to retrieve the proper protein names. The analyte names were then used to generate Sequence Retrieval System (LION Bioscience, Heidelberg, Germany) queries of the IPI database [20] using the SRS server at EBI (http://srs.ebi.ac.uk). The search parameters were as follows: protein name is in IPI AllText and OrganismName is Human. The data returned were Accession Number(s) and EntryName. The returned name from the IPI database was compared with the input analyte name, and records with the names not matching were discarded. IPI numbers corresponding to precursor forms of proteins were retained.

In the second search, the list of protein names against which the antibodies were raised was searched against the Human Protein Reference Database to identify all possible alternate names. These alternate names were further verified using the OMIM and Swiss-Prot databases. The IPI database was then searched using these names, and all IPI IDs, which corresponded to the protein name in question, were assigned to it. Each sequence corresponding to each IPI ID was further verified by conducting a BLASTP against the nr data set. The outputs were manually analyzed, and LocusLink identifiers were assigned to each sequence and cross-checked with those assigned in the IPI database. Alternate IPI IDs, as specified in the IPI data set, were also assigned so as to give all possible identifiers for each protein. Protein name and all alternate names were used to query the HUGO gene nomenclature committee's database, and the results verified using LocusLink identifiers. This allowed annotation of all entries with their gene name and gene symbol.

4.3 Results

4.3.1 Antibody-based measurements of the HUPO reference specimens

The PPP reference specimens were distributed to four different laboratories for immunoassay or antibody microarray analysis. Each of the four sites used a distinct technology for analyzing the specimens. The 39 immunoassays performed on DB clinical analyzers were based on immunonephelometric methods (33 tests) and sandwich-like enzyme immunoassays (6 tests) that use antibody-coated magnetic chromium dioxide particles. The GNF measured 88 different serum proteins using microarray-based sandwich assays detected by RLS. MSI used antibody microarrays to target 168 different proteins, mostly cytokines, using sandwich assays and detection by RCA [4, 21]. VARI measured 28 different serum proteins using TC-RCA detection on antibody microarrays [19]. The antibodies used by each site are listed in Supplemental Tab. 1. Each site independently designed their own experiments based on individual resources and experience, and the targeted proteins varied significantly between sites. The complete data sets are available at http://www.vai.org/vari/labs/haab.asp.

Two of the sites (MSI and GNF) ran the samples in triplicate, one in duplicate (VARI), and one had duplicate measurements for four of the samples (DB). The reproducibility of the replicate data is a good indicator of data quality. Replicate measurements showed good reproducibility for each data set, as depicted by the correlations of the different antibody measurements for the same sample in two separate experiments (Fig. 1). The average correlation coefficients between the different antibody measurements from replicate experiments were 0.99 for the DB set, 0.95 for the GNF set, 0.94 for the MSI set, and 0.96 for the VARI set. These high average correlations indicate that each data set is highly internally consistent.

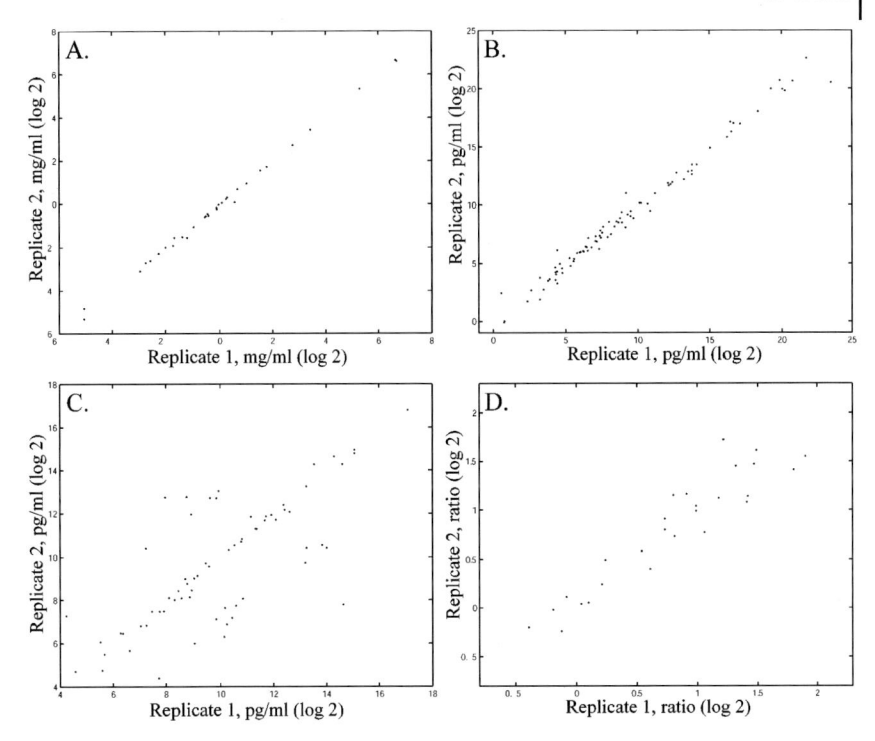

Fig. 1 Correlations between replicate measurements of one sample. Duplicate antibody measurements from a plasma sample were plotted against each other. Scatter plots from each of the four data sets are shown: (A) DB; (B) GNF; (C) MSI; (D) VARI. Correlations for each of the plots were 0.99, 0.95, 0.94, and 0.96, respectively. Plots are shown for the samples: (A) CAMS, citrate-plasma; (B) BDAA, citrate-plasma; (C) BDAF, citrate-plasma; and (D) BDCA, heparin-plasma.

Two of the data sets (GNF, MSI) used standard curves of purified antigens to calibrate the data and to calculate the concentrations of each of the measured proteins. DB analyzers used reference materials (standards, controls, and calibrators) that are based on IRMs and purified antigens for calibration and the determination of the concentrations of the analytes. The measured concentrations cover a broad range, from several mg/mL to below 1 pg/mL (Fig. 2). The GNF and MSI data sets, focusing on cytokine detection, account for most of the low-abundance measurements, while the DB and VARI sets focused on common mid-to-high-abundance serum proteins. Some overlap existed between the sets: six analytes were common between DB and GNF, three were common between DB and MSI, 11 were common between DB and VARI, 57 were common between GNF and MSI, 10 were common between GNF and VARI, and nine were common between MSI and VARI.

While the precision between replicates within each data set is good (Fig. 1), occasionally large differences were observed between platforms in the measured concentrations of common analytes. Of the 57 common analytes between GNF and

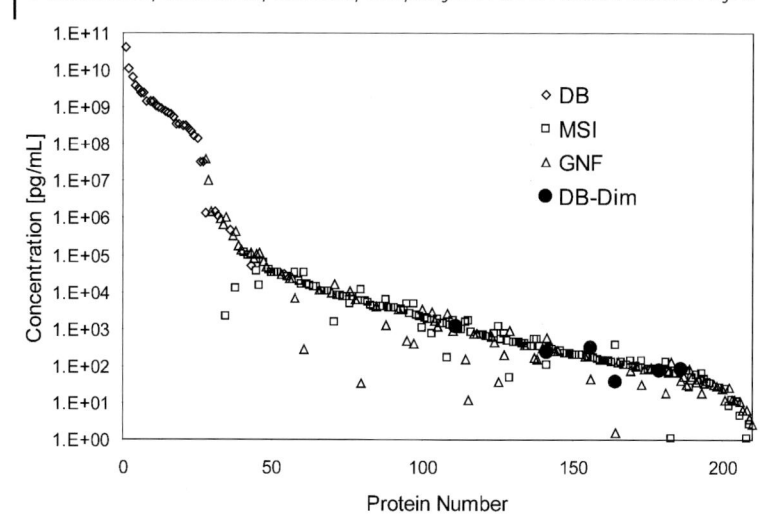

Fig. 2 Concentration range of the proteins measured in these studies. Geometric mean concentration over all the samples is plotted for each of 295 quantitative assays for 231 unique proteins. Set consists of quantitative measurements from the DB BN system (33 analytes), the DB Dimension system (6 analytes), MSI antibody microarrays (168 analytes), and GNF antibody microarrays (88 analytes). For analytes measured by more than one laboratory, the geometric mean concentration derived by each laboratory is displayed.

MSI, seven were measured more than ten-fold higher at GNF and eight were measured more than ten-fold higher at MSI. These deviations between assays in the measurement of common analytes can be seen in Fig. 2. Supplemental Tab. 2 provides the average measured concentrations of the analytes that were measured at more than one site. Interlaboratory variation is not uncommon and may be due to differences in the specificities of the antibodies used, the sample storage and treatment methods, and the calibration methods. The full exploration of the sources of variation between the laboratories was beyond the scope of this study, yet the existence of the occasional variation highlights the need for methods for calibration and validation across laboratories and platforms.

4.3.2
Systematic variation between the preparation methods of the PPP reference specimens

We investigated whether the blood preparation methods (serum, citrate-plasma, EDTA-plasma, heparin-plasma) introduced systematic bias into the abundances of all the proteins in general. A systematic bias in concentration would be evidenced by a consistent shift in the concentrations of analytes in one preparation method relative to the other methods. The protein abundances were compared between the samples that were prepared from the same starting material, *i.e.*, we compared the four preparations within the BDAA specimen set, the four preparations within the

Tab. 2 Concentrations and associated MS summary information. Information relating to 70 IPI numbers (66 unique analytes) that had a match between the analyte-derived lists and the MS-derived lists is presented. "Antibody name" = the name that was used in the searches for analyte-associated IPI numbers. "Name from analyte search" = the name in the IPI database that matched the antibody/analyte name. "Concentration" = the geometric mean concentration over all specimens as found by immunoassay or antibody microarray. "# Labs" = the number of laboratories (out of 18) that found a particular IPI number. "# Peptides" = the average number of different peptides found for that IPI number. "IPI set" = the analyte-associated list from which a match was found (see Section 2), either list 1, list 2, or both (1, 2, or B). In four instances, two different IPI numbers were associated with one analyte

Antibody name	Name from analyte search	Concentration, pg/mL	# Labs	# Peptides	Laboratory	IPI SET
Albumin	Albumin	4.0E + 10	17	201	DB	2
Transferrin	Transferrin	2.3E + 09	16	249	DB	2
Apolipoprotein A I	Apolipoprotein A-I	1.4E + 09	17	82	DB	2
α2-macroglobulin	Alpha-2-macroglobulin	1.4E + 09	17	211	DB	2
α1-antitrypsin	Serine (or cysteine) proteinase inhibitor, clade A (alpha-1 antiproteinase, antitrypsin), member 1	1.1E + 09	15	183	DB	2
C3c	Complement component 3	9.5E + 08	5	98	DB	2
Haptoglobin	Haptoglobin	8.8E + 08	18	113	DB	2
Hemopexin	Hemopexin	7.5E + 08	16	86	DB	2
Apolipoprotein B	Apolipoprotein B (including Ag(x) antigen)	7.2E + 08	13	328	DB	2
Fibrinogen	Fibrinogen, gamma polypeptide	6.7E + 08	16	66	DB	2
Fibrinogen	Fibrinogen, gamma polypeptide	6.7E + 08	12	51	DB	2
α1-acid-glycoprotein	Alpha-1-acid glycoprotein 2 precursor	6.1E + 08	14	24	DB	1
α1-acid-glycoprotein	Orosomucoid 1	6.1E + 08	16	45	DB	2
Antithrombin III	Serine (or cysteine) proteinase inhibitor, clade C (antithrombin), member 1	3.2E + 08	17	70	DB	2
Apolipoprotein A-II	Apolipoprotein A-II	3.0E + 08	15	18	DB	2
Prealbumin	Transthyretin (prealbumin, amyloidosis type I)	2.6E + 08	17	27	DB	2
Ceruloplasmin	Ceruloplasmin (ferroxidase)	2.1E + 08	15	134	DB	2

Tab. 2 Continued

Antibody name	Name from analyte search	Concentration, pg/mL	# Labs	# Peptides	Laboratory	IPI SET
C4	Complement C4 precursor [Contains: C4A anaphylatoxin]	1.7E + 08	17	157	DB	1
Plasminogen	Plasminogen	1.4E + 08	12	72	DB	2
Fibronectin	Fibronectin 1	1.1E + 08	1	86	DB	2
Apolipoprotein E	Apolipoprotein E	3.4E + 07	8	30	DB	2
vWF	Von Willebrand factor	1.3E + 06	2	46	GNF	2
β2Microglobulin	Beta 2-microglobulin protein	1.1E + 06	1	1	DB	1
β2Microglobulin	Beta-2-microglobulin	1.1E + 06	3	1	DB	2
sTfR	Transferrin receptor (p90, CD71)	5.8E + 05	1	2	DB	2
VAP-1	Amine oxidase, copper containing 3 (vascular adhesion protein 1)	1.2E + 05	2	6	MSI	2
Protein C	Mannose-binding lectin (protein C) 2, soluble (opsonic defect)	9.7E + 04	2	7	MSI	2
VCAM-I	Vascular cell adhesion molecule 1	9.4E + 04	3	9	MSI/GNF	2
TGFβ1	Transforming growth factor, beta 1 (Camurati-Engelmann disease)	7.5E + 04	2	2	GNF	2
IGF-BP3	Insulin-like growth factor binding protein 3	5.9E + 04	6	17	MSI/GNF	2
ICAM-1	Intercellular adhesion molecule 1 (CD54), human rhinovirus receptor	4.3E + 04	2	4	MSI/GNF	2
MMPg	Matrix metalloproteinase 9 (gelatinase B, 92 kDa gelatinase, 92 kDa type IV collagenase)	4.1E + 04	2	5	MSI/GNF	2
VE-cadherin	Cadherin 5, type 2, VE-cadherin (vascular epithelium)	3.0E + 04	3	11	MSI	2
M-CSF R	Colony stimulating factor 1 receptor, formerly McDonough feline sarcoma viral (v-fms) oncogene homolog	2.6E + 04	3	11	MSI	2
L-Selectin	Selectin L (lymphocyte adhesion molecule 1)	1.7E + 04	5	10	MSI	2
ALCAM	Activated leukocyte cell adhesion molecule	1.6E + 04	2	5	MSI	2
IGFBP2	Insulin-like growth factor binding protein 2, 36 kDa	1.5E + 04	1	3	MSI	2
TIMP1	Tissue inhibitor of metalloproteinase 1 (erythroid potentiating activity, collagenase inhibitor)	1.4E + 04	1	3	MSI/GNF	2

Tab. 2 Continued

Antibody name	Name from analyte search	Concentration, pg/mL	# Labs	# Peptides	Laboratory	IPI SET
EGF R1	Epidermal growth factor receptor (erythroblastic leukemia viral (v-erb-b) oncogene homolog, avian)	1.1E + 04	3	3	GNF	2
MMP2	Matrix metalloproteinase 2 (gelatinase A, 72 kDa gelatinase, 72 kDa type IV collagenase)	8.8E + 03	1	7	MSI/GNF	2
NAP-2	Nucleosome assembly protein 1-like 4	7.5E + 03	1	1	MSI	2
LIF Rα	Leukemia inhibitory factor receptor	5.0E + 03	2	4	MSI	2
PDGF-Rα	Platelet-derived growth factor receptor, alpha polypeptide	4.6E + 03	1	2	MSI	2
MMP1	Matrix metalloproteinase 1 (interstitial collagenase)	2.6E + 03	1	1	MSI/GNF	2
FasL	Tumor necrosis factor (ligand) superfamily, member 6	1.5E + 03	1	2	MSI/GNF	2
NSE	Enolase 2 (gamma, neuronal)	1.4E + 03	1	1	GNF	2
MMP8	Matrix metalloproteinase 8 (neutrophil collagenase)	9.0E + 02	1	1	MSI/GNF	2
VEGF-D	C-fos induced growth factor (vascular endothelial growth factor D)	5.0E + 02	1	1	MSI/GNF	2
ENA-78	Chemokine (C-X-C motif) ligand 5	3.4E + 02	1	1	MSI	2
CD30	Tumor necrosis factor receptor superfamily, member 8	3.3E + 02	1	2	MSI/GNF	2
MPIF-1	Chemokine (C-C motif) ligand 23	3.2E + 02	1	1	MSI	2
GROb	Chemokine (C-X-C motif) ligand 2	3.0E + 02	1	1	MSI	2
BDNF	Brain-derived neurotrophic factor	3.0E + 02	1	1	MSI	2
AFP	Alpha-fetoprotein	2.9E + 02	2	2	MSI/GNF	2
IGF-IR	Insulin-like growth factor 1 receptor	2.4E + 02	1	1	MSI/GNF	2
Calcitonin	Calcitonin/calcitonin-related polypeptide, alpha	1.9E + 02	1	1	GNF	2
Calcitonin	Calcitonin gene-related peptide type 1 receptor precursor	1.9E + 02	1	1	GNF	1
FGFβ	Fibroblast growth factor-20	1.6E + 02	1	1	GNF	1
IL-10Rβ	Interleukin 10 receptor, beta	1.5E + 02	1	1	MSI	2
Angp2	Angiopoietin 2	9.7E + 01	1	1	GNF	2
MCP-1	Splice isoform A of P15529	7.6E + 01	1	1	MSI/GNF	1

Tab. 2 Continued

Antibody name	Name from analyte search	Concentration, pg/mL	# Labs	# Peptides	Laboratory	IPI SET
SCF	KIT ligand	5.9E + 01	1	1	MSI/GNF	2
IFNγ	Interferon, gamma	5.4E + 01	1	1	MSI/GNF	2
OSM	Oncostatin M	4.8E + 01	1	1	MSI	2
IL1α	Interleukin 1, alpha	4.5E + 01	1	1	MSI/GNF	2
TNFα	Tumor necrosis factor (TNF superfamily, member 2)	3.7E + 01	2	1	MSI/GNF	2
AR	Androgen receptor (dihydrotestosterone receptor; testicular feminization; spinal and bulbar muscular atrophy; Kennedy disease)	2.6E + 01	1	1	MSI/GNF	2
I-TAC	Chemokine (C-X-C motif) ligand 11	2.3E + 01	1	1	MSI	2
CGB	Chorionic gonadotropin, beta polypeptide	1.9E + 01	1	1	MSI/GNF	2
IL7	Interleukin 7	7.0E + 00	1	1	MSI/GNF	2

BDAF specimen set, *etc.* For each preparation type (citrate-plasma, EDTA-plasma, *etc.*), the number of proteins that had a maximum concentration in that preparation was totaled. The number of proteins with minimum concentrations also was totaled for each preparation method. Those numbers were compared to the numbers of maxima or minima that would be expected by chance. Frequencies of maxima or minima much greater or lower than would be expected by chance could indicate systematic bias in the concentrations in a particular preparation method.

The results of that analysis are shown in Fig. 3. The proportion of proteins that had maxima (Fig. 3A) or minima (Fig. 3B) in each preparation type is indicated by the position on the *x*-axis of a different vertical line for each of the four data sets. The distribution of maxima and minima in each preparation method that would be expected by chance was calculated by permutation and is indicated by the histograms in each plot. As expected, the average frequency in the randomly permuted data is 0.25, since the maxima and minima are evenly distributed among the four preparation methods. All four data sets had a significantly lower frequency of maxima in citrate-plasma (Fig. 3A, top left), well below what is expected by chance. The GNF and MSI sets showed a high frequency of maxima in the EDTA-plasma samples (Fig. 3A, top right); the VARI measurements were often highest in heparin-plasma (Fig. 3A, lower left), and the DB measurements were frequently highest in the serum samples (Fig. 3A, lower right). For the minimum values, all methods showed a significantly frequent occurrence of minima for the citrate samples (Fig. 3B, top left), and the DB data were very seldom lowest using heparin-plasma or serum samples (Fig. 3B, lower left and lower right). The other frequencies are close to what might be expected by chance. These analyses show evidence for general biases in protein concentrations as a result of blood preparation method.

We examined the magnitudes of concentration differences between the sample types. For each protein, the concentration in each preparation method was divided by the maximum concentration found in that specimen set. For example, if a protein had a concentration of 100 pg/mL in citrate-plasma and 200 pg/mL in serum, citrate-plasma was given a 0.5 and serum was given a 1.0. The median concentration ratios for each preparation method are shown in Fig. 4 for each of the four data sets. Each data set shows the citrate-plasma preparation with the lowest average abundances, from about 85% of the maximum values (DB) to about 40% of the values (GNF and MSI). Consistent with the results from Fig. 3, serum had the highest concentrations in the DB set, EDTA-plasma in the MSI and GNF sets, and heparin-plasma in the VARI set. The variation between preparation methods is similar between the DB and VARI sets and between the GNF and MSI sets, and the GNF and MSI sets had broader variation in the relative abundances (larger error bars) than the other two sets. These relationships could be related to the similarity between the groups in the proteins measured; GNF and MSI measured mostly cytokines, while VARI and DB measured higher-abundance serum proteins.

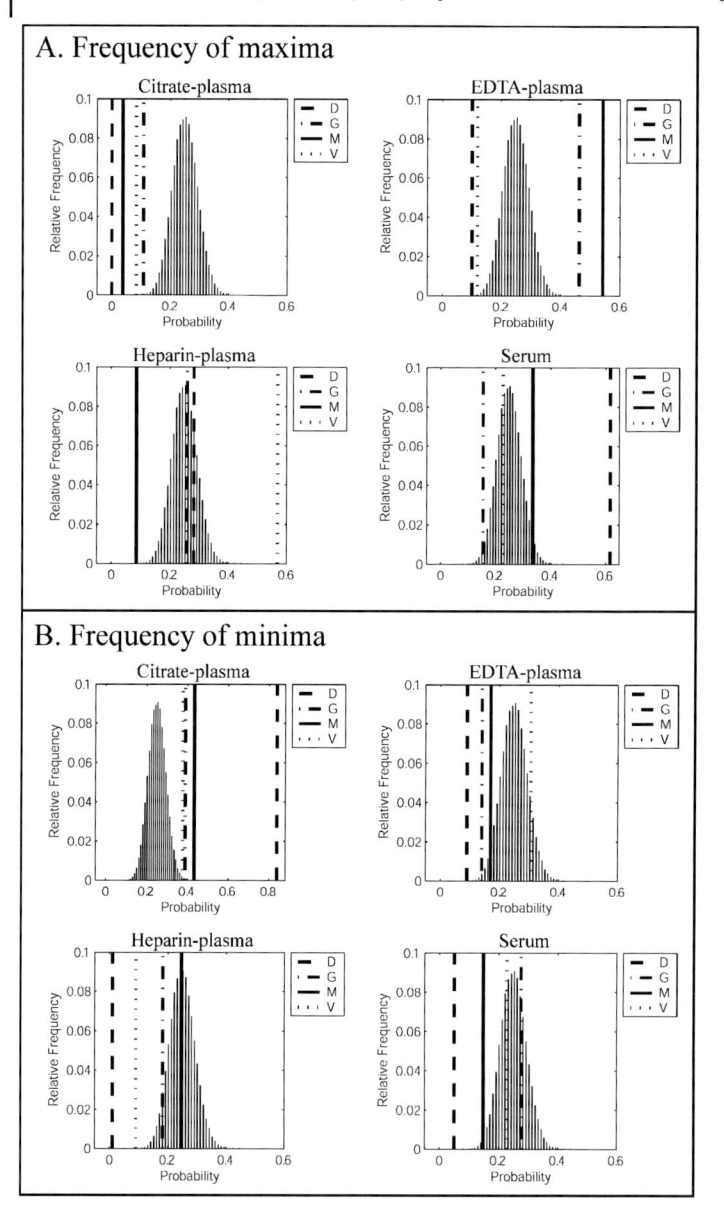

Fig. 3 Frequency of maxima (A) and minima (B) in each of the four preparation methods: citrate-plasma (upper left), EDTA-plasma (upper right), heparin-plasma (lower left), and serum (lower right). Position on the x-axis for the vertical lines in each plot indicate the frequency of maxima (A) or minima (B) in a given preparation method. Each line represents one of the four data sets, with the identities given in the legend: DB = dashed line, GNF = dotted/dashed line, MSI = solid line, VARI = dotted line. Distribution of randomly-occurring frequencies also is included in each graph.

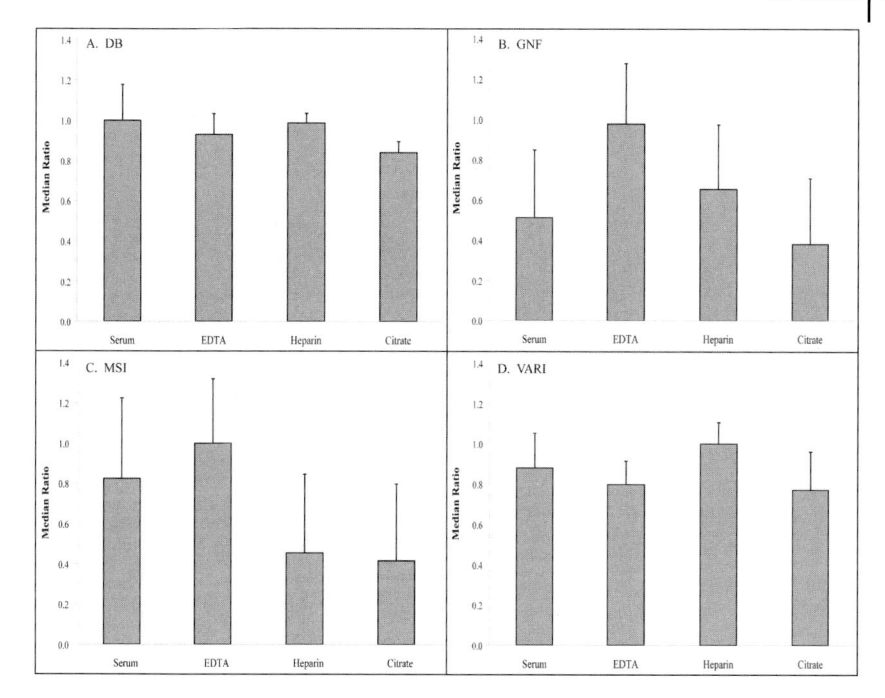

Fig. 4 Median relative concentration ratios in each sample type. Concentration of each protein in each preparation type was divided by the concentration of the preparation type that was highest for a given sample. Median relative change in concentration is depicted for each preparation type from the (A) DB, (B) GNF, (C) MSI, and (D) VARI data sets. Error bars represent the SD in relative concentration change over all the proteins.

4.3.3
Consistent alterations in specific protein abundances

We then examined whether specific proteins, as opposed to all the proteins in general, were consistently highest or lowest in a certain preparation type. Evidence for such a bias would be indicated by multiple specimen sets showing agreement in the alteration of the concentration of a specific protein, *e.g.*, if all three of the BD specimen sets showed a certain protein higher in a certain preparation method. We identified the proteins that always gave a highest value in one particular preparation method, in every specimen set, and in every replicate experiment. Such biases toward a particular preparation method are more than 99% likely not to have occurred by chance, as determined by a permutation test similar to that described above. A summary of these results is shown in Tab. 1. Many proteins were always highest in serum or in EDTA-plasma, particularly in the DB and GNF sets, respectively. Twenty-four proteins were always lowest in citrate-plasma in the DB set. These specific biases tend to follow the trends seen in the overall concentrations shown in Figs. 3, 4, but occasionally proteins are altered counter to those trends.

For example, in the GNF set, all the four preparation methods had proteins that were consistently elevated. A complete list of the proteins that seem to have concentrations systematically affected by preparation method, along with the magnitudes of the alterations, is provided in the Supplemental Tab. 3. The magnitude of the difference between preparation methods was usually below three-fold, but some proteins had much larger alterations (a ten-fold change or more) in certain preparation methods. The most consistent differences were in the DB set; the 24 proteins that were always lowest in citrate-plasma ranged from 73 to 88% of the maximum values, and the ten proteins that were always highest in serum ranged from 138 to 119% of the minimum values.

The bias for a particular preparation method in a specific protein is visually depicted in Fig. 5 for two representative proteins from each data set. The replicate measurements from each sample were plotted with respect to preparation method, with the solid lines representing the averages between the replicates. In each case shown, one preparation method is consistently highest in every sample and every replicate. EDTA-plasma, heparin-plasma, and serum each have examples in which the concentrations seem to be systematically elevated in one preparation method. Independently-collected ELISA data are plotted along with the microarray data for hemoglobin (Fig. 5G). The concordance between the microarray ELISA measurements are very good for each sample (0.94 over all the samples), validating the accuracy of the microarray measurements and the fact that the hemoglobin concentrations are highest in EDTA-plasma for these samples.

4.3.4
Linkage of MS data and antibody-based measurements

Another valuable use of these data for the PPP was to investigate relationships between the quantitative antibody-based measurements and the MS information derived from other work within the PPP. Based on the informatics integration methods (see Adamski et al., this issue), 9504 unique IPI proteins were included in the combined data (see http://www.bioinformatics.med.umich.edu/app1/test/). The link between the MS data and the antibody-based measurements was made through IPI numbers. Two different search methods were used to find IPI numbers that corresponded to the analytes measured in the quantitative antibody-based assays (see Section 2), generating two lists of analyte-associated IPI numbers. Seventy IPI numbers that were common between these lists and the MS summary data were identified and are presented in Tab. 2. In four cases, two IPI numbers were associated with the same analyte name. Tab. 2 also gives the average concentration (the geometric mean over all samples, including the NIBSC sample, and all data sets) of each analyte, the number of laboratories (out of 18) finding that IPI number, and the average number of peptides found for that IPI number. The relationships between the MS summary data and the average concentrations were examined (Fig. 6). Fig. 6A shows that individual laboratories made identifications in the 10–10 000 pg/mL range, with multiple laboratories finding the same IPI numbers above that range. Only single peptide identifications were made below

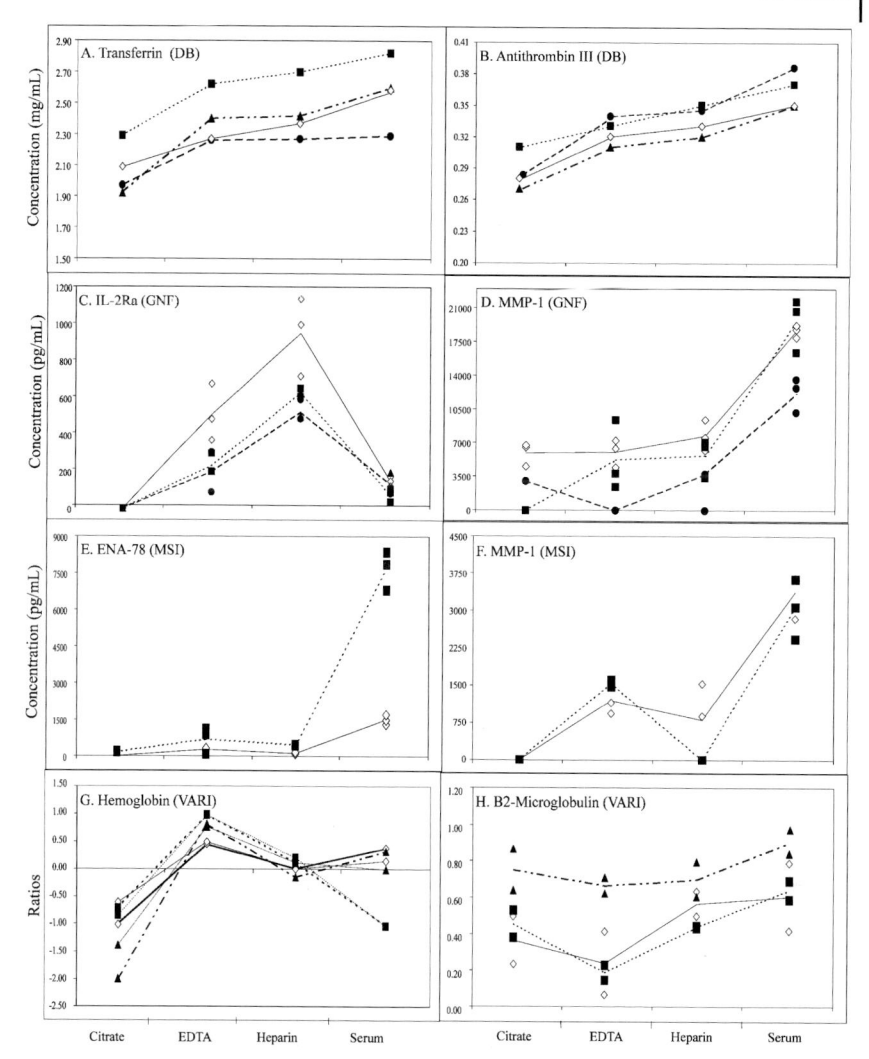

Fig. 5 Variation in the concentration of individual proteins across different preparation methods (see text for basis of selecting these proteins). Analyte and data set (in parentheses) are indicated in each plot. Replicate data from two to four different samples are plotted with respect to preparation method. Individual values for each sample are shown by the following: BD sample 1: open diamonds; BD sample 2: solid squares; BD sample 3: solid triangles; and CAMS: solid circles. Averages of the replicate data are shown by a solid line for BDAA, a dotted line for BDAF, two dots and a dash for BDCA, and a dashed line for the CAMS specimen. Graph G includes ELISA data that has been normalized to the same scale as the microarray data, represented by darker versions of the corresponding lines for the microarray data.

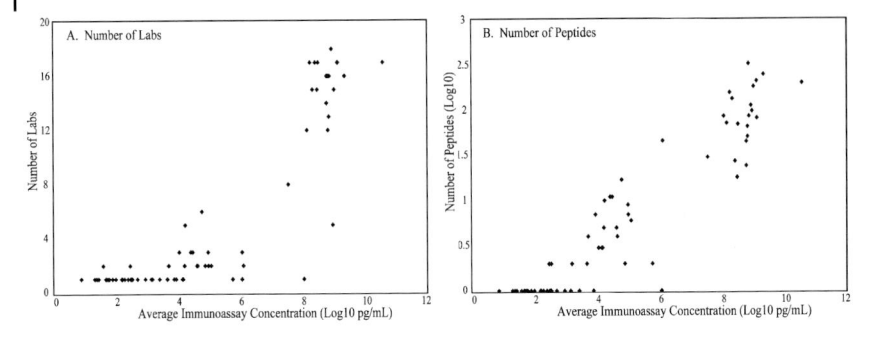

Fig. 6 MS summary data with respect to concentrations measured by immunoassays and antibody microarrays. Concentration is pg/mL (log base 10). (A) Number of laboratories finding a given protein, (B) number of peptides for each protein identification.

around 200 pg/mL, with a steadily increasing average number of peptides above that (Fig. 6B). Both metrics increased steadily with concentration. The lack of data points in the 1–100 µg/mL range is primarily due to the low number of immunoassay and antibody microarray measurements in that range, as shown in Fig. 2.

4.4
Discussion

The analysis of the HUPO PPP reference specimens by antibody-based methods provided a useful complement to the other studies of the PPP. This work examined the use of immunoassays and antibody microarray methods to investigate the systematic variation of specific proteins between the PPP's reference specimens' sample preparation methods and to provide insights into the concentration-dependence of protein discovery by MS methods. The use of four distinct methods from four independent laboratories gave a broad view of the capabilities of antibody-based methods. Each of the four data sets had highly internally reproducible data, as shown by the high average correlations between replicate data, although the values did not always agree in the measurements of common analytes. The occasional lack of concordance between the sets underscores the importance of the use of common IRMs for cross validation and calibration between laboratories and methods. An international reference standard for 15 abundant serum proteins, CRM 470 [14], has been developed; its use has significantly reduced interlaboratory variation in many protein assays in European quality assurance programs [22]. Of note, DB analyzers used standards, calibrators, and controls based on common IRMs that are generally applied in clinical chemistry. Antibody microarray measurements have not yet achieved the precision standard of clinical analyzers.

We investigated two aspects of the effect of sample preparation on protein concentration: systematic alterations of all proteins in general and consistent alterations in the concentrations of specific proteins. The most common general systematic alteration was a reduction of protein concentrations in the citrate-plasma preparation. This effect is attributable to the dilution of the plasma fraction of whole blood by the sodium citrate solution [23] and by the osmotic withdrawal of water from blood cells caused by the high salt concentration in the anticoagulant. When whole blood at a hematocrit of 0.4–0.5 is mixed with sodium citrate solution at a ratio of 9:1, the dilution of citrated plasma will be 15–19.5% (10% dilution from the citrate solution plus additional dilution from osmosis) [23]. The concentration reduction in citrate-plasma was the most consistent in the DB data and was explainable by the dilution factor, with 14 of the 17 consistently reduced proteins lower than the serum preparation by less than 20%. We might note that most of the DB analyses were not approved for use with citrate-plasma. The other data sets showed less consistent alterations in the citrate-plasma concentrations, perhaps due to lower precision in the measurements or other sources of variation besides dilution, as discussed below. Of great importance for proteomics analyses, the dilution in citrate-plasma did not seem to affect protein identification in PPP analyses using various fractionation and MS methods, as the citrate-plasma specimens gave similar numbers of proteins identified relative to the other specimen types and similar detection of low-abundance immunoassayed proteins (see Simpson *et al.*, and Omenn *et al.*, this issue).

The preparation method that generally gave the highest protein concentrations varied among the four data sets. The GNF and MSI sets had higher protein abundances in the EDTA-plasma preparation, the DB set had higher abundances in the serum preparation, and the VARI set had highest values in the serum and heparin-plasma. The GNF and MSI sets focused on cytokine detection, and the relatively higher concentration of the cytokines in EDTA-plasma could indicate a protective effect of EDTA on cytokine stability, perhaps through EDTA's role as a protease inhibitor. The more abundant, common serum proteins measured in the other two sets could be less susceptible to protease activity and therefore not necessarily higher in the EDTA-plasma preparation. Other sources of variation in concentration could be the anticoagulant-induced release of certain analytes by lymphocytes, such as the release of tumor M2-PK in heparin-plasma but not in EDTA-plasma [24], interference in certain assays by anticoagulants, or variability in protease activity or protein stability due to the presence or absence of certain anticoagulants.

The analysis of specific proteins showed that certain proteins were always highest or always lowest in certain preparation methods. The fact that some of these alterations were counter to the overall trends noted above shows that blood preparation methods can have variable effects on specific proteins or antibodies. Anticoagulants may in some cases specifically interact with certain proteins or specifically affect the stability of certain proteins. Such effects have been seen in previous studies. In one study, the levels of several hormones were either elevated or reduced between matched serum and EDTA-plasma and between matched serum and citrate-plasma samples [25]. Another study showed that parathyroid hormone is more stable in EDTA-plasma than in serum [26]. The levels of the cytokines IL-6, TNF-α, and leptin were found to be highly variable in citrate-anticoagulated- and heparin-anticoagulated-plasma but not in EDTA-anticoagulated-plasma or serum [27]. In some cases, an anticoagulant might actually bind to specific proteins. For example, EDTA binds to hemoglobin [28], which might be related to the observed consistent elevation of the hemoglobin measurements in the EDTA-plasma samples.

Based on the above observations, it is clear that comparisons between samples are only accurate when using samples that were collected with precisely the same method. Which preparation method to use in every case, however, is less obvious. No single preparation method is optimal for every analyte – the use of certain anticoagulants may interfere with some assays, and the activation of the clotting cascade may be detrimental for other assays. Therefore, the development of assays for individual proteins needs to be evaluated and optimized on a case-by-case basis. The information contained in Supplemental Tab. 3 could be used as a starting point for identifying potential anticoagulant-protein interactions that could affect an assay. Although assays for individual proteins must be individually optimized, it would be advantageous to use a single preparation method for proteomics methods and highly-multiplexed assays. Additional studies with an appropriate number of samples of each blood preparation method have to be performed to address the optimal blood preparation method for proteomics and highly-multiplexed studies, perhaps focusing on the consistency and stability of analytes rather than simply on concentration.

The final part of this study investigated the use of the antibody measurements to determine the concentration dependence of MS protein identification, using summary data from 18 different laboratories. A clear dependence on concentration was observed for both the number of laboratories finding certain proteins and the number of peptides found for each protein. It is encouraging that a precipitous decline in identifications at lower concentrations was not observed, but rather a steady decrease through most of the concentration range. Although the likelihood of identifying a protein and the quality of the identifications drop significantly for lower-abundance analytes, identifications were still made in the pg/mL range. Continued refinements and improvements in the technologies should make the identification of low-abundance proteins more common.

These studies demonstrate the benefits of high-throughput, high-precision, and high-sensitivity antibody-based analytical methods. We identified general and specific alterations in the protein concentrations that are related to the blood preparation method. In general, it appears that many cytokines are more stable in EDTA-plasma, specific interactions may occur in some cases with each anticoagulant, and a general dilution occurs with the use of citrate as an anticoagulant. The antibody-based methods also were useful for providing insights in the performance of MS-based protein identifications, showing that low concentration protein identifications are less frequent but still possible. In the continuing projects of the PPP, immunoassays and antibody microarrays will be useful in further studying these and other topics, such as characterizing the variation of many proteins in large populations of samples. Calibration using certified reference standards will be needed to reduce variation between laboratories and platforms.

F.V. thanks Herbert Schwarz, Harald Ackermann, and Lori Sokoll for excellent technical assistance and Carsten Schelp, Mary Lou Gantzer, Fritz Behrens, Manfred Lammers, Alex Rai, and Daniel Chan for helpful discussions. B.B.H. thanks John Marcus for excellent technical work. We thank Marcin Adamski and Raji Menon for assistance with the PPP data sets and Akhilesh Pandey for helpful discussions on the IPI protein searches. B.B.H.

acknowledges support from the Michigan Proteome Consortium, the Michigan Economic Development Corporation (GR-356), and the Van Andel Research Institute. G.M. was supported by the NIH (grant 1P41RR018627-01) and the NSF (grant IIS-9988085). B.H.G. thanks Professor Peter Schultz and the Genomics Institute of the Novartis Research Foundation for continuing support. G.S.O. was supported by the Michigan Economic Development Corporation (GR-356). We acknowledge trans-NIH (NCI grant supplement NCI-84982) and corporate support for the overall PPP (see www.hupo.org/ppp for detailed acknowledgements).-MSI and DB generously contributed resources to the study.

4.5 References

[1] Haab, B. B., Dunham, M. J., Brown, P. O., *Genome Biol.* 2001, *2*, 1–13.

[2] MacBeath, G., Schreiber, S. L., *Science* 2000, *289*, 1760–1763.

[3] Miller, J. C., Zhou, H., Kwekel, J., Cavallo, R., Burke, J., Butler, E. B., Teh, B. S., Haab, B. B., *Proteomics* 2003, *3*, 56–63.

[4] Schweitzer, B., Roberts, S., Grimwade, B., Shao, W., Wang, M., Fu, Q., Shu, Q., Laroche, I. *et al.*, *Nat. Biotechnol.* 2002, *20*, 359–365.

[5] Huang, R.-P., Huang, R., Fan, Y., Lin, Y., *Anal. Biochem.* 2001, *294*, 55–62.

[6] Nielsen, U. B., Geierstanger, B. H., *J. Immunol. Methods* 2004, *290*, 107–120.

[7] Haab, B. B., *Proteomics* 2003, *3*, 2116–2122.

[8] Omenn, G. S., *Proteomics* 2004, *4*, 1235–1240.

[9] Birkmeyer, R. C., Diaco, R., Hutson, D. K., Lau, H. P., Miller, W. K., Neelkantan, N. V., Pankratz, T. J., Tseng, S. Y. *et al.*, *Clin. Chem.* 1987, *33*, 1543–1547.

[10] Lammers, M., Gressner, A. M., *J. Clin. Chem. Clin. Biochem.* 1987, *25*, 363–367.

[11] Cuka, S., Dvornik, S., Drazenovic, K., Mihic, J., *Clin. Lab.* 2001, *47*, 35–40.

[12] Baudner, S., Haupt, H., Hubner, R., *J. Clin. Lab. Anal.* 1994, *8*, 177–190.

[13] Steinmetz, J., Tarallo, P., Fournier, B., Caces, E., Siest, G., *Eur. J. Clin. Chem. Clin. Biochem.* 1995, *33*, 337–342.

[14] Whicher, J. T., *Clin. Biochem.* 1998, *31*, 459–465.

[15] Finney, H., Newman, D. J., Gruber, W., Merle, P., Price, C. P., *Clin. Chem.* 1997, *43*, 1016–1022.

[16] Yguerabide, J., Yguerabide, E. E., *Anal. Biochem.* 1998, *262*, 157–176.

[17] Yguerabide, J., Yguerabide, E. E., *Anal. Biochem.* 1998, *262*, 137–156.

[18] Saviranta, P., Okon, R., Brinker, A., Warashina, M., Eppinger, J., Geierstanger, B. H., *Clin. Chem.* 2004, *50*, 1907–1920.

[19] Zhou, H., Bouwman, K., Schotanus, M., Verweij, C., Marrero, J. A., Dillon, D., Costa, J., Lizardi, P. M. *et al.*, *Genome Biol.* 2004, *5*, R28.

[20] Kersey, P. J., Duarte, J., Williams, A., Karavidopoulou, Y., Birney, E., Apweiler, R., *Proteomics* 2004, *4*, 1985–1988.

[21] Schweitzer, B., Wiltshire, S., Lambert, J., O'Malley, S., Kukanskis, K., Zhu, Z., Kingsmore, S. F., Lizardi, P. M. *et al.*, *Proc. Natl. Acad. Sci. USA* 2000, *97*, 10113–10119.

[22] Johnson, A. M., Whicher, J. T., *Clin. Chem. Lab. Med.* 2001, *39*, 1123–1128.

[23] Lammers, M., *Eur. J. Clin. Chem. Clin. Biochem.* 1996, *34*, 369.

[24] Hugo, F., Fischer, G., Eigenbrodt, E., *Anticancer Res.* 1999, *19*, 2753–2757.

[25] Kohek, M., Leme, C., Nakamura, I. T., De Oliveira, S. A., Lando, V., Mendonca, B. B., *BMC Clin. Pathol.* 2002, *2*, 2.

[26] Glendenning, P., Laffer, L. L., Weber, H. K., Musk, A. A., Vasikaran, S. D., *Clin. Chem.* 2002, *48*, 766–767.

[27] Flower, L., Ahuja, R. H., Humphries, S. E., Mohamed-Ali, V., *Cytokine* 2000, *12*, 1712–1716.

[28] Thillet, J., Chu, A. H., Romeo, P., Tsapis, A., Ackers, G. K., *Hemoglobin* 1983, *7*, 141–157.

5
Depletion of multiple high-abundance proteins improves protein profiling capacities of human serum and plasma*

Lynn A. Echan, Hsin-Yao Tang, Nadeem Ali-Khan, KiBeom Lee and David W. Speicher

Systematic detection of low-abundance proteins in human blood that may be putative disease biomarkers is complicated by an extremely wide range of protein abundances. Hence, depletion of major proteins is one potential strategy for enhancing detection sensitivity in serum or plasma. This study compared a recently commercialized HPLC column containing antibodies to six of the most abundant blood proteins ("Top-6 depletion") with either older Cibacron blue/Protein A or G depletion methods or no depletion. In addition, a prototype spin column version of the HPLC column and an alternative prototype two antibody spin column were evaluated. The HPLC polyclonal antibody column and its spin column version are very promising methods for substantially simplifying human serum or plasma samples. These columns show the lowest nonspecific binding of the depletion methods tested. In contrast other affinity methods, particularly dye-based resins, yielded many proteins in the bound fractions in addition to the targeted proteins. Depletion of six abundant proteins removed about 85% of the total protein from human serum or plasma, and this enabled 10- to 20-fold higher amounts of depleted serum or plasma samples to be applied to 2-D gels or alternative protein profiling methods such as protein array pixelation. However, the number of new spots detected on 2-D gels was modest, and most newly visualized spots were minor forms of relatively abundant proteins. The inability to detect low-abundance proteins near expected 2-D staining limits was probably due to both the highly heterogeneous nature of most plasma or serum proteins and masking of many low-abundance proteins by the next series of most abundant proteins. Hence, non2-D methods such as protein array pixelation are more promising strategies for detecting lower abundance proteins after depleting the six abundant proteins.

* Originally published in Proteomics 2005, 13, 3292–3303.

Exploring the Human Plasma Proteome. Edited by Gilbert S. Omenn
Copyright © 2006 WILEY-VCH Verlag GmbH & Co. KGaA, Weinheim
ISBN: 3-527-31757-9

5.1
Introduction

A major challenge of proteome research is detecting disease biomarkers in biological fluids. Disease markers, described by Adkins as "proteins that undergo a change in concentration or state in association with a biological process or disease," can be key factors for early diagnosis, monitoring response to therapy, and detection of relapse of most types of cancers as well as of other diseases [1]. However, disease biomarkers are usually present at relatively low concentrations (ng/mL or less). Serum and plasma offer particularly promising resources for biomarker discovery because collection of these samples is minimally invasive and the blood is thought to contain the majority of protein constituents found in the body [2–4].

The complex nature of serum and plasma, and the presence of a modest number of proteins at mg/mL levels, e.g., 0.1–40$^+$ mg/mL, make detection of low-abundance disease biomarkers very challenging. Although protein-rich plasma and serum have been used as diagnostic tools for decades, there are still fewer than 1000 distinct proteins identified, and only a small portion of these known proteins have been shown to have diagnostic potential [5, 6]. Traditional 2-D gel methods are most commonly used for most quantitative proteome analyses. However, when applied to analysis of serum or plasma, sample load capacity of 2-D gels is severely limited by the presence of high-abundance proteins in addition to other well-known limitations. Consequently, prefractionation methods as well as alternative non2-D methods are being used to divide proteomes into smaller subsets to identify as many proteins, or patterns of proteins, as possible and detect low-abundance disease biomarkers [3, 7, 8].

A highly promising first step for most analysis strategies of serum or plasma is to deplete as many of the major proteins as possible. A range of methods to deplete high-abundance proteins have been evaluated in the past, specifically Cibacron blue, – a chlorotriazine dye which has a high affinity for albumin [9–11], – as well as Protein A or G to deplete immunoglobulins [12–13]. While dye-based kits bind the majority of albumin, they often also bind a large number of nonspecific proteins, resulting in potential losses. This nonspecific binding probably includes both proteins that bind to the dye as well as some minor proteins that bind albumin. Cibacron blue and other dye-based methods are known to bind proteins with nucleated binding domains as well as *via* ionic and hydrophobic interactions [9–11]. In addition, the buffers typically used with these kits are unlikely to dissociate minor proteins that are complexed with albumin [14].

Conversely, Protein A and G may not bind all of the immunoglobulin subgroups, thereby leaving a portion of these very heterogeneous abundant proteins behind. Individual antibody methods have proven to be more specific in depleting targeted proteins and give more complete removal of abundant proteins. Monoclonal antibodies are a promising choice for their high specificity, but they may not recognize all forms of the targeted protein, including proteolytic fragments and PTM forms of the antigen [15]. Polyclonal antibodies, on the other hand, are more likely to deplete multiple structural forms of a protein. Ideally, for biomarker discovery it is desirable to deplete as many high-abundance proteins as possible while minimizing incidental losses of nontargeted proteins. Thus, a recently developed depletion method that mixes six high-specificity polyclonal antibodies to rapidly and efficiently deplete multiple proteins in a single purification step is particularly promising [16]. A commercial version of this method, multiple affinity removal system (MARS), recently became available and was used for these experiments.

In this study, we compared depletion of six abundant human blood proteins using a polyclonal HPLC column with: older Cibacron blue/Protein A or G depletion methods, two prototype antibody spin columns, and no depletion. The most critical considerations for major protein depletion are the extent to which undesired nonspecific losses of proteins occur during major protein depletion, and the potential positive impact that major protein depletion has on detection of lower abundance proteins. Results using normal human serum and plasma show that the HPLC column containing polyclonal antibodies to six abundant human proteins can efficiently and reproducibly deplete about 85% of the total protein. Although this depletion allows larger amounts of serum or plasma to be analyzed, the next most abundant proteins subsequently interfere with detection of very low-abundance proteins by masking major regions of the gel. In addition, most blood proteins are structurally heterogeneous due to physiological and/or artifactual proteolysis, and varying degrees of PTMs. As a result, most proteins are separated into many spots on 2-D gels, making detection more difficult. Hence, non2-D gel analysis methods are more likely to detect low-abundance proteins.

5.2
Materials and methods

5.2.1
Serum/plasma collection

Human serum (BDCA02-SERUM) and plasma (BDCA02-HEP) were provided from the HUPO Specimen Collection and Handling Committee (BD Preanalytical Solutions, Franklin Lakes, NJ, USA). Sample collection and handling were performed as previously described [17]. Additional normal human serum and plasma were collected with full participant consent from a single healthy donor at the Wistar Institute, Philadelphia, PA. For serum samples, blood was collected by venipuncture into BD Vacutainer serum separation tubes (SST), allowed to clot on ice for 30 min, centrifuged at $1300 \times g$ for 10 min at 4°C after which serum was pooled, aliquoted, and snap-frozen within 60 min of collection, and stored at −80°C. Plasma samples were collected into BD Vacutainer plasma separation tubes (PST) coated with lithium heparin anticoagulant, centrifuged as above within 30 min of collection, pooled, aliquoted, and frozen for storage at −80°C. The total protein concentrations of plasma and serum samples were estimated using a BCA Protein Assay (Pierce Chemical, Rockford, IL, USA).

5.2.2
MARS

For removal of the top six most abundant proteins, plasma or serum was applied to either a single MARS HPLC column (Agilent Technologies, Wilmington, DE, USA) (4.6 mm × 50 mm) or a two-column configuration where the 50 mm column was connected in tandem to a second, longer 4.6 mm × 100 mm column for a total column length of 150 mm. These HPLC columns contain polyclonal antibodies to human albumin, transferrin, haptoglobin, α-1-antitrypsin, IgG, and IgA. Typically, when the two-column configuration was used, 40 µL of human plasma or serum was diluted five-fold with the manufacturer's equilibration buffer, filtered through a 0.22 µm microcentrifuge filter tube, and injected onto the antibody column. The flow-through fractions containing unbound proteins from sequential injections were collected, pooled, and concentrated using a 5K molecular weight cutoff (MWCO) spin concentrator. The concentrated pool was either used immediately or aliquoted and stored at −80°C for future use. Affinity-bound major proteins were eluted with the manufacturer's elution buffer, neutralized with 1 M NaOH, and pooled for analysis on 1- and 2-D gels.

5.2.3
Multiple affinity removal spin cartridge

Typically, 10 µL of human plasma or serum was diluted 20-fold with the manufacturer's equilibration buffer, filtered through a 0.22 µm microcentrifuge filtration tube, and loaded onto the antibody resin in a microcentrifuge spin column. The sample was either passed into the column at slow speed (1.5 min, 100 × g, RT) or incubated at 4°C for 15 min. In both cases, the column was washed twice with the equilibration buffer and centrifuged (2.5 min, 100 × g) to collect the total unbound plus wash fractions. The bound fraction was collected with the manufacturer's elution buffer and neutralized, and both the flow-through fraction containing unbound proteins and the bound fractions were concentrated with a 5K MWCO spin concentrator for immediate analysis or aliquoted and stored at −80°C for future use.

5.2.4
Microscale solution IEF (MicroSol IEF) (ZOOM™-IEF) fractionation

Prior to microSol IEF, varying amounts of serum or plasma up to 250 µL, with or without protein depletion, were diluted to 1 mL in 8 M urea, 10 mM glycine. Samples were reduced with 10 mM DTT for 1 h at RT, concentrated to 200 µL using a 5K MWCO spin concentrator, rediluted to 1 mL with 8 M urea, 10 mM glycine (to dilute reducing reagents), and alkylated with 25 mM iodoacetamide (IAM) for 2 h at RT in the dark. Alkylation was quenched with 1% DTT for 15 min at RT. Following in-solution reduction and alkylation, depleted plasma or serum samples were prefractionated by microscale solution IEF as previously described [18], using a ZOOM-IEF fractionator (Invitrogen, Carlsbad, CA, USA). Proteins were separated into seven small volume (∼700 µL) pools where the separation chambers were defined by immobilized gel membranes having pH values of 3.0, 4.4, 4.9, 5.4, 5.9, 6.4, 8.1, and 10.0, respectively. In these experiments the pH 3.0 and 10.0 membranes were commercially available (Invitrogen), while the remaining membranes were prepared as described previously [18].

For protein array pixelation, ZOOM-IEF fractions were further separated by 1-D PAGE in individual lanes on short minigels. In general, the highest possible protein amounts that did not cause extensive band distortion were loaded into 10-well 10% NuPAGE™ (Invitrogen) SDS gels and electrophoresed using MOPS running buffer until the tracking dye had migrated 4 cm. Proteins were visualized by staining with Colloidal Blue (Invitrogen). Each lane was subsequently cut into uniform 1 mm slices with the MEF-1.5 Gel Cutter (The Gel Company, San Francisco, CA, USA). Two gel slices were combined *per* digestion tube and the gel slices were in-gel trypsin-digested as previously described [19], and analyzed by LC-MS/MS.

5.2.5
2-DE

IPG strips were purchased from Amersham Biosciences (San Francisco, CA, USA) and proteins were isoelectrofocused using the PROTEAN IEF Cell™ (Bio-Rad, Hercules, CA, USA) IEF system, essentially as described by Go″rg *et al.* [20]. Briefly, samples were thawed and diluted into IEF sample buffer containing 9 M urea, 2 M thiourea, 4% CHAPS, 0.1 M DTT, 0.8% pH 3–10 linear carrier ampholyte buffer to yield the desired protein amount in a volume that could be absorbed by the IPG strip used (typically 400 µL for an 18 cm IPG strip). IPG strips (typically 18 cm) were rehydrated with sample buffer containing serum proteins at 50 V for 12 h and then the applied voltage was increased in a linear fashion to a maximum of 10 000 V until a total of 60 000 Vh was reached. The IPG strips were equilibrated and applied to 18 cm × 19 cm, 1.0 mm thick second dimension 10% Tris-tricine polyacrylamide gels [21], cast without stacking gels and with sodium thiosulfate added to reduce silver stain background, as described by Hochstrasser *et al.* [22]. Gels were run at 100 mA constant current with external cooling (~4 h) until the tracking dye migrated to within 1 cm of the bottom of the gel. Either colloidal CBB or Silver Quest (Invitrogen) was used to visualize proteins after 2-DE. For protein identification, protein spots were excised using biopsy punches (Miltex Instruments, Lake Success, NY, USA) and subjected to in-gel trypsin digestion. Protein identifications of the 2-D gel spots were performed using an LTQ linear IT mass spectrometer (Thermo Electron, Marietta, OH, USA). The MS/MS spectra were searched against the NCBI non-redundant database, and each assignment was manually validated.

5.2.6
LC-MS/MS

Protein identification using tryptic peptides for protein array pixelation was performed using an LCQ Deca XP + IT mass spectrometer (Thermo Electron) interfaced with an Eldex MicroPro pump and an autosampler. Tryptic peptides were separated by

RP-HPLC on a C18 nanocapillary column (0.75 µm id × 100 mm packed with 5 µm Magic particles, Michrom BioResources). Solvent A was 0.58% acetic acid in Milli-Q water, and solvent B was 0.58% acetic acid in ACN. Peptides were eluted into the mass spectrometer at 200 nL/min using an ACN gradient. In some experiments with complex samples, gradient lengths were increased to enhance detection of additional proteins. The mass spectrometer was set to repetitively scan m/z from 375 to 1600 followed by data-dependent MS/MS scans on the three most abundant ions with dynamic exclusion enabled. The MS/MS spectra were searched against the International Protein Index (IPI, version 2.21) for protein identifications using SEQUEST. Consistent with the HUPO recommendations, peptide identifications were filtered using the following criteria: $X_{corr} > 1.9$ for charge state 1+, $X_{corr} > 2.2$ for charge state 2+, or $X_{corr} > 3.75$ for charge state 3+; $\Delta C_n > 0.1$; $R_{sp} < 4$.

5.3
Results

5.3.1
Depletion of major proteins to enhance detection of lower abundance proteins

The effectiveness of major protein depletion using several different methods is summarized in Fig. 1. As noted above, prior to the availability of the MARS polyclonal antibody column, the primary method for depleting major plasma or serum proteins was to use Cibacron blue for albumin and Protein A or G for IgG. As part of this study, five commercial kits were tested that used either a blue-dye alone, or in association with Protein A or G to deplete albumin and IgG. There were only minor differences among the kits (data not shown). The best of the dye-based products tested appeared to be the Proteoprep Blue Albumin Depletion kit (Sigma-Aldrich, St. Louis, MO, USA), which is shown here as representative of the dye-based/Protein G spin columns (Fig. 1A, dye-based affinity). While this method depletes considerable amounts of albumin and IgG (Fig. 1, panel A), substantial amounts of these proteins remain in the unbound fractions. In addition, there are a number of high molecular weight bands that nonspecifically bind along with the targeted proteins. In contrast, a dual anti-albumin and IgG antibody column (Fig. 1B), a β-test product subsequently commercialized as the Proteoprep Immunoaffinity Albumin and IgG Depletion Kit (Sigma-Aldrich), shows more complete depletion of albumin and IgG. However, several high molecular weight proteins are evident on 1-D gels. Some, but not all of these high molecular weight bands are cross-linked or incompletely reduced albumin or IgG. Not surprisingly, the most efficient enrichment of lower abundance proteins is achieved using the MARS HPLC column. The observed protein bands in the bound fraction on 1-D gels are the six targeted proteins, with no apparent evidence of nonspecific binding, at least using this low-sensitivity, low-resolution detection method (Fig. 1, panel C). In addition, a more convenient β-test spin column version of the MARS was evaluated in preliminary experiments. Initial 1-D gel results suggested depletion similar to that obtained with the HPLC column (Fig. 1, panel D).

5.3 Results

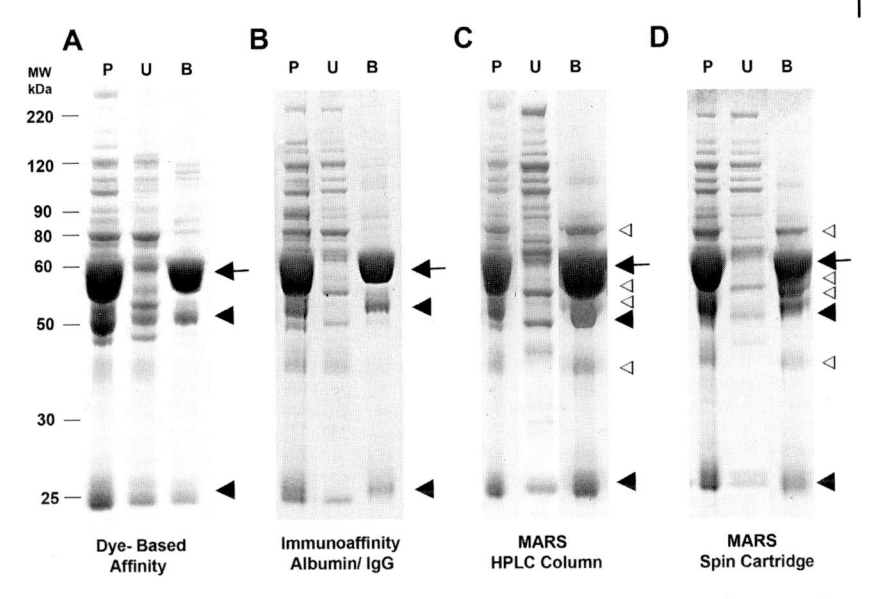

Fig. 1 Comparison of major protein depletion techniques. (A) Human plasma (50 μL) was processed using a representative blue-dye/Protein G based kit. While a significant amount of albumin (arrows) and IgG (solid arrowheads) is depleted from the plasma, there is nonspecific binding of other proteins. (B) Human plasma (25 μL) depleted using the prototype Proteoprep Immunoaffinity Albumin and IgG depletion kit. Majority of albumin and IgG was successfully depleted using this column; however, other major proteins still limit sample loads and detection of low-abundance proteins, and nontargeted bands appear in the bound fraction. (C, D) Plasma (15 and 10 μL) were separated using the Multiple Affinity Removal HPLC (4.6 cm × 50 mm) and spin column, respectively. Six most abundant plasma proteins were effectively removed. Albumin and IgG are indicated as above; transferrin, haptoglobin, IgA, and α-1-antitrypsin are indicated by open arrowheads. Samples were separated on 10% Bis-Tris 1-D gels and stained with colloidal CBB; all samples loaded onto the gel were volume-normalized to the undepleted sample (10 μg); P, undepleted plasma, U, unbound fraction, B, bound fraction.

5.3.2
Evaluation of high-abundance protein removal using 2-DE

More detailed comparisons of the alternative depletion methods were conducted using 2-D gels (Fig. 2). The incomplete depletion of albumin and IgG using the dye-based spin column was confirmed by analysis of the flow-through fraction at two different loads (Fig. 2, top panels). Even at low loads (100 μg) both "depleted" proteins could be readily detected. Nonetheless, as a result of the reduced levels of these two most abundant proteins, sample loads could be increased up to five-fold compared with unfractionated serum before excessive horizontal streaking was observed with a high-sensitivity silver stain. The more efficient removal of albumin and IgG with the two-protein antibody affinity spin column was readily evident

Dye-Based Affinity

Not Depleted - 100μg   Depleted From 100μg   Depleted From 500μg

Immunoaffinity Albumin/IgG Depletion Kit

Not Depleted - 100μg   Depleted From 100μg   Depleted From 500μg

Multiple Affinity Removal System (HPLC)

Not Depleted - 100μg   Depleted From 100μg   Depleted from 2mg

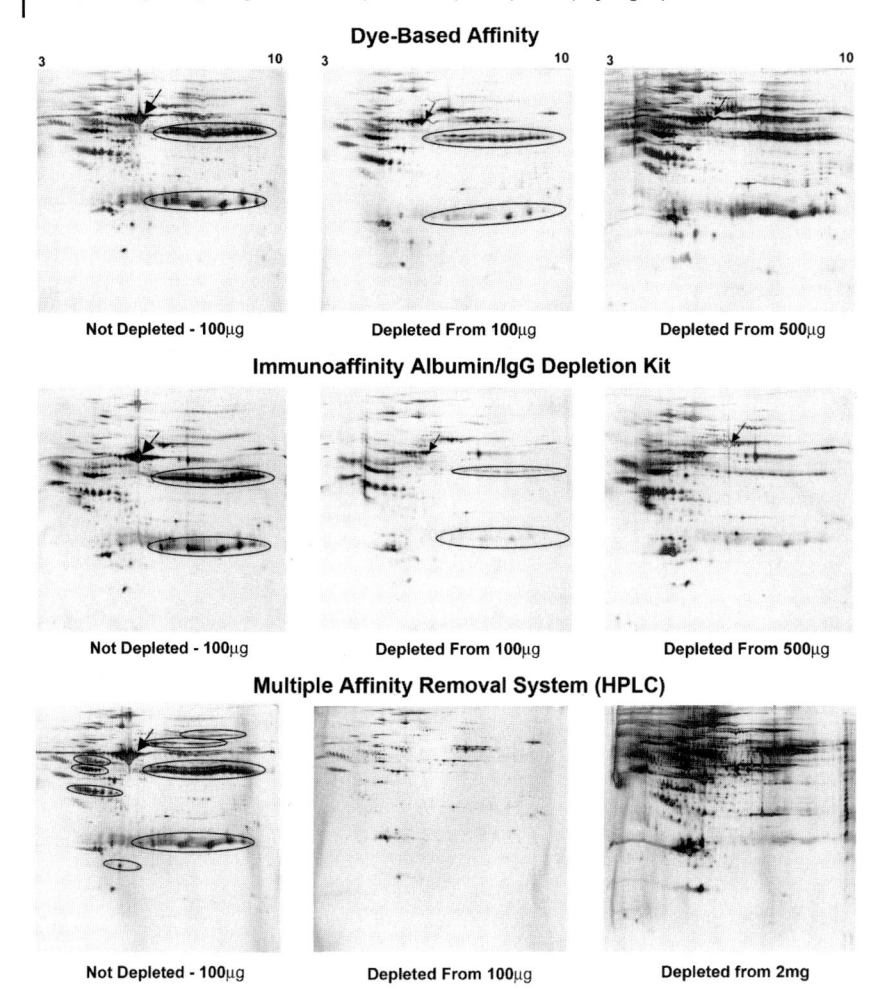

Fig. 2 Evaluation of major protein depletion on 2-D gels. Top panels, separation of human plasma before and after depletion using a dye affinity method shows incomplete removal of albumin (indicated with arrow) and IgG (ellipses) in the depleted fractions. Dual-antibody depletion shows slightly better removal of the two proteins from a serum sample of the same blood donor (middle panels). Bottom panels, separation of plasma on the MARS HPLC column (albumin indicated with arrow; the other five proteins, including major proteolytic products and aggregates are enclosed in ellipses). Top-6 depletion effectively removes ~85% of total protein content (based on manufacturer product claims as well as BCA assay results), allowing 10- to 20-fold higher protein loads, at which point other abundant proteins become problematic.

when depleted fractions from 100 and 500 μg of serum were analyzed (Fig. 2, middle panels). Specifically, the unbound fraction from the albumin/IgG immunoaffinity resin resulted in less horizontal streaking at a five-fold increased load, and reduced amounts of albumin and IgG were observed compared with the dye-based method. Finally, as expected, depletion of six abundant proteins on the

MARS HPLC column instead of only two proteins resulted in further improvements in protein loading capacity on 2-D gels (Fig. 2, bottom panels). Under these conditions, the HPLC column removed about 85% of the total protein in human serum or plasma samples and this reduced complexity enabled 10- to 20-fold higher loads of depleted serum or plasma (1.0–2.0 mg) on silver-stained gels compared with unfractionated samples (about 100 µg). In addition, 2-D gels of pools of the bound fraction from replicate runs of the same sample are essentially identical, indicating that the same set of proteins are continually removed (data not shown).

5.3.3
Specificity of major protein depletion

The specificity of alternative methods of affinity depletion was evaluated by analyzing the bound fractions on 2-D gels at a load equivalent to that of the initial serum or plasma samples; i.e., bound fractions derived from 100 µg. These results show that the dye-based affinity resin contained the largest number of non-specifically bound proteins, since this fraction should contain only albumin and IgG (Fig. 3A). Similarly, the 2-D gel of the bound fraction from the albumin/IgG immunoaffinity spin column showed extensive nonspecific removal of additional proteins (Fig. 3B). This fraction contained nearly as many proteins as the dye assay bound fraction, suggesting that improved purification buffers are needed to minimize nonspecific binding. In contrast, all major protein spots observed in the bound fraction from the MARS HPLC column were the six targeted proteins (Fig. 3C). Furthermore, most of the moderate- and low-intensity spots were tentatively identified as proteolytic fragments of these six proteins based upon comparisons to the 2-D plasma protein map on the Swiss-Prot website (http://us.expasy.org/cgi-bin/map2/def?PLASMA_HUMAN). The more specific removal of targeted proteins by the MARS HPLC column compared with the albumin/IgG immunoaffinity spin column cannot be due to clonality, since both are polyclonal antibodies. However, there may be differences in affinity of the antibodies that affects specificity and, in addition, the methods for purifying the polyclonal antibodies may differ. Another factor is the type of resin and cross-linker used and a final consideration is the purification buffers used for protein depletion. The proprietary purification buffers associated with the MARS HPLC column purification scheme have apparently been well optimized for disruption of weak nonspecific interactions. However, these buffers are apparently not necessarily universally suited for major protein depletion using immunoaffinity resins because when we used the MARS purification buffers with the prototype albumin/IgG immunoaffinity column, the binding capacity was greatly reduced and most proteins, including the majority of albumin and IgG, passed through the column. Subsequent analysis of the bound fractions on 2-D gels showed little binding of albumin and IgG, and many other nonspecific proteins adhering to the column resin (data not shown).

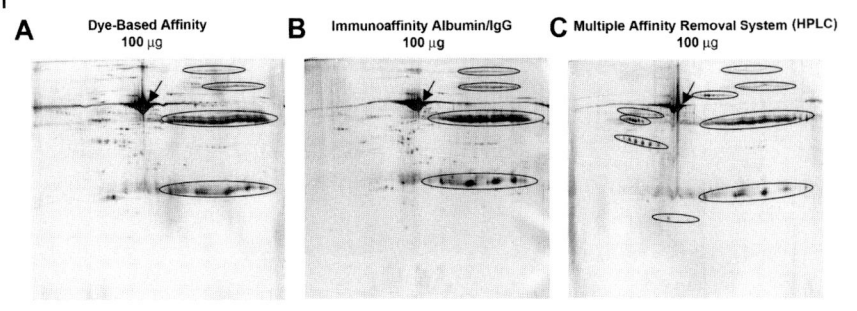

Fig. 3 Evaluation of bound fractions after depletion. (A) Cibacron blue/Protein G bound fraction after column stripping. While this column is meant to deplete albumin (arrow) and IgG (ellipses) only, there are a number of other low-abundance proteins comigrating with the major proteins. (B) Dual-antibody immunoaffinity (albumin and IgG) bound fraction displays a similar degree of nonspecific binding of proteins other than albumin and IgG as that observed with the dye-based affinity system. (C) MARS HPLC antibody column shows a much cleaner bound fraction; the six targeted proteins (arrow for albumin, enclosed in ellipses for other proteins) can be seen with few other nonspecific proteins.

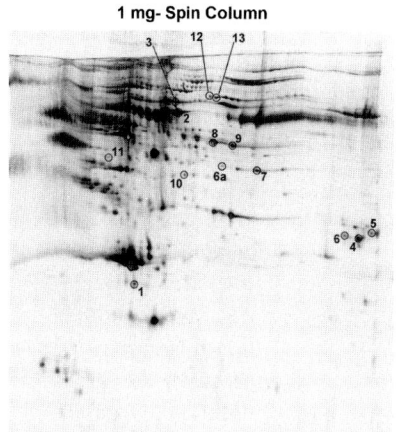

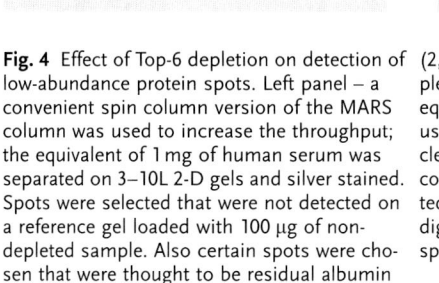

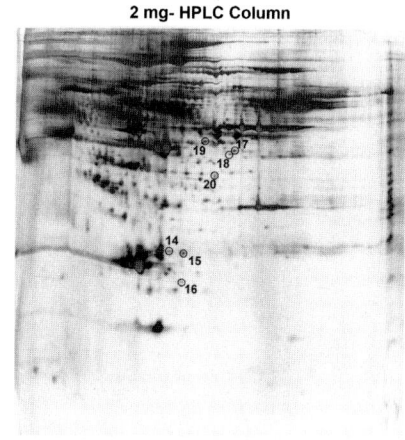

Fig. 4 Effect of Top-6 depletion on detection of low-abundance protein spots. Left panel – a convenient spin column version of the MARS column was used to increase the throughput; the equivalent of 1 mg of human serum was separated on 3–10L 2-D gels and silver stained. Spots were selected that were not detected on a reference gel loaded with 100 µg of nondepleted sample. Also certain spots were chosen that were thought to be residual albumin (2, 3) and transferrin (12, 13) to verify completeness of depletion. Right panel – the equivalent of 2 mg of human plasma depleted using the MARS HPLC column. New spots (circles with numbers) selected at this higher load compared with undepleted sample were selected. All spots selected were in-gel trypsin-digested and analyzed using a linear IT mass spectrometer.

5.3.4
Impact of Top-6 protein depletion on detection of lower abundance proteins using 2-D gels

The increased volume of serum or plasma that can be loaded onto 2-D gels after depleting six of the most abundant proteins should result in improved capacity to detect lower abundance proteins by this method. To evaluate the extent to which lower abundance proteins could actually be detected, silver-stained 2-D gels of high loads of depleted serum and plasma (Fig. 4) were compared to a 100 µg load of unfractionated serum or plasma. Many of the moderate and low-intensity protein spots that were detected on the depleted sample gels were actually detectable on the unfractionated sample gel. However, a few new spots that were below the detection limit on the unfractionated gel could be seen on the more heavily loaded post-depletion gels. These spots as well as several spots from the MARS spin column unbound fraction that were tentatively identified as incompletely depleted albumin and transferrin were excised (Fig. 4), digested with trypsin, and identified using LC-MS/MS. As anticipated, the spots in the serum albumin region (spots 2, 3) and the transferrin region (spots 12, 13) were identified as these proteins, which indicated they were incompletely depleted when the prototype spin column was used. Since the same antibodies were used by the same manufacturer to produce both products, it is most likely that this difference in efficiency was due to either an overloading of the spin column or, more likely, simply a difference in purification format, i.e., it is well known that batch purifications tend to be less effective than column chromatography using the same resin.

Although most of the analyzed samples were moderate or very faint silver-stained spots, more than one protein was identified for some spots (Tab. 1) due to the high-sensitivity of the linear IT mass spectrometer. Surprisingly, most of the new "low-abundance" spots that appeared at high protein loads were apparently minor forms of major proteins such as ceruloplasmin and complement proteins. The lowest abundance proteins detected were amyloid P serum component, at an expected protein concentration of 28 µg/mL [23], and carboxypeptidase N, with an expected plasma concentration of 30 µg/mL [24], indicating that the combination of depleting six major proteins and high-resolution broad-range 2-D gels are not sufficient for detection of proteins in the ng/mL range.

5.3.5
Combining Top-6 protein depletion with microSol IEF prefractionation and narrow pH range gels

We previously showed that prefractionating complex proteomes including undepleted serum using microSol IEF followed by analysis of fractions on very narrow pH range gels could isolate albumin in a single very narrow pH range fraction and could substantially expand the number of protein spots that could be reproducibly resolved [25]. Hence, in this study we evaluated the utility of combining Top-6 protein depletion with subsequent microSol IEF fractionation and analysis of each fraction on very narrow pH range gels. However, only a moderate further increase

Tab. 1 Proteins identified from silver-stained 2-D gels after MARS depletion

Spot no.	Primary ID[a]	Other IDs
1	Ceruloplasmin (4557485, 1)[b]	
2	Serum albumin (28592, 39)	Complement C8-α chain precursor (25757820, 10) Peptidoglycan recognition protein L precursor (21361845, 8) Heparin cofactor II (123055, 7) α-2-macroglobulin (177872, 4) Complement component 3 (40786791, 4) Hemopexin (11321561, 1) Carboxypeptidase B2 (13937897, 2)
3	Albumin precursor (4502027, 16)	Histidine-rich glycoprotein precursor (4504489, 10) Complement C4 binding protein α (4502503, 8) Inter-α-trypsin inhibitor heavy chain H1 precursor (2851501, 9)
4	Complement component 1q, B chain (11038662, 7)	
5	Complement component 1q, A chain (7705753, 3)	
6	Complement component 1q, B chain (11038662, 7)	Complement component 3 (40786791, 1) Desmoglein 1 preproprotein (4503401, 1) Migration inhibitory factor-related protein 14 variant E (7417329, 1)
6a	Complement component 3 (40786791, 6)	
7	Complement component 3 (40786791, 15)	
8	β-2-glycoprotein I precursor (4557327, 18)	Kallistatin precursor (1708609, 2) α-2-macroglobulin (177872, 1)
9	β-2-glycoprotein I precursor (4557327, 8)	
10	Complement component 4A preproprotein (14577919, 7)	
11	Ceruloplasmin (1620909, 1)	
12	Transferrin (4557871, 38)	α-2-macroglobulin (224053, 4) Ig heavy chain (106378, 2) Coagulation factor XII precursor (182292, 1)

Tab. 1 Continued

Spot no.	Primary ID[a]	Other IDs
13	Transferrin (4557871, 37)	α-2-macroglobulin (224053, 3) Ig heavy chain (106378, 3) Coagulation factor XII precursor (182292, 2)
14	Preserum amyloid P component (337758, 4)[c]	Serum albumin precursor (4502027, 2) Ig heavy chain V-region (197102, 1)
15	Preserum amyloid P component (337758, 5)[c]	Antipneumococcal antibody 7C5 light chain variable region (21311315, 2)
16	Apolipoprotein M (22091452, 7)	Apolipoprotein A-I precursor (4557321, 3) Fibrinogen-α A (223918, 2)
17	Hemopexin (11321561, 11)	
18	Carboxypeptidase N (4503011, 3)[c]	
19	Fibrinogen-β chain precursor (399492, 17)	Hemopexin (11321561, 1) Complement factor I precursor (116133, 1)
20	Complement C4A precursor (2144577, 3)	Complement factor H-related protein 2 (1064908, 2)

a) For spots where multiple proteins were identified, primary identifications were defined by the protein with the highest ion current for matched peptides.
b) Numbers in parentheses refer to the NCBI accession number followed by the number of unique peptides identified. MS/MS spectra for all reported peptides identified have been manually validated.
c) Indicates proteins with expected concentrations of <0.1 mg/mL.

```
Top 6 Protein Depletion
         ↓
ZOOM IEF Fractionation
         ↓
    1-D SDS gel
         ↓
In-gel Trypsin Digestion
         ↓
LC-MS/MS or LC/LC-MS/MS
         ↓
  Protein Identification
```

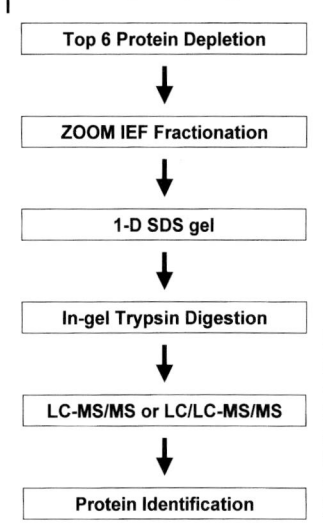

Fig. 5 Scheme for detection of low-abundance proteins using major protein depletion followed by a multi-dimensional downstream separation strategy. To identify the maximum number of low-abundance proteins in plasma or serum samples, major protein depletion should be coupled with multiple downstream fractionation techniques, such as solution IEF and 1-D SDS-PAGE prior to LC-MS/MS to increase detection sensitivity.

in the number of reproducibly resolved spots was obtained when Top-6 depleted plasma was fractionated into seven pH ranges. Apparently due to the very wide dynamic range of concentrations even after depleting six of the most abundant spots, only about 2000–3000 total spots were resolved on a series of seven slightly overlapping narrow pH range 2-D gels (data not shown). Since this strategy increased the number of 2-D gels required to survey a complete proteome, this modest improvement in resolution did not justify the much higher workload. In contrast, microSol IEF fractionation followed by narrow pH range gels could reproducibly resolve more than 8000 protein spots when working with human cancer cell lysates due to the smaller dynamic range of protein abundance (data not shown).

5.3.6
Analysis of Top-6 depleted serum and plasma using protein array pixelation

The effectiveness of major protein depletion on enhancing detection of lower abundance proteins by non2-D gel methods was also assessed. Fig. 5 summarizes a scheme for a promising new protein profiling method that combines multiple dimensions of protein separations with one or more dimensions of peptide separations prior to MS/MS. After major protein depletion, the unbound fraction is alkylated and further fractionated by microSol IEF. Each microSol IEF fraction is then separated by 1-D SDS-PAGE and each lane is cut into uniform slices. The result is a 2-D protein array where each point or pixel on the array contains a group of proteins within a specific pI and M_r range.

To evaluate the effectiveness of protein depletion on detection of proteins by protein array pixelation, duplicate aliquots of the HUPO BDCA02-HEP human plasma were either not depleted or depleted of the top six proteins using the MARS HPLC column. Both samples were then separated into seven pH range fractions on a ZOOM-IEF fractionator, and fraction 3, pH 4.9–5.4, from each of the two samples

Tab. 2 Effects of Top-6 depletion on identification of proteins using the protein array pixelation method[a]

Number of peptide/protein[b]	Not depleted	Depleted
>4	37	54
4	4	1
3	7	12
2	11	8
1	50	85
Total	109	160

a) Number of protein identifications obtained from pixelation of a 1-D gel lane containing fraction 3 (pH 4.9–5.4). Redundant protein identifications within each gel lane have been deleted.
b) Peptides were filtered using the following criteria: $X_{corr} > 1.9$ ($z = 1$), 2.2 ($z = 2$), 3.75 ($z = 3$) and $\Delta C_n > 0.1$, and $R_{sp} < 4$.

was separated on 1-D SDS-PAGE until the tracking dye had migrated 4 cm on a minigel. The lane was sliced into 20 uniform slices prior to trypsin digestion and LC-MS/MS analysis on a Thermo LCQ-XP + IT mass spectrometer. The results comparing depletion with nondepletion for fraction 3 are summarized in Tab. 2. Substantially more unique proteins were identified after major protein depletion. Further analysis of the 66 proteins common to both samples indicates that 39 proteins were identified with more unique peptides in the depleted sample, 12 proteins have more peptides in the undepleted sample, and 15 proteins did not show any changes in the number of peptides identified. Of the 12 proteins identified by more peptides in the undepleted sample, three proteins were among those depleted by the MARS column, i.e., haptoglobin, albumin, and α-1-antitrypsin. The remainder of the 12 proteins from the undepleted sample had only one or two additional peptides being identified. In contrast, 30 of the 39 proteins identified by more peptides in the depleted sample had at least three additional matched peptides compared with the nondepleted sample. Hence, by depleting the major proteins, we increased the proportion of lesser abundance proteins identified in this single ZOOM-IEF fraction, leading to substantially more proteins identified and increased sequence coverage for identified proteins. However, even after major protein depletion, most of the high-quality MS/MS spectra data acquired matched high- or medium-abundance blood proteins. Since comparisons in this experiment were based on only one of seven fractions, the above differences could be extrapolated by about sevenfold if all fractions from depleted and nondepleted samples were compared.

Fig. 6 shows a representative gel from microSol IEF separation of ~200 μL depleted or nondepleted human plasma. Because Top-6 protein depletion reduced the total protein by about seven-fold, volumes of depleted sample fractions loaded on the gels were increased approximately seven-fold relative to nondepleted samples for comparison. Due to the sample simplification that occurred after major protein de-

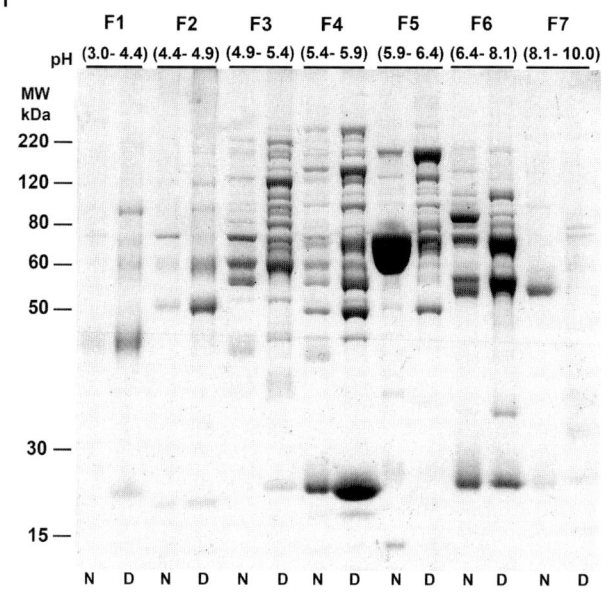

Fig. 6 ZOOM-IEF fractionation of plasma proteins. Nondepleted (N), or depleted (D) BDCA02 human plasma (about 200 μL each) representing 16.5 or 2.4 mg total protein, respectively, were separated using the ZOOM-IEF fractionator into seven discrete pH pools (~700–800 μL final fraction volumes), run on 10% Bis-Tris 1-D gels and stained with colloidal CBB. All aliquots loaded to the gel (~1.7 μL for nondepleted sample, and ~12 μL for depleted sample) were equivalent to about 35 μg of the original plasma protein.

pletion, much larger volumes of some fractions could be applied to 1-D SDS gels before band distortion due to overloading occurred. However, the next most abundant proteins limited 1-D gel loads and therefore limited detection of even lower abundance proteins. When protein array pixelation was applied to a complete human plasma sample proteome after depletion, the most abundant proteins detected as estimated based on sequence coverage are summarized in Tab. 3.

5.4
Discussion

Interest in developing improved methods for major protein depletion from serum, plasma, and other biological fluids has recently increased as proteomics technologies are sought that can aid the discovery of disease-related biomarkers. For serum or plasma this ideally requires the ability to routinely detect proteins present at ng/mL – pg/mL levels in samples that contain a modest number of proteins at the 0.1–40^+ mg/mL level. A logical solution would be to eliminate as many high-abundance proteins as possible so that lower abundance proteins could be detected. However, major protein depletion strategies have their critics, primarily due to the risk that proteins of interest could be removed along with the targeted proteins. A further complication is that major proteins such as albumin apparently function as carriers or molecular sponges that bind other potentially important proteins and peptides. For example, in one study when dye-based albumin depletion was used with very mild wash conditions designed to preserve noncovalent interactions, up to 63 other proteins were identified in the albumin fraction by LC-MS/MS [14].

Tab. 3 Next most abundant proteins after depletion[a]

Serum	% Coverage[b]	Plasma	% Coverage
Apolipoprotein A-1	86.1	Fibrinogen-γ	72.4
Complement C3	86.1	Apolipoprotein A-1	65.5
Transthyretin	80.3	Hemopexin	59.3
Plasma retinol binding protein	76.1	Fibrinogen-β	57.6
Apolipoprotein E	72.9	Plasminogen	55.3
Plasminogen	72.6	Albumin	53.7
Ribosomal protein C5	72.6	α-2-macroglobulin	53.1
Vitamin-D binding protein	72.0	Complement C3	52.4
Hemopexin	71.9	Complement C3	51.1
Apolipoprotein M	70.2	Prothrombin	51.1
Apolipoprotein A-IV	68.9	Ceruloplasmin	50
α-2-macroglobulin	68.3	Ig-α-I chain C	49
Prothrombin	65	Vitamin-D binding protein	48.9
Ceruloplasmin	64	Apolipoprotein A-IV	48.7
Tetranectin	63.4	Complement H	46.9

a) Major protein-depleted serum and plasma were fractionated using ZOOM-IEF and proteins were run on 1-D SDS-PAGE for analysis on LC-MS/MS.
b) Abundant proteins are ranked according to protein coverage as a crude estimate of abundance.

These results are consistent with the current study, since we show that there are large differences in the amount of nontargeted protein losses among types of affinity media and with different wash buffers. Specifically, among the resins analyzed in this study, the most extensive losses occurred with all dye-based affinity resins, while the MARS HPLC column with six polyclonal antibodies had the lowest level of nontargeted protein losses based on gel analysis.

Despite high nonspecific protein binding, dye-based affinity columns have several advantages relative to antibody-based methods. Due to their relatively low cost, they are disposable, which avoids any danger of cross-contamination between different samples. In addition, these resins are usually in a spin column format that does not require expensive HPLC equipment and that facilitates processing of samples in a cold room to minimize proteolysis. The spin column format has the added advantage that many samples or multiple aliquots of a single sample can be readily processed in parallel. Furthermore, the high nonspecific binding of dye-based affinity methods can be turned into an advantage by using the method as a fractionation step rather than a depletion step. That is, both bound and unbound fractions can be analyzed, which allows a more comprehensive analysis of the serum/plasma proteome.

The major advantage of the MARS antibody column is that it can efficiently deplete six high-abundance proteins including different molecular forms and many proteolytic products of these proteins with low nonspecific losses of other proteins. This demonstrates that it is feasible to efficiently deplete multiple major proteins in a single step with minimal losses of other proteins. Because about 85% of the total protein content of serum or plasma has been removed, this is clearly an advantage in terms of the volume equivalent of serum or plasma that can be introduced into diverse downstream analysis methods. However, the degree to which this increased load capacity contributes to detection of low-abundance proteins is highly dependent upon the downstream analysis method used. The major disadvantages of antibody-based depletion resins are those features inherent to working with antibodies, namely, relatively high cost and low sample capacity. Fortunately, antibodies are highly robust proteins, and based on many years of experience using mAb and polyclonal antibody affinity matrices for protein purification, it seems likely that such columns will last for many purification cycles if appropriate care is taken to minimize proteolysis and column clogging. In addition, as illustrated by the preliminary results described here with the prototype MARS spin column, the polyclonal antibody resin also works effectively in the spin column format, which allows parallel processing of multiple samples or aliquots and does not require complex instrumentation.

The efficient performance of the MARS column with minimal losses of target proteins apparently reflects an excellent match of specific antibodies and a specific, well-designed, proprietary binding buffer (MARS Buffer A). In contrast, other antibody affinity matrices that we tested, including the two antibody antialbumin/IgG spin column (Figs. 1–3), an antihuman albumin monoclonal, and various chicken antibodies to human plasma proteins, showed much higher nonspecific binding (data not shown). However, the MARS Buffer A is not a universal immunoaffinity binding buffer because it was not compatible with other antibody resins we tested (see above).

This study showed that effective depletion of six abundant proteins resulted in the ability to load larger equivalent amounts of serum or plasma into downstream separation modes including 2-D gels. But, even when 10 to 20 times more Top-6 depleted serum or plasma was applied to 2-D gels, only a modest number of new spots were detected and most of these spots were minor forms of major proteins. This was quite surprising because the silver stain we used should have a detection threshold of 0.5–1.0 ng or less, and the protein load in Fig. 4B was derived from about 25 µL. This suggests that proteins present in serum or plasma at about 20–40 ng/mL or higher should be detectable if they are recovered in good yield as a single spot. But the lowest abundance proteins detected in this study are known to be present in serum at about 30 µg/mL (see above). The fact that all observed proteins were about 1000-fold more abundant than the theoretical detection limit is probably due primarily to the extensive heterogeneity of high- and medium-abundance proteins caused by extensive, variable PTMs, and physiological as well as artifactual proteolysis and oxidative damage. As a result, these abundant proteins obscure most of the 2-D gel image. For example, in the heavily loaded silver-stained gel shown in Fig. 4B, about 50% of the available separation area is heavily stained and any new low-abundance proteins that would appear in these areas could not be readily detected. Actually the only substantially open area in this gel is the low-molecular weight region but serum and plasma contain very few proteins that are less than 30 kDa, so this is a minimally useful region of the gel. Hence, the extensive heterogeneity of high- and medium-abundance proteins severely limits the utility of 2-D gels for detection of low-abundance proteins, which may often also be structurally heterogeneous and will be spread among many spots.

Even after depleting six abundant proteins, serum and plasma are extremely complex and still have a very wide range of protein abundances. To effectively mine the low-abundance regions of these proteomes, multiple high-resolution separation modes must be effectively integrated. One particularly promising approach is a new method that we are developing, which incorporates high-resolution protein and peptide separations into a 4- or 5-D protein profiling strategy (Fig. 5 [26]). This method uses three sequential separation modes to separate proteins: Top-6 protein depletion, microSol IEF, and 1-D SDS-PAGE. The 1-D gel lanes from each microSol IEF fraction are then cut into uniform slices and these latter two modes define a 2-D protein array where each point or pixel in the array contains a group of proteins with a range of known pIs and a narrow range of molecular weights. Each of these gel slices is then digested with trypsin and analyzed by LC-MS/MS or LC/LC-MS/MS. Although the data shown above was obtained on the Thermo LCQ XP+, subsequent analyses on a higher sensitivity linear IT mass spectrometer (Thermo LTQ) showed dramatic increases in the number of proteins that can be identified without increasing total analysis time. Comprehensive analysis of all pixels from a human serum sample on the LTQ IT mass spectrometer using the protein array pixelation method resulted in identification of about 2400 proteins that passed the HUPO-defined stringency filter for SEQUEST data (see Section 2). Most importantly, a number of proteins at the low nanogram *per* mL level could be identified [26]. Of course as with most complex peptide mixture analyses, the majority of identifications are based on single peptide hits and better informatics tools are needed to more reliably distinguish false positives from true positives. A recently published analysis of cerebrospinal fluid using the MARS antibody depletion column prior to shotgun MS analysis also found that this column was specific for the targeted proteins and their removal enhanced detection of lower abundance proteins [27].

In summary, efficient depletion of six abundant proteins from human serum or plasma enables the detection of more proteins with greater protein coverage when a multidimensional protein-peptide separation strategy is used. However, the next most abundant proteins rapidly become limiting both in terms of sample loading capacities and because most of the mass spectrometer time is spent identifying remaining high- and medium-abundance proteins. Hence, ideally a highly specific polyclonal antibody column that can deplete at least 18–22 of the most abundant proteins, which comprise 98–99% of total serum protein content, would be desirable. In this regard, during preparation of this manuscript, an immunoaffinity resin containing 12 polyclonal chicken antihuman antibodies, the Seppro™ Mixed 12 spin column (Genway Biotech, San Diego, CA, USA) became commercially available and may be a promising method of further simplifying serum and plasma for biomarker discovery.

This work was supported in part by the National Institutes of Health Grants CA94360 and CA77048 to D.W.S., and institutional grants to the Wistar Institute including an NCI Cancer Core Grant (CA10815), and the Commonwealth Universal Research Enhancement Program, Pennsylvania Department of Health. The authors would also like to thank Agilent Technologies and Sigma-Aldrich for graciously supplying prototype products for β-test studies.

5.5
References

[1] Adkins, J. N., Varnum, S. M., Auberry, K. J., Moore, R. J., Angell, N. H., Smith, R. J., Springer, D. L., Pounds, J. G., *Mol. Cell. Proteomics* 2002, *1.12*, 947–952.

[2] Anderson, N. L., Anderson, N. G., *Mol. Cell. Proteomics* 2002, *1.11*, 845–867.

[3] Zhang, R., Barker, L., Pinchev, D., Marshall, J., Rasamoelisolo, M., Smith, C., Kupchak, P. et al., *Proteomics* 2004, *4*, 244–256.

[4] Tirumalai, R. S., Chan, K. C., Prieto, D. A., Issaq, H. J., Conrads, T. P., Veenstra, T. D., *Mol. Cell. Proteomics* 2003, *2.10*, 1096–1103.

[5] Lathrop, J. T., Anderson, N. L., Anderson, N. G., Hammond, D. J., *Curr. Opin. Mol. Ther.* 2003, *5*, 250–257.

[6] Pieper, R., Gatlin, C. L., Makusky, A. J., Russo, P. S., Schatz, C. R., Miller, S. S., Su, Q. et al., *Proteomics* 2003, *3*, 1345–1364.

[7] Adam, B. L., Vlahou, A., Semmes, O. J., Wright, G. L., *Proteomics* 2001, *1*, 1264–1270.

[8] Chertov, O., Biragyn, A., Kwak, L. W., Simpson, J. T., Boronina, T., Hoang, V. M., Prieto, D. A. et al., *Proteomics* 2004, *4*, 1195–1203.

[9] Travis, J., Bowen, J., Tewksbury, D., Johnson, D., Pannell, R., *Biochem. J.* 1976, *157*, 301–306.

[10] Gianazza, E., Arnaud, P., *Biochem. J.* 1982, *201*, 129–136.

[11] Ahmed, N., Barker, G., Oliva, K., Garfin, D., Talmadge, K., Georgiou, H., Quinn, M., Rice, G., *Proteomics* 2003, *3*, 1980–1987.

[12] Govorukhina, N. I., Keizer-Gunnink, A., van der Zee, A. G. J., de Jong, S., de Bruijn, H. W. A., Bischoff, R., *J. Chromatogr. A* 2003, *1009*, 171–178.

[13] Wang, Y. Y., Cheng, P., Chan, D. W., *Proteomics* 2003, *3*, 243–248.

[14] Zhou, M., Lucas, D. A., Chan, K. C., Issaq, H. J., Petricoin, E. F., Liotta, L. A., Veenstra, T. D., Conrads, T. P., *Proteomics* 2004, *25*, 1289–1298.

[15] Steel, L. F., Trotter, M. G., Nakajama, P. B., Mattu, T. S., Gonye, G., Block, T., *Mol. Cell. Proteomics* 2003, *2.4*, 262–270.

[16] Pieper, R., Su, Q., Gatlin, C. L., Huang, S. T., Anderson, N. L., Steiner, S., *Proteomics* 2003, *3*, 422–432.

[17] Omenn, G. S., *Proteomics* 2004, *4*, 1235–1240.

[18] Zuo, X., Speicher, D. W., *Anal. Biochem.* 2000, *284*, 266–278.

[19] Speicher, K. D., Kolbas, O., Harper, S., Speicher, D. W., *J. Biomol. Tech.* 2000, *11*, 74–86.

[20] Go″rg, A., Obermaier, C., Boguth, G., Weiss, W., *Electrophoresis* 1999, *20*, 712–717.

[21] Schagger, H., von Jagow, G., *Anal. Biochem.* 1987, *166*, 368–379.

[22] Hochstrasser, D. F., Patchornik, A., Merril, C. R., *Anal. Biochem.* 1988, *173*, 412–423.

[23] Bijl, M., Bootsma, H., van der Geld, Y., Limburg, P. C., Kallenberg, C. G. M., van Rijswijk, M. H., *Ann. Rheum. Dis.* 2004, *63*, 831–835.

[24] Matthews, K. W., Mueller-Oritz, S. L., Wetsel, R. A., *Mol. Immunol.* 2004, *40*, 785–793.

[25] Zuo, X., Speicher, D. W., *Proteomics* 2002, *2*, 58–68.

[26] Tang, H. Y., Ali-Khan, N., Echan, L. A., Levenkova, N., Rux, J., Speicher, D. W., *Proteomics* 2005, *5*, this issue.

[27] Maccarrone, G., Milfay, D., Birg, I., Rosenhagen, M., Holsboer, F., Grimm, R., Bailey, J. et al., *Electrophoresis* 2004, *25*, 2402–2412.

6
A novel four-dimensional strategy combining protein and peptide separation methods enables detection of low-abundance proteins in human plasma and serum proteomes*

Hsin-Yao Tang, Nadeem Ali-Khan, Lynn A. Echan, Natasha Levenkova, John J. Rux and David W. Speicher

A novel strategy, termed protein array pixelation, is described for comprehensive profiling of human plasma and serum proteomes. This strategy consists of three sequential high-resolution protein prefractionation methods (major protein depletion, solution isoelectrofocusing, and 1-DE) followed by nanocapillary RP tryptic peptide separation prior to MS/MS analysis. The analysis generates a 2-D protein array where each pixel in the array contains a group of proteins with known pI and molecular weight range. Analysis of the HUPO samples using this strategy resulted in 575 and 2890 protein identifications from plasma and serum, respectively, based on HUPO-approved criteria for high-confidence protein assignments. Most importantly, a substantial number of low-abundance proteins (low ng/mL – pg/mL range) were identified. Although larger volumes were used in initial prefractionation steps, the protein identifications were derived from fractions equivalent to approximately 0.6 μL (45 μg) of plasma and 2.4 μL (204 μg) of serum. The time required for analyzing the entire protein array for each sample is comparable to some published shotgun analyses of plasma and serum proteomes. Therefore, protein array pixelation is a highly sensitive method capable of detecting proteins differing in abundance by up to nine orders of magnitude. With further refinement, this method has the potential for even higher capacity and higher throughput.

6.1
Introduction

There is considerable interest in systematically analyzing the human plasma proteome to identify novel biomarkers that can be used for improved early diag-

---

* Originally published in Proteomics 2005, 13, 3329–3342

Exploring the Human Plasma Proteome. Edited by Gilbert S. Omenn
Copyright © 2006 WILEY-VCH Verlag GmbH & Co. KGaA, Weinheim
ISBN: 3-527-31757-0

nosis of a wide range of diseases. Plasma or serum is easily and widely collected and its proteome contains thousands of proteins including proteins secreted or shed by most cells and tissues as well as proteins that leak into the blood from damaged tissue [1]. The presence or change in concentration of blood proteins is likely to reflect the state of health of an individual. A number of proteins discovered through targeted studies are currently being used as diagnostic markers for diseases such as acute myocardial infarction (creatine kinase MB, myoglobin, and troponin T [2]), prostate cancer (prostate-specific antigen [3]), and ovarian cancer (CA125 [4]). However, it is likely that the blood contains many additional disease biomarkers that will have greater diagnostic value than the handful of biomarkers discovered.

While the human plasma proteome potentially contains many different important biomarkers for most human diseases, several factors make it difficult to characterize. Plasma proteins are present in a very wide dynamic range, varying by a factor of at least 10^{10} in abundance, and many of these proteins have a high degree of heterogeneous PTMs [1]. The ability to identify low-abundance plasma proteins is particularly severely limited by several major proteins that are present at >1 mg/mL. For example, albumin together with immunoglobulins contributes to more than 80% of the total plasma proteins at about 40 and 12 mg/mL, respectively [1, 5]. In contrast, many bioactive proteins and potential biomarkers of disease are low-abundance proteins that are typically found at ng/mL – pg/mL levels or less.

The strategies that have been most frequently used to overcome the dynamic range problem of plasma proteins are to fractionate the plasma proteome into smaller subsets, and/or to deplete one or more of the major proteins, particularly albumin and immunoglobulins [5–12]. Numerous dye-based and immunoaffinity methods for major protein depletion have been described and are available commercially. Immunoaffinity methods are preferred, as they provide the most efficient depletion of targeted major proteins with reduced nonspecific binding of other proteins [7–9]. Alternatively, albumin can be efficiently separated based on its pI by microscale solution IEF (MicroSol-IEF) into a single fraction [10, 11]. Both major protein depletion and MicroSol-IEF methods have resulted in increased detection of lower abundance proteins when analyzed by 2-DE [8, 9, 11]. While removal of major proteins is beneficial, multiple orthogonal fractionation steps have been used to further facilitate detection of low-abundance proteins [8, 12].

A popular alternative to 2-DE is the shotgun or multidimension protein identification technology (MudPIT) approach which involves proteolytically digesting complex protein mixtures into peptides that are further subjected to multidimensional separations prior to analysis by ESI-MS/MS [13]. The most common form of multidimensional separations involves strong cation exchange (SCX) chromatography followed by RP-LC [12, 14, 15]. Alternate peptide separation strategies, such as ampholyte-free liquid-phase IEF [16] and CZE [17], have also been used in the analysis of human serum proteome.

Compared to 2-DE, the MudPIT approach has the potential of higher throughput and is capable of identifying more proteins from the plasma proteome. In a 2-DE study, 325 proteins were identified from human serum after 3-D fractionation using immunodepletion of nine abundant proteins, anion-exchange, and SEC [8]. In comparison, 490 proteins were identified with the MudPIT technique using immunoglobulin depletion and 2-D peptide separations by SCX and RP-LC [12]. While the 2-DE technology is relatively mature, the MudPIT method is constantly improving due to technological advances mainly to the LC and MS components of the system. In a recent study using ultra-high-performance SCX/RP-LC coupled to MS/MS, at least 800 proteins (depending on the criteria used) were identified from human plasma proteome [15]. These proteins were identified without prior depletion of major proteins, indicating that the improvement to the LC system and the longer gradient used were capable of overcoming the dynamic range problem of the plasma proteome to a certain degree. However, since immunoglobulins, which contain highly variable regions, were not depleted in the study, many of the proteins identified (up to 38%) belong to the immunoglobulin group [15].

The total number of proteins in the human plasma proteome is unknown but has been estimated to contain up to 10 000 proteins [18]. A recent analysis of the human plasma proteome by combining four separate sources of protein identification, including a 2-DE and two separate MudPIT experiments, has resulted in a conservative nonredundant list of 1175 proteins [19]. Interestingly, only 46 proteins are common to all four sources. This indicates that current methodologies cannot consistently provide comprehensive coverage of the human plasma proteome. Clearly, further reduction in the complexity of the human plasma proteome by additional more efficient fractionation steps is required to effectively mine the lower abundance proteins that have potential to be the next generation of disease biomarkers. Realizing the need for better methodology to analyze the human plasma proteome, HUPO has established the Plasma Proteome Project (PPP), and one of its aims is to determine the best technology platform for comprehensive profiling of the human plasma and serum proteomes [20].

As a participant of the HUPO PPP, in this report we describe a novel 4-D separation strategy to analyze the human plasma and serum proteomes that combines many of the benefits of 2-DE and MudPIT approaches. This strategy, termed protein array pixelation, consists of three sequential protein fractionation methods (major protein depletion, MicroSol-IEF fractionation, and SDS-PAGE). The result is a 2-D array of pixels or gel slices that is conceptually equivalent to a low-resolution 2-D gel. That is, each pixel in the array contains a group of proteins in a gel slice with a known pI and molecular weight (MW) range. Each pixel is then digested with trypsin followed by RP-LC peptide separation prior to ESI-MS/MS analysis. Using HUPO plasma and serum samples, we demonstrate that the protein array pixelation strategy is a highly sensitive method capable of detecting proteins that differ in abundance up to nine orders of magnitude.

6.2
Materials and methods

6.2.1
Materials

Human plasma (Caucasian American Sample Set; Lot # BDCA02-Heparin) and serum (Caucasian American Sample Set; Lot # BDCA02-Serum) were obtained from the HUPO Specimen Collection and Handling Committee [20]. The total protein concentration of plasma and serum was estimated using a BCA Protein Assay (Pierce Chemical, Rockford, IL, USA). Trypsin digest was performed with porcine sequencing grade modified trypsin (Promega, Madison, WI, USA). HPLC-grade ACN was obtained from J. T. Baker (Phillipsburg, NJ, USA). All reagents and buffers were prepared with Milli-Q water (Millipore, Bedford, MA, USA).

6.2.2
Top six protein depletion

Removal of the six most abundant proteins in human plasma was achieved with a single 4.6 × 50 mm multiple affinity removal system (MARS) HPLC column (Agilent Technologies, Wilmington, DE, USA). The MARS column contains polyclonal antibodies to human albumin, transferrin, haptoglobin, α-1 antitrypsin, IgG, and IgA. Typically, plasma was diluted five-fold with the manufacturer's equilibration buffer and filtered through a 0.22 μm microcentrifuge filter tube, and aliquots containing ∼1 mg total protein were injected onto the antibody column. A total of 193 μL (14.5 mg) of plasma was depleted. The flow-through fractions from sequential injections were collected, pooled, and concentrated to 200 μL (2.4 mg) using a 5 K MWCO spin concentrator (Millipore). Affinity-bound major proteins were eluted with the manufacturer's elution buffer, neutralized with 1 M NaOH, concentrated as above, and stored at −70°C. The concentrated unbound fraction (depleted plasma) was reduced with 10 mM DTT for 1 h at 23°C in 1 mL of buffer (final volume) containing 8 M urea, 10 mM glycine, pH 8.5. The reaction volume was subsequently reduced to 200 μL using a 5 K MWCO spin concentrator, and alkylated with 25 mM iodoacetamide in 1 mL of buffer (final volume) containing 8 M urea, 10 mM glycine, pH 8.5, for 2 h at 23°C. Reaction was quenched by adding DTT to 1% final concentration. Prior to MicroSol-IEF, salts and reagents were removed by buffer exchange using a 5 K MWCO spin concentrator.

For the analysis of human serum, the major proteins from 415 μL (35.3 mg) of serum were depleted using two MARS columns, where the 50 mm column was connected in tandem to another 4.6 × 100 mm column. All buffers used in the depletion were supplemented with protease inhibitors (1 mM DFP, 1 μg/mL leupeptin, 1 μg/mL pepstatin, 5 mM EDTA). Each injection contained ∼200 μL of five-fold diluted serum. The unbound fractions were pooled and concentrated to 240 μL (4.3 mg). Proteins were reduced and alkylated in the presence of 100 mM Tris-Cl,

8 M urea, pH 8.3 with 20 mM DTT, and 60 mM iodoacetamide for 1 h at 37°C each. Reaction was terminated with 60 mM DTT for 15 min at 37°C. Salts and reagents were removed by precipitation with 9 vol of acetone.

6.2.3
MicroSol-IEF fractionation

MicroSol-IEF was performed using a ZOOM IEF Fractionator (Invitrogen, Carlsbad, CA, USA). Depleted plasma was fractionated on a seven-chamber device, separated by immobiline gel membranes having pH values of 3.0, 4.4, 4.9, 5.4, 5.9, 6.4, 8.1, and 10.0. The pH 3.0 and 10.0 membranes were obtained commercially (Invitrogen), while the remaining membranes were prepared as described previously [21]. Reduced and alkylated plasma was diluted to 3.5 mL and adjusted to the same constituent concentrations as the MicroSol buffer, *i.e.*, 8 M urea, 2 M thiourea, 4% CHAPS, 1% DTT, 0.2% carrier ampholytes, pH 3–10L. Aliquots (700 µL) of the sample were loaded into the inner five chambers, and the remaining two outer chambers were filled with MicroSol buffer without sample. Depleted serum was fractionated on a five-chamber device, with pH 3.0, 4.6, 5.4, 6.2, 7.0, and 10.0 membranes obtained commercially (Invitrogen). Acetone-precipitated serum was dissolved in 700 µL of MicroSol buffer and was loaded into the central chamber of the device only.

6.2.4
Protein array pixelation

Following MicroSol-IEF, the fractions were separated by 1-D PAGE in individual lanes on short minigels. In some cases, proteins were extracted from the membrane partitions by two sequential incubations with 400 µL of MicroSol buffer. Each fraction was run on separate gels to avoid possible cross-contamination from other fractions. The highest possible protein amounts that did not cause extensive band distortion were loaded into 10-well 10% NuPAGE (Invitrogen) SDS gels and electrophoresed using MOPS running buffer until the tracking dye had migrated 4 cm (plasma analysis) or 6 cm (serum analysis) into the gels. In fractions containing very low amounts of proteins (F1 and M1 of serum analysis), the proteins were concentrated by precipitation with 9 vol of acetone and electrophoresed for only 2 cm. Proteins were visualized by staining with Colloidal Blue (Invitrogen). Each lane was subsequently cut into uniform 1 mm slices with the MEF-1.5 Gel Cutter (The Gel Company, San Francisco, CA, USA). Generally, two gel slices were combined (*i.e.*, 2 mm pixels) *per* digestion tube and the pixels were digested in-gel with trypsin as previously described [22], and analyzed by LC-ESI-MS/MS. In the serum analysis, the gel lanes were pixelated in a variable manner depending on the band intensity. Intense bands were digested as 1 mm pixels, while regions of the lane without much staining were digested as 4 mm pixels.

6.2.5
LC-ESI-MS/MS methods

Tryptic peptides from pixelation of the fractionated plasma sample were analyzed on an LCQ Deca XP+ IT mass spectrometer (Thermo Electron, San Jose, CA, USA) interfaced with a MicroPro pump (Eldex, Napa, CA, USA) and an autosampler. Serum tryptic digests were analyzed on an LTQ linear IT mass spectrometer (Thermo Electron) coupled with a NanoLC pump (Eksigent Technologies, Livermore, CA, USA) and autosampler. For each pixel, 5 µL (plasma samples) or 7 µL (serum samples) of the tryptic digest (total ∼30 µL) was analyzed. Tryptic peptides were separated by RP-HPLC on a nanocapillary column, 75 µm id × 20 cm PicoFrit (New Objective, Woburn, MA, USA), packed with MAGIC C18 resin, 5 µm particle size (Michrom BioResources, Auburn, CA, USA). In some of the initial optimization experiments, POROS R2 C18 resin, 10 µm particle size (Applied Biosystems, Foster City, CA, USA) was used. Solvent A was 0.58% acetic acid in Milli-Q water, and solvent B was 0.58% acetic acid in ACN. Peptides were eluted into the mass spectrometer at 200 nL/min using an ACN gradient. Each RP-LC run consisted of a 10 min sample load at 1% B; a 75 min total gradient consisting of 1–28% B over 50 min, 28–50% B over 14 min, 50–80% B over 5 min, 80% B for 5 min before returning to 1% B in 1 min. To minimize carryover, a 36 min blank cycle was run between each sample. Hence, the total sample-to-sample cycle time was 121 min. In some optimization experiments, a 49 min gradient (1–28% B over 27 min, 28–50% B over 11 min, 50–80% B over 5 min, 80% B for 5 min before returning to 1% B in 1 min) was used instead of the 75 min gradient.

The mass spectrometers were set to repetitively scan m/z from 375 to 1600 followed by data-dependent MS/MS scans on the three most intense (LCQ Deca XP+) or the ten most abundant (LTQ) ions with dynamic exclusion enabled. In some experiments, gas-phase fractionation using different m/z ranges was performed as described in Fig. 2B.

6.2.6
Data analysis

Proteins from each pixel were identified from the MS/MS spectra using the SEQUEST Browser program (Thermo Electron). DTA files were generated from MS/MS spectra using an intensity threshold of 500 000 (Deca XP+ data) or 5000 (LTQ data), and minimum ion count of 30. The DTA files generated were processed by the ZSA, CorrectIon, and IonQuest algorithms of the SEQUEST Browser program, and searched against the International Protein Index (IPI) human protein database [23] version 2.21 (July, 2003) containing 56 530 entries as requested by HUPO. In some of the optimization experiments (data sets for Fig. 2A), the National Center for Biotechnology Information nonredundant database (01/15/2004) was also used. To reduce database search time, the databases were indexed with the following parameters: average mass range of 500–3500, length of 6–100, tryptic cleavages with 1 (for LTQ analysis) or 2 (for LCQ Deca XP+ analysis) internal missed cleavage sites, static modification of Cys by carboxamidomethylation, and dynamic modification of Met to methionine sulfoxide (+16 Da). The DTA files were searched with a 2.5 Da peptide mass tolerance and 0 Da fragment ion mass tolerance. Other search parameters were identical to those used for database indexing. For each pixel, the peptides identified were assembled into the minimum number of unique proteins using SEQUEST SUMMARY with a depth of 3. Perl programs were developed for parsing, storing, analyzing, and retrieving SEQUEST results. Data from SEQUEST SUMMARY were stored in a relational database (Oracle 9i) with Perl Object layer.

The peptides from each protein were initially filtered using the following criteria: $X_{Corr} \geq 1.9$ ($z = 1$), 2.3 ($z = 2$), 3.75 ($z = 3$) and $\Delta C_n \geq 0.1$; or $S_f \geq 0.7$. Further data analysis used the HUPO defined criteria where peptides were filtered using $X_{Corr} \geq 1.9$ ($z = 1$), 2.2 ($z = 2$), 3.75 ($z = 3$) and $\Delta C_n \geq 0.1$; and $R_{Sp} \leq 4$. For both criteria, redundant peptides with the same accession number were removed and different forms (charge states and modification) of the same peptide were counted as a single-peptide hit. Keratins were also excluded from all data sets.

6.3 Results and discussion

6.3.1 Protein array pixelation strategy

We previously showed that MicroSol-IEF is capable of providing high-resolution fractionation of serum samples, resulting in albumin being confined into a single fraction [10, 11]. This fractionation approach has substantially expanded the number of proteins that can be detected by 2-DE since higher protein loads can be analyzed, in most fractions, without interference from the highly abundant albumin. We have also examined a number of commercially available methodologies for depleting abundant proteins from human plasma/serum and found that the Agilent MARS column is highly efficient in depleting the six most abundant proteins (albumin, transferrin, haptoglobin, α-1 antitrypsin, IgG, and IgA) with minimal nonspecific binding of other proteins [9]. Removal of the major proteins allowed higher amounts of serum or plasma to be loaded onto 2-D gels. However, when the minor protein spots were analyzed by LC-ESI-MS/MS, most of these proteins turned out to be proteolytic products of major proteins [9]. To further enhance detection of lower abundance proteins, the depleted plasma/serum was subjected to MicroSol-IEF fractionation followed by 2-DE analysis. This very time-consuming series of 2-D gels only moderately increased the number of protein spots detected (data not shown). Hence, 2-DE is not an efficient method for detecting lower abundance proteins (<μg/mL) of the human plasma/serum proteome.

To overcome these limitations of 2-DE, we developed the protein array pixelation strategy for comprehensive profiling of the human plasma proteome (Fig. 1). The first step is major protein depletion using the Agilent MARS column. Following

Fig. 1 Diagram of the protein array pixelation strategy used for analysis of the HUPO plasma sample (BDCA02-Heparin). Strategy consists of four separation steps (major protein depletion, MicroSol-IEF fractionation, 1-D gel separation, and RP-LC separation of peptides) followed by MS/MS analysis.

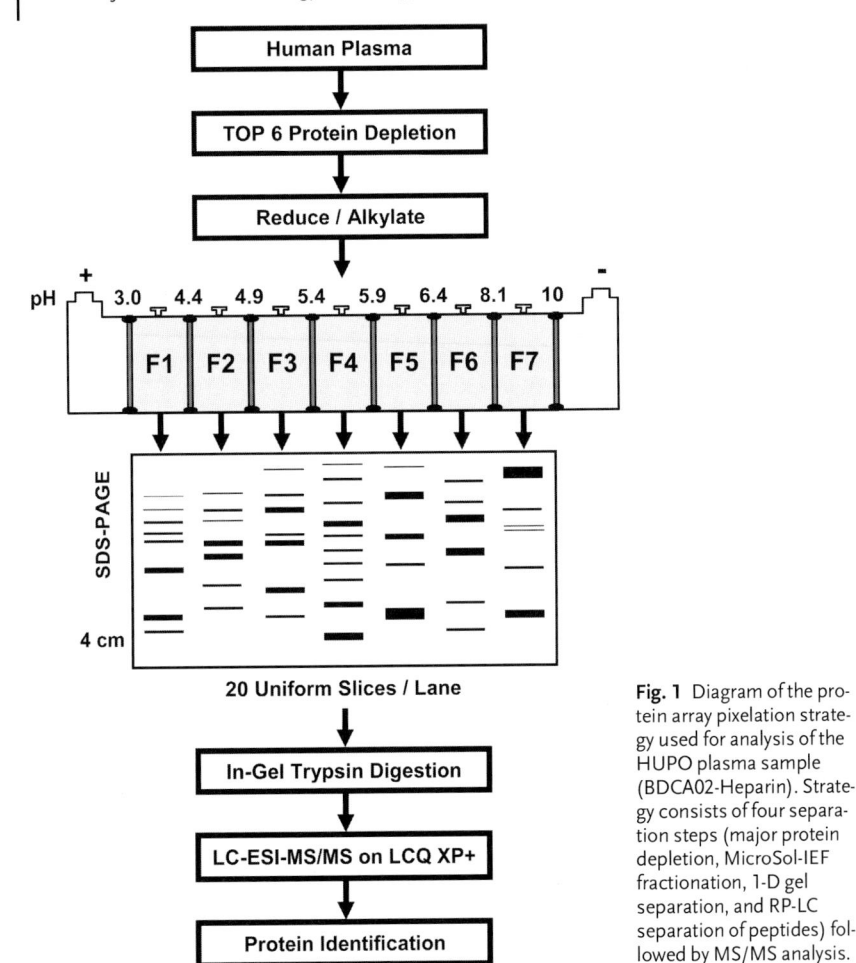

reduction and alkylation of the unbound (depleted) proteins, MicroSol-IEF is used as the second fractionation step to further reduce the complexity of the plasma proteome. Each fraction is subsequently electrophoresed on 1-D gels, sliced into pixels, digested individually with trypsin, and analyzed by LC-ESI-MS/MS. In initial analysis of the data, proteins were identified from peptides that passed the stringent criteria of $X_{Corr} \geq 1.9$ ($z = 1$), 2.3 ($z = 2$), 3.75 ($z = 3$) and $\Delta C_n \geq 0.1$, which is based on a commonly used relatively stringent published criteria [13]. Due to the concern that the strict $X_{Corr}/\Delta C_n$ used may eliminate some correctly identified low-level proteins, we also incorporated an additional scoring scheme, S_f (final score), which was developed by William Lane at Harvard University and is available in the commercial version of SEQUEST Browser. The S_f score examines the X_{Corr}, ΔC_n, S_p, R_{Sp}, and Ions scores of SEQUEST using a neural network and combines them into a single score that reflects the strength of peptide assignment on a scale of 0–1. Peptides with S_f score ≥ 0.7 were considered to have a high probability of being correct (William Lane, personal communication). Therefore, peptide assignments by SEQUEST were also considered positive if they had an S_f value of ≥ 0.7, regardless of the $X_{Corr}/\Delta C_n$ scores.

In this study, emphasis is given to proteins identified by multiple peptides (≥ 2 peptides) because the chance multipeptide proteins are false positives decreases exponentially with each additional peptide identified [24]. Since multiple peptides with lower X_{Corr} values can provide the same confidence as a single peptide with a high X_{Corr} value [24], the inclusion of the S_f score in our analysis should not generate a significant increase in false identifications of multipeptide proteins.

6.3.2
Optimization of protein array pixelation

A number of parameters that could affect the performance of the protein array pixelation were examined to optimize the method (Fig. 2). The first consideration is the size of the pixel used for tryptic digestion. The smaller the pixel size, the more total samples will need to be analyzed by LC-ESI-MS/MS, thereby substantially increasing the total time needed to completely analyze a proteome. To determine the effect of pixel size, a test sample of nondepleted serum was loaded on multiple lanes of a 1-D gel and electrophoresed for the full distance. When the same 4 mm region of a gel lane was examined with pixel size of 1 mm (four pixels total), 2 mm (two pixels total), and 4 mm (one pixel total), the largest number of nonredundant proteins was identified from the four 1 mm pixels analyzed separately. When 2 mm pixels were used the number of identified proteins decreased moderately but when a 4 mm pixel was used the decrease was dramatic (Fig. 2A, columns 1–3). Even though the total analysis time decreased four-fold with the single 4 mm pixel analysis *versus* four 1 mm pixels, the 58% decrease in high-confidence protein identifications (≥ 2 peptides) is clearly unacceptable. The 2 mm pixel size is a good compromise between the total analysis time and the number of proteins detected, since compared with 1 mm pixels, the analysis time was reduced by 50% and the high-confidence identifications were reduced by only 15% (Fig. 2A).

In these analyses, the number of protein identifications could be improved by increasing the sample injection volume from 2 to 4 µL (21% increase in high-confidence proteins; Fig. 2A, columns 3 and 4). Although this is a modest increase, it does not increase the analysis time and is therefore a positive factor. Increasing the RP-LC gradient time increased the high-confidence protein identification by 36% and the analysis time by 27% (Fig. 2A, columns 3 and 5). Hence, this change had a marginal advantage. A greater increase was observed when 10 µm C18 particle size POROS R2 resin was replaced with 5 µm MAGIC C18 particle size resin where a 57% increase in high-confidence proteins was obtained for a constant analysis time (Fig. 2A, columns 6 and 7). Extending the column length from 10 to 20 cm did not appreciably increase the number of proteins identified (6% increase in high-confidence proteins), but substantially improved protein coverage, since a 30% increase in the number of proteins with ≥ 3 peptides was observed with the 20 cm column (Fig. 2A, columns 7 and 8).

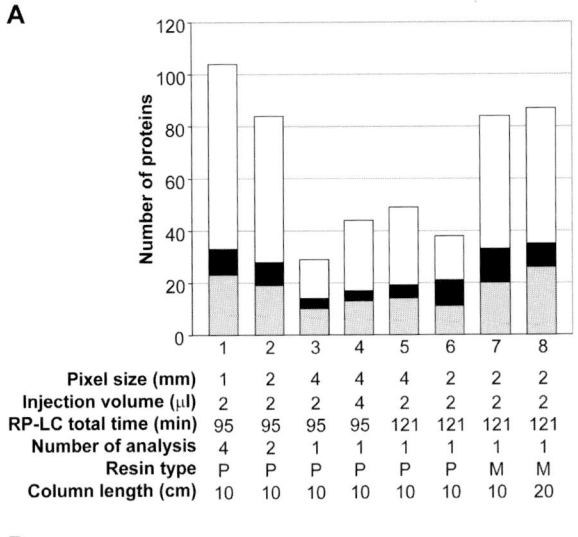

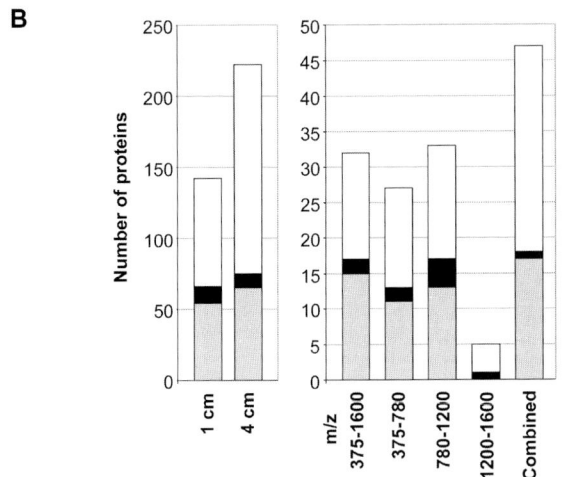

Fig. 2 Parameters affecting the efficiency of the protein array pixelation strategy. (A) Bar chart displaying the effect of pixel size (columns 1–3), sample injection volume (columns 3 and 4), RP-LC time (columns 4 and 5), type of C18 resin (columns 6 and 7), and column length (columns 7 and 8) on the number of nonredundant proteins identified. P, POROS R2 C18 10 μm; M, MAGIC C18 5 μm. (B) Bar charts showing the effect of gel separation distance and gas-phase fractionation on the number of nonredundant proteins identified. Number of proteins identified from the human plasma F3 MicroSol-IEF fraction electro-phoresed for 1 cm (10 × 1 mm size pixel) and 4 cm (20 × 2 mm size pixel) on 1-D gels are shown. Gas-phase fractionation of the F3–7 pixel from the human plasma sample was analyzed using the full m/z range of 375–1600, or with three separate m/z ranges as indicated. Last column shows the combined number of nonredundant proteins identified from the three separate m/z ranges. Number of proteins identified by 1, 2, and ≥3 unique peptides are indicated by the white, black, and gray bars, respectively.

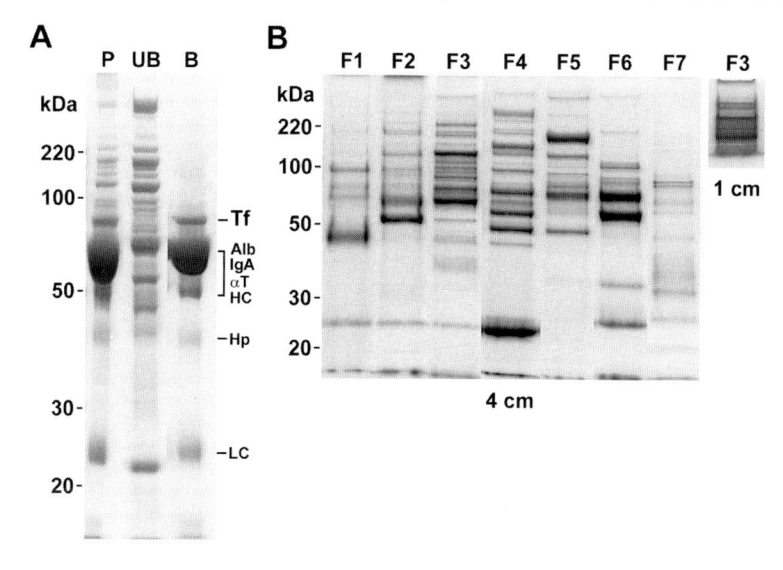

Fig. 3 Major protein depletion and MicroSol-IEF separation of human plasma proteins. (A) 1-D gel showing the plasma proteins [P] before depletion, and unbound [UB] and bound [B] proteins from the MARS antibody column. Tr, transferrin; Alb, albumin; αT, antitrypsin; HC, Ig heavy chain; Hp, haptoglobin; LC, Ig light chain. (B) Seven MicroSol-IEF fractions of the depleted plasma proteins were subjected to 1-D gel separations for a total distance of 4 cm from the bottom of the wells. Separation of F3 fraction for 1 cm is shown in the right panel. Proteins were separated on 10% bis-Tris NuPage gels using MOPS buffer, and stained with Colloidal blue.

We also examined the effect of 1-D gel separation distance on the number of proteins identified. For this analysis, a MicroSol-IEF fraction of the major protein-depleted plasma sample (F3, see below) was electrophoresed for a total distance of 4 or 1 cm (see Fig. 3B). The 4 cm lane was divided into 2 mm pixels for a total of 20 pixels, whereas the 1 cm lane was analyzed as 1 mm pixels for a total of 10 pixels. The total number of nonredundant proteins identified from the 4 cm lane was 56% greater than the 1 cm lane and the high-confidence identifications increased by 14% (Fig. 2B). Because longer gel separation distances are likely to increase the total number of analyses *per* proteome, the benefits of increased SDS gel separation distance are ambiguous. If a substantial number of the identifications based on one peptide are correct, the increased analysis time may be worthwhile.

A well-known major factor that limits peptide identification capability of complex peptide mixtures using LC-ESI-MS/MS is coelution of more peptides from the RP column than the mass spectrometer can analyze. One method of addressing this problem is gas-phase fractionation, where a single sample is repeatedly analyzed using different segments of the full m/z range in each run [25]. To test the utility of gas-phase fractionation in the current method, a 2 mm pixel (F3 pixel 7, depleted plasma sample; see below) was analyzed using the unsegmented m/z of 375–1600 approach and compared with gas-phase fractionation using three separate m/z segments of 375–780, 780–1200, and 1200–1600 (Fig. 2B). In the gas-phase fractionation experiment, most proteins were identified using the m/z range of 780–1200. In contrast, least proteins were identified using m/z of 1200–1600, and all proteins identified in this segment were also found in the other two segments (data not shown). However, peptides identified in the m/z 1200–1600 segment are important because they increased sequence coverage of many proteins. By combining the three m/z segments, the total number of nonredundant proteins identified increased by 47% compared to the single unsegmented analysis. However, the number of high-confidence proteins increased by a marginal 6%. Taking into consideration the three-fold increase in the analysis time, the segmented approach does not appear to be an efficient strategy for comprehensive proteome analysis using the protein array pixelation strategy.

6.3.3
Total analysis time for protein array pixelation of human plasma proteome

As emphasized in the above discussion of separation parameters, a major consideration for any comprehensive proteome analysis strategy is the total time required to analyze an entire proteome. Some improvements such as increased injection volume, smaller resin size, and longer column length can be implemented without affecting the total run time and therefore even modest improvements in protein coverage or number of proteins identified are considered positive improvements. However, increasing the number of pixels *per* SDS gel lane, increasing the RP-LC gradient time for better peptide separation, and increasing gel separation distances will increase total proteome analysis time as well as increase the number of proteins identified. Therefore, a practical compromise between improved number of proteins detected and increased analysis time has to be achieved. We felt that a generally acceptable time frame for complete proteome analysis should be similar to the time required to perform a MudPIT analysis of the human plasma/serum proteome [12, 15]. Therefore, based upon the optimization results discussed in Section 3.2, we decided to fractionate the depleted plasma using MicroSol-IEF into seven fractions, followed by 1-DE of each fraction for a total distance of 4 cm (Fig. 1). Each gel lane is sliced into 2 mm size pixels to produce a total of 140 pixels. Following tryptic digestion, each pixel is analyzed by LC-ESI-MS/MS with an RP-LC gradient time of 75 min. However, to minimize carryover from the previous run especially with increased sample injection volume, a short blank gradient is run after each analytical run. Hence, the total RP-LC run time from sample to sample is 121 min (see Section 2.5). In total, 11.8 days will be required to complete the analysis of the 140 pixels from the plasma proteome. This compares favorably with ~9.8 days to analyze the 135 SCX fractions of human serum proteome [12], and ~13.9 days for 77 LC-ESI-MS/MS runs from two cycles of SCX-LC of the human plasma proteome [15].

6.3.4
Systematic protein array pixelation of the human plasma proteome

The profiling of human plasma proteome began with major protein depletion from a total of 193 µL (14.5 mg) plasma (BDCA02-Heparin) using the MARS antibody column (Fig. 3A). Following depletion, 2.4 mg of unbound proteins were recovered, indicating that the six targeted proteins constituted approximately 83% of plasma proteins in this sample. Analysis of the bound fraction by 2-DE showed that the bound proteins were the six targeted proteins with no apparent evidence of other proteins [9].

The depleted plasma was then fractionated by MicroSol-IEF into seven fractions (Fig. 3B). Based on the 1-D gel analysis of the MicroSol-IEF fractions, the plasma proteins were well distributed throughout the seven fractions although the terminal fractions (F1 and F7) have the least amount of proteins as judged by the staining intensity (Fig. 3B). Many protein bands (including nondepleted abundant proteins) were present only in a specific fraction, indicating that MicroSol-IEF effectively separated proteins based on their pI. For example, a major protein with apparent MW of approximately 25 kDa was located almost exclusively in F4 (pH 5.4–5.9) with minor amounts found in more acidic fractions as observed by 1-D gel (Fig. 3B). Subsequent analysis by MS/MS identified this protein as apolipoprotein A-I precursor with calculated MW of 30.8 kDa and pI of 5.6. The good agreement with the observed values, confirmed the effectiveness of the pI and MW separation in this strategy. Hence, MicroSol-IEF not only further reduced the complexity of the plasma proteome, but also confined most remaining abundant proteins into specific fractions. This allows higher amounts of samples to be analyzed in downstream processes and permits us to dig deeper into the plasma proteome for lower abundance proteins.

Following MicroSol-IEF fractionation, the seven fractions were further separated by 1-D SDS-PAGE for a total distance of 4 cm (Fig. 3B). The gel lanes were then sliced and analyzed as uniform 2 mm pixels for a total of 140 pixels. Each pixel was digested in-gel with trypsin and analyzed by LC-ESI-MS/MS on an LCQ Deca XP+ mass spectrometer. In order to obtain a better correlation between the observed and the calculated MW of the proteins identified, the amount of sample loaded on the gel was limited to avoid overloading and to provide the optimal resolution of protein bands. In addition, the edge of the gel lanes where some degree of vertical smearing is frequently observed was excluded when the lane was cut. Depending upon protein concentration, between 1.3 and 2.8% (average 1.9%) of each Micro-Sol-IEF fraction was loaded onto 1-D gels used for pixelation. This average amount is equivalent to approximately 3.7 µL (278 µg) of the original plasma sample. Following tryptic digestion of the pixels, only 16.7% of each digestion mixture was analyzed by LC-ESI-MS/MS analysis. Hence, an amount equivalent to 0.6 µL (45 µg) of the original plasma sample was actually consumed in the final analysis.

From the LC-ESI-MS/MS analysis of the 140 pixels, a 2-D array of the human plasma proteome was generated (Fig. 4A). Each pixel in the array has a distinct range of MW and pI as shown and contains a group of identified proteins (from 3 to

Tab. 1 Number of nonredundant proteins identified from human plasma/serum using different filters

Sample	Filter	Number of nonredundant proteins[c]			
		Total	≥3	2	1
Plasma	S_f[a]	744	140	45	559
Serum	S_f[a]	4377	365	387	3625
Plasma	HUPO[b]	575	138	36	401
Serum	HUPO[b]	2890	297	223	2370
Plasmacommon[d]	HUPO[b]	319	132	29	158
Serumcommon[e]	HUPO[b]	319	178	33	108
Combined[f]	HUPO[b]	3146	316	251	2579
Without Ig[g]	HUPO[b]	3104	308	241	2555

a) Filter used: $X_{Corr} \geq 1.9$ ($z = 1$), 2.3 ($z = 2$), 3.75 ($z = 3$) and $\Delta C_n \geq 0.1$; or $S_f \geq 0.7$
b) Filter used: $X_{Corr} \geq 1.9$ ($z = 1$), 2.2 ($z = 2$), 3.75 ($z = 3$) and $\Delta C_n \geq 0.1$; and $R_{Sp} \leq 4$
c) ≥3, 2, and 1 indicate the number of unique peptides *per* protein
d) Proteins in plasma that are also identified in the serum data set
e) Proteins in serum that are also identified in the plasma data set
f) Both plasma and serum data sets were combined for analysis.
g) Combined data set with immunoglobulin entries removed

36 proteins) defined by one or more peptides that passed the $X_{Corr}/\Delta C_n/S_f$ criteria. Each pixel was assigned a name in the format Fx-y, where x is the MicroSol-IEF fraction (1–7), and y is the MW fraction from 1 (largest) to 20 (smallest). In general, the number of proteins identified in the pixels corresponds roughly to the staining density of the gel (Figs. 3B, 4A). A total of 744 nonredundant proteins defined by 3235 nonredundant peptides were identified from all the 140 pixels. Of these, 185 proteins (24.9%) were identified by at least two different peptides (high-confidence) whereas the majority (75.1%) was single-peptide proteins (Tab. 1).

A unique feature of this method is that the 2-D array can also be used to display the distributions of specific proteins that provide insight into their MW, p*I*, and the presence of alternate forms of each protein such as alternate splices and proteolytic fragments (Fig. 4B–D). Of course due to the fact that many plasma proteins are heterogeneously modified such as by glycosylation and proteolytic processing, the observed MW and p*I* are not expected to closely match the values derived from amino acid sequences. Due to the high sensitivity of the mass spectrometer, the high- and moderate-abundance proteins (mg/mL – µg/mL) were commonly found in more than one pixel. Since the relative abundance of a specific protein can be roughly determined from the number of unique peptides identified [26], the primary position of an abundant protein is determined by the pixel containing the maximum number of peptides. The distribution of three proteins with varying abundance (apolipoprotein B-100, 720 µg/mL; ceruloplasmin, 210 µg/mL; metalloproteinase inhibitor 1, 14 ng/mL [27]) is shown in Fig. 4. The distribution of

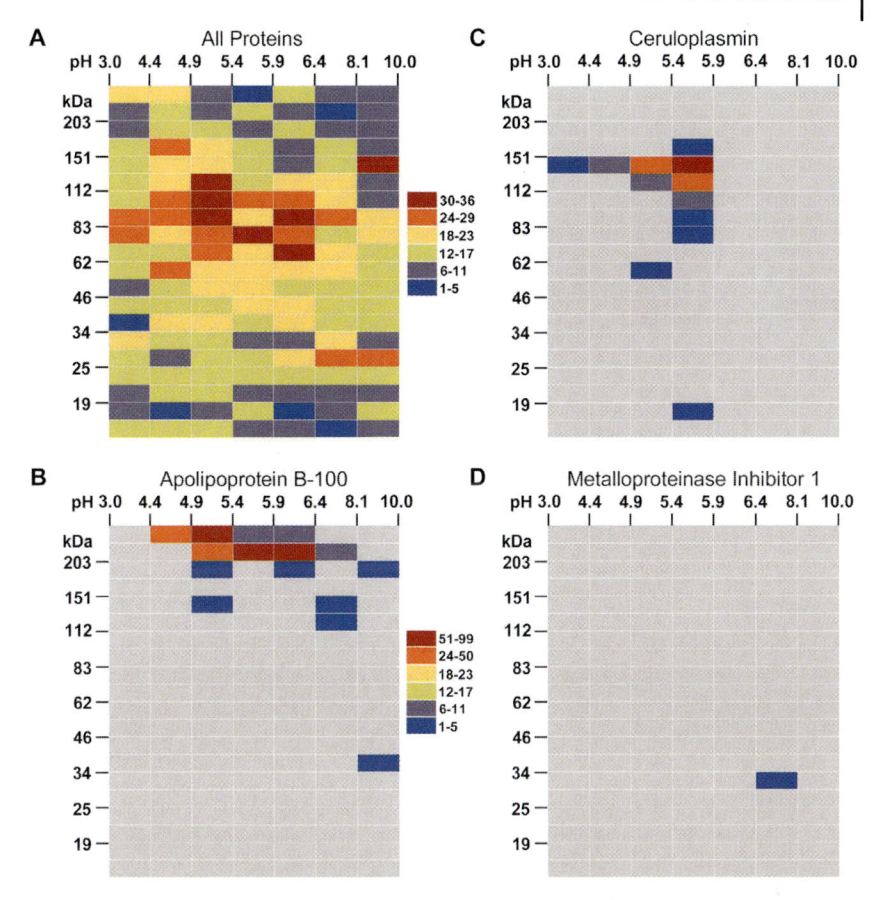

Fig. 4 Distributions of identified plasma proteins in the 2-D protein array. (A) Heat map showing the number of proteins identified that passed the $X_{Corr}/\Delta C_n/S_f$ criteria for each pixel in the analysis of the human plasma sample. Redundant proteins among pixels were not eliminated in this data set. Total number of proteins identified was 2255. Total number of nonredundant proteins was 744, which were defined by 3235 nonredundant peptides. (B–D) Heat maps showing the distributions of peptides identified for apolipoprotein B-100, ceruloplasmin, and metalloproteinase inhibitor 1.

apolipoprotein B-100 in the array indicated that the protein was present in at least two major forms (Fig. 4B). Both major forms were larger than 200 kDa; the smaller form (in F4–2 and F5–2) had a pI in the range of pH 5.4–6.4, whereas the pI of the larger form (in F3–1) is between pH 4.9 and 5.4. The observed MW of the protein and the multiple forms observed are consistent with the calculated MW of 515.6 kDa, and the reported forms of the protein such as B-74, B-48, and B-26 with apparent MW of 400, 259, and 140 kDa, respectively [28]. The observed pI of the protein is slightly lower than the theoretical pI of 6.6, which could be caused by heterogeneous modifications such as glycosylation [28]. The moderate-abundance protein, ceruloplasmin, was found mainly in F4–5 which is consistent with the calculated MW of 122.2 kDa and the theoretical p*I* of 5.4 (Fig. 4C). Unlike high- and moderate-abundance proteins, low-abundance proteins were usually identified by a single peptide that was found in only one or two pixels. For example, metalloproteinase inhibitor 1 precursor identified by the single-peptide GFQALGDAADIR is found only in pixel F6–15 at ~30 kDa and p*I* between 6.4 and 8.1 (Fig. 4D). These values are close to the expected MW of 23.2 kDa and p*I* of 8.5 for the protein. The MS/MS spectrum of this peptide was verified by manual inspection (see also Fig. 7). Hence, the MW and p*I* values derived from the 2-D array can be used to reinforce the protein identifications made by SEQUEST, especially for proteins identified by a single peptide, which is the group of proteins that predominates in most shotgun proteomics approaches.

6.3.5
Systematic protein array pixelation of the human serum proteome

Protein array pixelation of the HUPO serum sample, BDCA02-Serum, was performed using a method similar to that used for the plasma sample, except this method incorporated several refinements to further improve coverage of the proteome (Fig. 5A). The major protein depletion was performed on 415 µL (35.3 mg) of serum using a dual MARS column. A total of 4.3 mg of unbound proteins were recovered, indicating that approximately 88% of the total serum protein content was removed in this sample compared with the removal of 83% of total plasma proteins. This difference was at least partially due to more effective removal of targeted major proteins using the dual column compared with the single MARS column depletion of the plasma. This is consistent with the number of albumin peptides observed in both samples after depletion, where 9.2% sequence coverage of albumin was obtained from the depleted serum compared to 59.1% sequence coverage from the depleted plasma sample. Similarly, serotransferrin was identified with 40.7% sequence coverage in the depleted plasma sample but was not detected in the depleted serum sample. Since minor amounts of major proteins such as albumin (~40 mg/mL concentration) are still major components of the sample, they will still interfere with the overall analysis. Hence, it is better to use a longer antibody column and under-load the column to ensure the most effective depletion of targeted proteins as possible.

Following reduction and alkylation of the depleted serum proteins, the sample was fractionated into five pH fractions by MicroSol-IEF (Fig. 5B). The fractionation was performed using the commercially available pH membrane partitions, which greatly simplify the MicroSol-IEF procedure. To compensate for the reduced MicroSol-IEF fractions, the majority of the fractions (F2–F5) were separated on 1-D gels for a total distance of 6 cm. Compared to the 4 cm separation of the plasma sample, the longer separation distance should allow for increased sample loading (up to 50%) without overloading the gels. Due to the lower amount of proteins in F1, this fraction was concentrated by acetone precipitation and electrophoresed for only 2 cm to minimize empty regions in the gel lane (Fig. 5B). In addition, we also

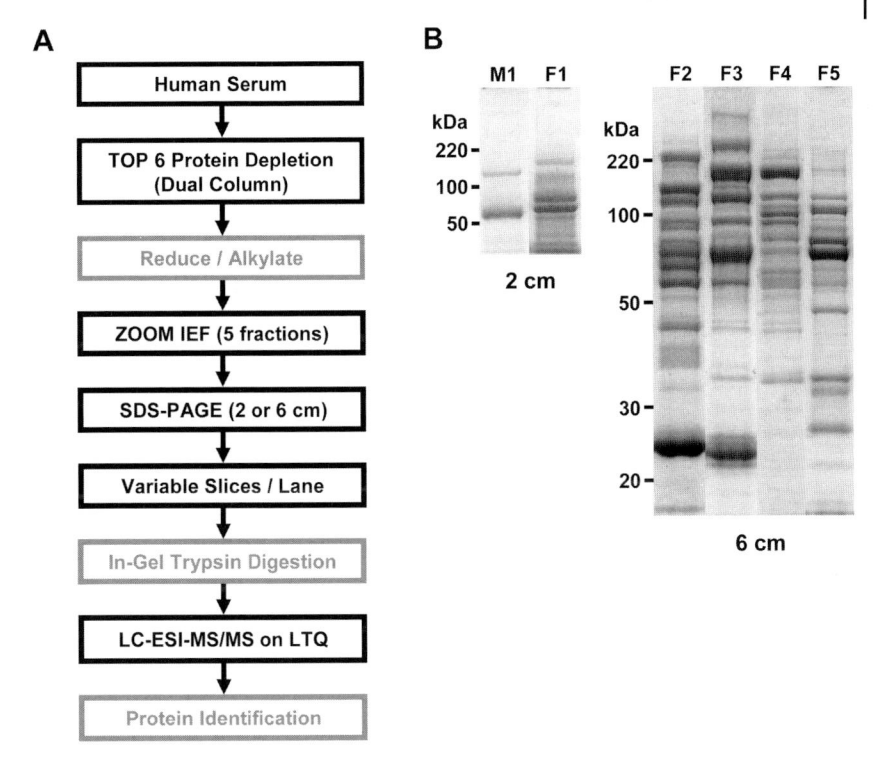

Fig. 5 Protein array pixelation of the HUPO serum sample (BDCA02-Serum). (A) Diagram showing the improved methodologies for analysis of the human serum sample. Steps identical to those used for analysis of the plasma sample are shown in gray. (B) 1-D gel showing the five MicroSol-IEF fractions (F1–F5) of major extracted proteins from the membrane partition (M1) between the anode buffer and F1 to detect proteins that might be trapped in the membrane. The M1 fraction was also concentrated by acetone precipitation prior to 1-D gel separation for 2 cm (Fig. 5B).

protein-depleted human serum. M1 shows protein extracted from the pH 3.0 membrane. Separation distances for each fraction are as indicated. Proteins were separated on 10% bis-Tris NuPage gels using MOPS buffer, and stained with Colloidal blue.

Following gel electrophoresis, the gel lanes containing M1 and F1 were analyzed as 2 mm size pixels. In the initial optimization studies presented above, more unique proteins were identified from four 1 mm size pixels than a single 4 mm pixel (Fig. 2A). To potentially increase the number of proteins identified, gel lanes containing the F2-F5 fractions were pixelated in a variable manner (1–4 mm size pixel) depending on the band intensity. Regions of the gel with intense staining were analyzed as 1 mm pixels, and regions without much staining were analyzed as 4 mm pixels (Figs. 5B, 6A). For a direct comparison of pixel size using current methods, fraction F3 was also reanalyzed as uniform 2 mm pixels. In total, 159 pixels were generated for tryptic digestion and analyzed by LC-ESI-MS/MS

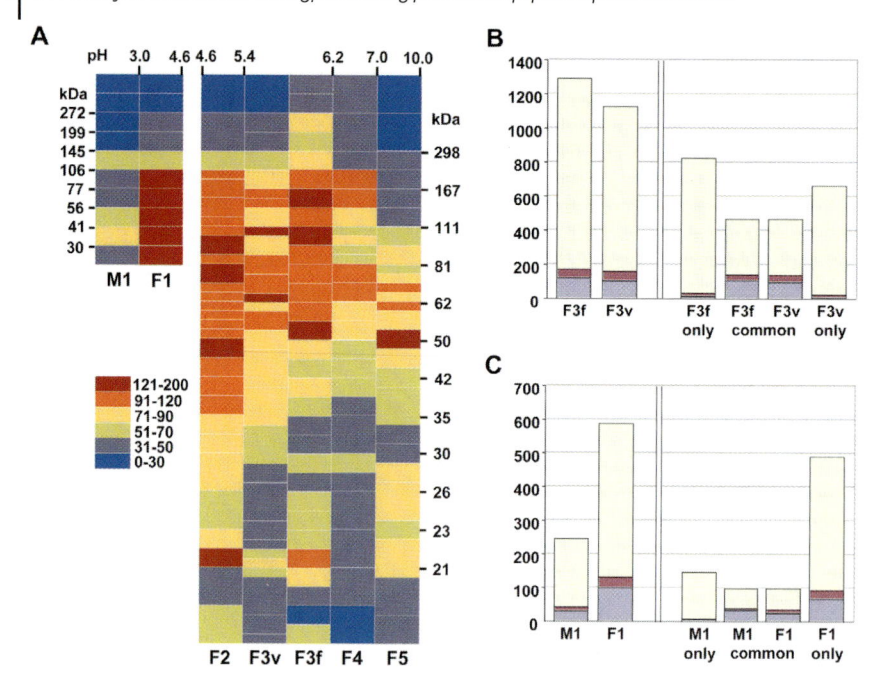

Fig. 6 Result from analysis of the human serum proteome. (A) Heat map showing distribution of the identified serum proteins in the 2-D protein array. Redundant proteins among pixels were not eliminated in this data set. Total number of proteins identified was 11 656. Total number of nonredundant proteins was 4377, which were defined by 9393 nonredundant peptides. (B) Comparison of the number of nonredundant proteins identified from the F3 fraction using the fixed pixelation strategy (F3f) and variable pixelation strategy (F3v). (C) Comparison of the number of nonredundant proteins identified from the M1 and F1 fractions. Proteins unique to a single data set are indicated in the "only" columns, and proteins present in both data sets are indicated in the "common" columns. Number of proteins identified by 1, 2, and $\geq$3 unique peptides are indicated by the yellow, purple, and blue bars, respectively.

with a total analysis time of 13.4 days (121 min RP-LC total run time *per* sample). Depending upon protein concentration, between 1.1 and 5.2% (2.5% average) of each soluble MicroSol-IEF fraction was used for gel pixelation. The average is equivalent to 10.4 µL (885 µg) of the original serum sample. After tryptic digestion, 23.3% of the digested material was injected and analyzed by LC-ESI-MS/MS. Therefore, protein identification was performed using an amount equivalent to approximately 2.4 µL (204 µg) of the original serum sample.

All samples from this serum analysis were analyzed using a Thermo Electron linear IT LTQ mass spectrometer, which is more sensitive and has a faster scan rate than the LCQ Deca XP+ [29]. The number of proteins identified for all 159 pixels is shown as a heat map in the 2-D array (Fig. 6A). Each pixel contained between 13 and 199 proteins that pass the $X_{Corr}/\Delta C_n/S_f$ criteria defined above. Comparison of the uniform (F3f) and variable (F3v) pixelation methods of the F3 fraction indicated that the variable pixelation method did not offer any improvement over the uniform pixelation method (Fig. 6B). In fact, uniform pixelation identified 6.9% more high-confidence proteins compared to the variable pixelation method. In addition, the uniform pixelation method is easier and quicker to perform, as there is no need to correlate the pixel size with band intensity. The total number of nonredundant proteins identified from the 159 pixels was 4377 from a total of 9393 nonredundant peptides. Of these, 752 proteins (17.2%) were identified as high-confidence and the majority (82.8%) was single-peptide proteins (Tab. 1). Therefore, the overall improvements due to further refinement of the method and, most importantly, use of the highly sensitive LTQ mass spectrometer, resulted in about fourfold increase in the number of high-confidence proteins identified in serum compared to the plasma analysis.

The establishment of the pH gradient during MicroSol-IEF is dependent on the membrane partitions between each fraction [21]. During MicroSol-IEF, some proteins can be partially or completely trapped in the membrane partitions and are therefore excluded from the soluble fractions. To investigate this possibility, proteins were extracted from the five membrane partitions and analyzed on 1-D gels (data not shown). Except for M1 (pH 3.0 membrane partition between anode buffer and F1), all protein bands from the other membrane partitions appeared to have corresponding protein bands in adjacent soluble fractions. Furthermore, the protein bands from the membrane partitions are much less intensely stained than their soluble fractions counterparts. The only exception is apolipoprotein B-100 which is found predominately in the membrane partition between F3 and F4, presumably due to its large size of ~540 kDa [28]. The tentative conclusion that most proteins trapped in membrane partitions were also partially recovered in adjacent fractions was further supported by parallel analysis of membrane and soluble fractions from similar serum separations on 2-D gels (data not shown). In M1, however, two apparently unique protein bands were observed in the 1-D gel analysis (Fig. 5B shows the concentrated M1 proteins). Pixelation of the M1 fraction and comparison with F1 fraction indicated that 42 high-confidence proteins were identified in M1, but only 7 were not identified in F1 (Fig. 6C). Out of these seven high-confidence M1 proteins, six were not identified elsewhere in the entire serum proteome analysis. Hence, a few acidic proteins were found exclusively in the pH 3 membrane partition, and it is likely that a small number of unique proteins are in other membrane partitions and could be detected if higher sensitivity methods like LC-ESI-MS/MS are used instead of 1- or 2-D gels. Therefore, it may be advantageous to include membrane partition extracts in the analyses to provide more comprehensive coverage of proteins.

6.3.6
Analyses of human plasma and serum proteomes using HUPO filter criteria

All the data described above were analyzed using the $X_{Corr}/\Delta C_n/S_f$ criteria. From experience, we know that a substantial number of the proteins identified by a single peptide using these criteria are incorrect and the probability of a protein being correctly identified increases with the number of unique peptides identified. However, for biomarker discovery it is better to have a less stringent filter so that potentially interesting low-abundance proteins will not be excluded from the analysis. This, however, will inevitably increase the number of false positives and will require more efforts to verify the data generated. As the aim of the HUPO PPP is to provide the most accurate description of the human serum/plasma proteome possible, a more stringent filter ($X_{Corr} \geq 1.9$ ($z = 1$), 2.2 ($z = 2$), 3.75 ($z = 3$) and $\Delta C_n \geq 0.1$; and $R_{Sp} \leq 4$) was selected by HUPO to minimize false identifications. The analysis of our plasma and serum data sets using the HUPO criteria is summarized in Tab. 1. With the more stringent HUPO criteria, the number of nonredundant proteins identified by ≥ 2 peptides from the smaller plasma data set was reduced by only 5.9% while the single-peptide proteins were reduced by 28.3%. A larger reduction was observed with the serum data set, where proteins identified by ≥ 2 peptides and a single peptide were reduced by 30.9 and 34.6%, respectively. In both data sets, the reduction in proteins identified by ≥ 2 peptides was mainly contributed by the two-peptide proteins (Tab. 1). This is consistent with the expectation that identification errors are more likely to happen for single-peptide proteins followed by two-peptide proteins, and least likely for proteins identified by more than two peptides.

In this study, nonredundant proteins are defined as proteins with different accession number. The in-house software used for the analysis of these datasets dit not eliminate potential redundancy caused by SEQUEST assignment of the same peptide in different pixel to different but homologous database entries. To address this issue, the plasma and serum datasets were reanalyzed using the DTASelect program [30] that is capable of grouping redundant identifications. The program was used to filter peptides using the HUPO high stringency criteria. Proteins that were subsets of others of contained the description 'keratin' were removed. A total of 576 and 2725 nonredundant proteins were reported for the plasma and serum datasets, respectively, using the DTASelect program compared with 575 and 2890 proteins using our in-house software. Therefore, the redundancy in our analysis is very minimal.

When both data sets were combined, a total of 3146 nonredundant proteins were identified using the HUPO criteria (Tab. 1). Of these, 567 (18.0%) were identified by ≥ 2 unique peptides and 82.0% were single-peptide proteins. Since immunoglobulins were depleted in both samples, they constituted only 1.3% of the total nonredundant proteins (or 3.2% of proteins with ≥ 2 peptides) identified in the combined data set (Tab. 1). The number of proteins that are common to both data sets is only 319, and is limited by the lower sensitivity method used in the plasma analysis (Tab. 1). In addition, 92.5% of the proteins with ≥ 2 peptides in plasma were identified in the serum analysis, but only 40.6% of serum proteins with ≥ 2 peptides were identified in the plasma analysis. However, 49.5% of the common proteins identified in plasma are single-peptide proteins. Since these single-peptide proteins were identified using a different instrument and sample, it is likely that a large percentage of the single-peptide proteins identified in plasma are probably correct. In support of this, many of the single-peptide proteins identified in plasma, as well as in serum, have rich MS/MS fragmentation patterns that agree

Tab. 2 Examples of low-abundance proteins (<100 ng/mL) and the corresponding peptides identified in the human plasma and serum samples

Sample	Name	ng/mL[a)]	Sequence	z	X_{Corr}	ΔC_n	R_{Sp}
Plasma	Vascular endothelial-cadherin	30	VHDVNDNWPVFTHR	3	4.08	0.52	1
Serum	Vascular endothelial-cadherin	30	DTGENLETPSSFTIK	2	4.32	0.43	1
			EYFAIDNSGR	2	2.59	0.55	1
			KPLIGTVLAM*DPDAAR	3	3.76	0.32	1
			VDAETGDVFAIER	2	3.75	0.61	1
			VHDVNDNWPVFTHR	2	3.45	0.51	1
			YEIVVEAR	2	2.39	0.27	1
Plasma	L-selectin	17	NKEDCVEIYIK	2	3.56	0.38	1
Serum	L-selectin	17	NKEDCVEIYIK	2	4.29	0.34	1
			SLTEEAENWGDGEPNNK	2	4.38	0.50	1
			SLTEEAENWGDGEPNNKK	2	5.22	0.61	1
			SYYWIGIR	2	2.65	0.35	1
			TICESSGIWSNPSPICQK	2	5.33	0.50	1
Plasma	Metalloproteinase inhibitor 1	14	GFQALGDAADIR	2	3.72	0.45	1
Serum	Metalloproteinase inhibitor 1	14	GFQALGDAADIR	2	3.05	0.26	1
			HLACLPR	2	2.38	0.11	1
			LQSGTHCLWTDQLLQGSEK	3	5.32	0.52	1
Serum	Vascular endothelial growth factor D	0.500	SEQQIRAASSLEELLR	2	2.47	0.13	2
Serum	Calcitonin	0.190	SALESSPADPATLSEDEAR	2	2.24	0.15	3
Serum	Tumor necrosis factor (TNF-a)	0.041	PWYEPIYLGGVFQLEK	2	2.87	0.30	1

[a)] Concentration values were obtained from [27]
* Indicates methionine oxidation

well with peptide sequences assigned by SEQUEST. Examples of the MS/MS spectra for single-peptide proteins identified in both data sets are shown in Fig. 7. All of the major peaks in both MS/MS spectra can be accounted for by fragment ions from the predicted peptide sequences, indicating that the peptide assignment is correct. Of particular interest is the protein creatine kinase M which, in the MB isoform, is an important serum marker for myocardial infarction [31]. Therefore, even though the single-peptide protein category contains the most false positives, it also contains many important correct entries that cannot be ignored.

Examples of low-abundance proteins identified in the plasma and serum samples using HUPO criteria are shown in Tab. 2. The list provides an estimate of the detection limit of the protein array pixelation strategy. Some proteins in the low ng/mL can be detected from 45 µg of plasma using the LCQ Deca XP+, whereas some proteins in the pg/mL can be detected from the 204 µg of serum analyzed using the LTQ mass spectrometer. Not surprisingly, the ability to detect low abundance proteins decreases with protein abundance. For example, among the low abundance proteins described by Haab et al. [27] in their Tab. 2, 14 out of the 20 proteins in the

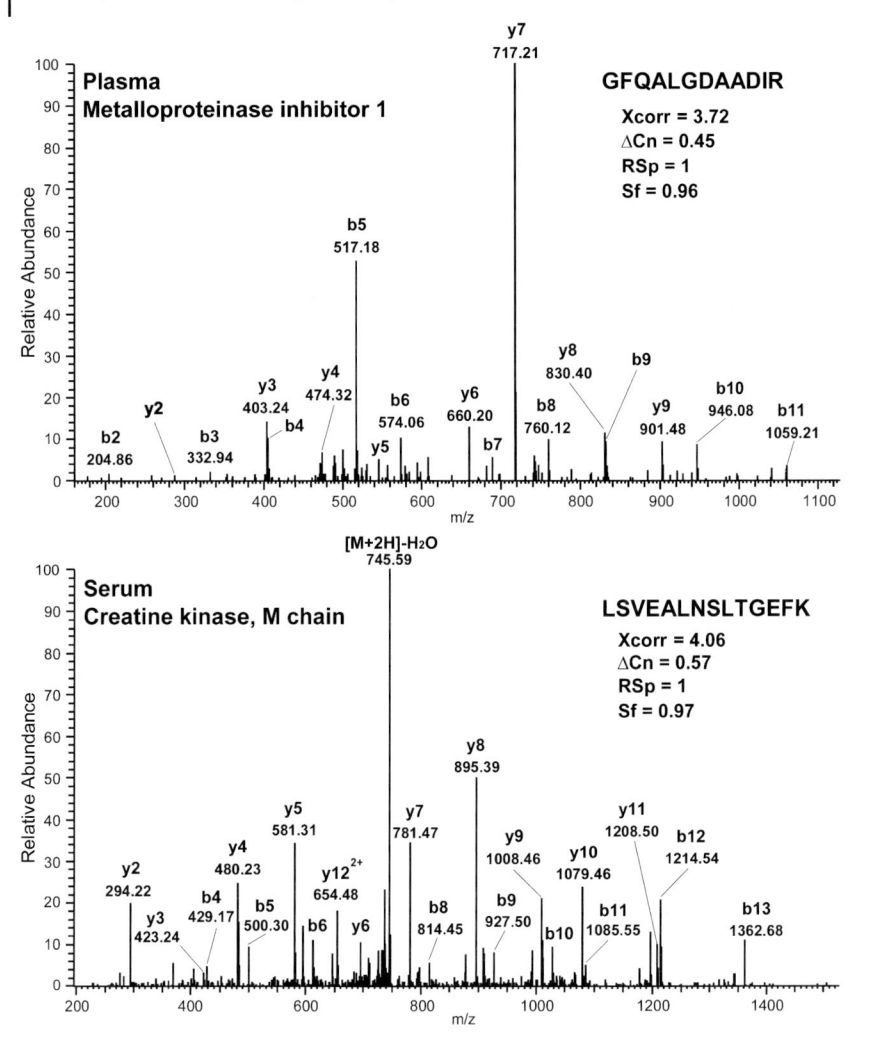

Fig. 7 Representative MS/MS spectra of low-abundance proteins identified by single-peptide matches. MS/MS spectra of the doubly charged ions with m/z 617.45 (GFQALGDAADIR) and m/z 755.22 (LSVEALNSLTGEFK) are shown.

1 to 100 ng/mL concentration range were detected in our serum analysis, whereas only 2 out of 19 proteins at concentrations below 1 ng/mL were detected. In addition, most of the lower abundance proteins identified in plasma are single-peptide proteins whereas the same proteins were identified with multiple peptides in the serum analysis using the more sensitive linear IT mass spectrometer. This indicates that the use of the highly sensitive LTQ mass spectrometer coupled with our optimized method allows detection of proteins up to a concentration range of 10^9.

6.4 Concluding remarks

This study demonstrates the utility of a novel 4-D protein profiling strategy, protein array pixelation, for comprehensive profiling of human plasma and serum proteomes. The four separations used in this strategy greatly reduce plasma/serum complexity, allowing access to proteins differing in abundance by up to nine orders of magnitude. Using HUPO criteria for high-confidence protein identifications, this strategy has detected a total of 3104 nonredundant proteins, after excluding keratins and immunoglobulins. Although larger amounts of sample are used for early steps, the final LC-ESI-MS/MS analyses are based on very low amounts of sample (45 µg of plasma and 204 µg of serum). Of these identified proteins, 549 were identified with two or more unique peptides. The total time required for analyzing each sample was similar to MudPIT approaches described by others [12, 15]. Analysis of the HUPO serum sample (BDCA02) using the highly sensitive LTQ mass spectrometer and an optimized method produced a very rich data set that contained >90% of the proteins with two or more peptides identified in the plasma sample. Most importantly, many low-abundance proteins (<100 ng/mL – pg/mL levels) were identified in this data set. In conclusion, the protein array pixelation strategy is a powerful method for comprehensive protein profiling and for protein biomarker discovery.

We would like to thank Brian Haab for sharing data prior to publication. We are also grateful to John Yates, III form The Scripps Research Institute for providing the DTASelect software. This work was supported in part by the National Institutes of Health Grants CA94360 and CA77048 to D.W.S., and institutional grants to the Wistar Institute including an NCI Cancer Core Grant (CA10815), and the Commonwealth Universal Research Enhancement Program, Pennsylvania Department of Health.

6.5 References

[1] Anderson, N. L., Anderson, N. G., *Mol. Cell. Proteomics* 2002, *1*, 845–867.

[2] Lindahl, B., Venge, P., Wallentin, L., *Coron. Artery Dis.* 1995, *6*, 321–328.

[3] Stamey, T. A., Yang, N., Hay, A. R., McNeal, J. E., Freiha, F. S., Redwine, E., *N. Engl. J. Med.* 1987, *317*, 909–916.

[4] Bast, R. C., Klug, T. L., St. John, E., Jenison, E., Niloff, J. M., Lazarus, H., Berkowitz, R. S. *et al.*, *N. Engl. J. Med.* 1983, *309*, 883–887.

[5] Greenough, C., Jenkins, R. E., Kitteringham, N. R., Pirmohamed, M. Park, B. K., Pennington, S. R., *Proteomics* 2004, *4*, 3107–3111.

[6] Marshall, J., Jankowski, A., Furesz, S., Kireeva, I. Barker, L., Dombrovsky, M., Zhu, W., *et al.*, *J. Proteome Res.* 2004, *3*, 364–382.

[7] Pieper, R., Su, Q., Gatlin, C. L., Huang, S. T. Anderson, N. L., Steiners, S., *Proteomics* 2003, *3*, 422–432.

[8] Pieper, R., Gatlin, C. L., Makusky, A. J., Russo, P. S., Schatz, C. R., Miller, S. S., Su, Q. *et al.*, *Proteomics* 2003, *3*, 1345–1364.

[9] Echan, L. A., Tang, H.-Y., Ali-Khan, N., Lee, K., Speicher, D. W., *Proteomics* 2005 *22*, this issue.

[10] Zuo, X., Echan, L., Hembach, P., Tang, H.-Y., Speicher, K. D., Santoli, D., Speicher, D. W., *Electrophoresis* 2001, *22*, 1603–1615.

[11] Zuo, X., Speicher, D. W., *Proteomics* 2002, *2*, 58–68.
[12] Adkins, J. N., Varnum, S. M., Auberry, K. J., Moore, R. J., Angell, N. H., Smith, R. D., Springer, D. L., Pounds, J. G., *Mol. Cell. Proteomics* 2002, *1*, 947–955.
[13] Washburn, M. P., Wolters, D., Yates, J. R., *Nat. Biotechnol.* 2001, *19*, 242–247.
[14] Fujii, K., Nakano, T., Kawamura, T., Usui, F., Bando, Y., Wang, R., Nishimura, T., *J. Proteome Res.* 2004, *3*, 712–718.
[15] Shen, Y., Jacobs, J. M., Camp, D. G. II, Fang, R., Moore, R. J., Smith, R. D., Xiao, W. *et al.*, *Anal. Chem.* 2004, *76*, 1134–1144.
[16] Xiao, Z., Conrads, T. P., Lucas, D. A., Janini, G. M., Schaefer, C. F., Buetow, K. H., Issaq, H. J., Veenstra, T. D., *Electrophoresis* 2004, *25*, 128–133.
[17] Janini, G. M., Chan, K. C., Conrads, T. P., Issaq, H. J., Veenstra, T. D., *Electrophoresis* 2004, *25*, 1973–1980.
[18] Wrotnowski, C., *Genet. Eng. News* 1998, *18*, 14.
[19] Anderson, N. L., Polanski, M., Pieper, R., Gatlin, T., Tirumalai, R. S., Conrads, T. P., Veenstra, T. D. *et al.*, *Mol. Cell. Proteomics* 2004, *3*, 311–326.
[20] Omenn, G. S., *Proteomics* 2004, *4*, 1235–1240.
[21] Zuo, X., Speicher, D. W., *Anal. Biochem.* 2000, *284*, 266–278.
[22] Speicher, K. D., Kolbas, O., Harper, S., Speicher, D. W., *J. Biomol. Tech.* 2000, *11*, 74–86.
[23] Kersey, P. J., Duarte, J., Williams, A., Karavidopoulou, Y., Birney, E., Apweiler, R., *Proteomics* 2004, *4*, 1985–1988.
[24] MacCoss, M. J., Wu, C. C., Yates, J. R., *Anal. Chem.* 2002, *74*, 5593–5599.
[25] Yi, E. C., Marelli, M., Lee, H., Purvine, S. O., Aebersold, J. D., Goodlett, D. R., *Electrophoresis* 2002, *23*, 3205–3216.
[26] Yu, L.-R., Conrads, T. P., Uo, T., Kinoshita, Y., Morrison, R. S., Lucas, D. A., Chan, K. C. *et al.*, *Mol. Cell. Proteomics* 2004, *3*, 896–907.
[27] Haab, B. B., Geierstanger, B. H., Michailidis, G., Vitzthum, F., Forrester, S., Okon, R., Saviranta, P. *et al.*, *Proteomics*, 2005, *5*. DOI 10.1002/pmic.200400470
[28] Kane, J. P., *Annu. Rev. Physiol.* 1983, *45*, 637–650.
[29] Schwartz, J. C., Senko, M. W., Syka, J. E., *J. Am. Soc. Mass Spectrom.* 2002, *13*, 659–669.
[30] Tabb, D. L., McDonald, W. H., Yates, J. R. III, *J. Proteome Res.* 2002, *1*, 21–26.
[31] Mair, J., Artner-Dworzak, E., Dienstl, A., Lechleitner, P., Morass, B., Smidt, J., Wagner, I. *et al.*, *Am. J. Cardiol.* 1991, *68*, 1545–1550.

7
A study of glycoproteins in human serum and plasma reference standards (HUPO) using multilectin affinity chromatography coupled with RPLC-MS/MS*

Ziping Yang, William S. Hancock, Tori Richmond Chew and Leo Bonilla

The glycoproteome is a major subproteome present in human plasma. In this study, we isolated and characterized approximately 150 glycoproteins from the human plasma and serum samples provided by HUPO using a multilectin affinity column. The corresponding tryptic digest was separated by RP-HPLC coupled to an IT mass spectrometer (3-D LCQ). Also in this study, a new system, namely an Ettan MDLC system coupled to a linear ITLTQ, was compared with the previous LCQ platform and gave a greater number of protein identifications, as well as better quality. When we compared the composition of the glycoproteomes for the plasma and serum samples there was a close correlation between the samples, except for the absence of fibrinogen from the identified-protein list in the latter sample, which was presumably as a result of the clotting process. In addition, the analysis of the samples from three ethnic specimens, Caucasian American, Asian American, and African American, were very similar but showed a higher angiotensinogen plasma level and a lower histidine-rich glycoprotein level in Caucasian American samples, and a lower vitronectin level in African American blood samples.

7.1
Introduction

The glycoproteome is one of the major subproteomes of human plasma, as many proteins are secreted from the tissues, such as the liver, in a glycosylated form [1–4]. It is proposed from literature studies that about 50% of all plasma proteins are glycosylated [5], which was confirmed by lectin capture experiments, if one excludes albumin [6]. The plasma glycoproteome has important clinical value, as many biomarkers are glycosylated, such as the breast cancer biomarkers CA125 and ERBB [7,

---

* Originally published in Proteomics 2005, 13, 3353–3366

Exploring the Human Plasma Proteome. Edited by Gilbert S. Omenn
Copyright © 2006 WILEY-VCH Verlag GmbH & Co. KGaA, Weinheim
ISBN: 3-527-31757-0

8]. Zhang *et al.* [9] have developed a method to specifically enrich glycoproteins from human serum by capturing N-linked glycoproteins using hydrazide chemistry. In this method, the captured proteins were digested and the N-linked glycopeptides were then isolated from the complex. The glycopeptides were treated with PNGase F (to release glycans) and identified using MS/MS. This method could also be used for comparative quantification if coupled with isotope labeling. Using lectin affinity to enrich glycoproteins is another approach, which avoids chemical derivatization and the potential for side reactions. In order to comprehensively study the serum glycoproteome, we have developed a multilectin affinity system to efficiently and specifically enrich glycoproteins from human serum [6]. In that study we demonstrated that the use of a set of lectins (optimized to a given sample) overcame the broad specificity and lack of complete glycoprotein capture that is typically achieved with a single lectin. In addition, we demonstrated that this approach was specific to glycoproteins, gave good recovery and was reproducible.

The challenge of a comprehensive study of the human plasma proteome is its wide dynamic range. To better identify low-abundance proteins in plasma, the removal of the most abundant protein(s) using an immunoaffinity approach has been proven to be effective [10]. However, with the depletion of these high-abundance proteins, such as albumin, it has been suggested that some interesting proteins are also lost due to protein complex formation. The glycoprotein enrichment process described here automatically improves the dynamic range of serum protein analysis, since the nonglycosylated albumin is largely removed. Although some albumin is retained in the affinity systems with its associated glycoproteins, this approach minimizes nonspecific losses.

In this research, we analyzed HUPO human plasma and serum samples from different ethnic groups using multilectin affinity chromatography followed by trypsin digestion and LC-MS/MS. The results of a comparison of the plasma and serum glycoproteomes are reported here, as well as the results obtained with samples collected from different ethnic groups.

7.2
Materials and methods

7.2.1
Materials

Human plasma and serum samples were provided by HUPO. These samples (total of 12) were from the pools of three ethnic groups including Caucasian American, African American and Asian American, and the plasma was treated with sodium citrate, lithium heparin, or K_2EDTA (Tab. 1). Agarose-bound lectins (Con A, wheat germ agglutinin (WGA), and Jacalin lectin (JAC) were purchased from Vector Laboratories (Burlingame, CA, USA).

Tab. 1 Number of proteins identified in each plasma/serum sample

Sample type	Ethnic group	Caucasian American CA	Asian American AA	African American AFA
Sodium citrate (CIT) plasma		76	63	69
K₂EDTA (EDTA) plasma		71	84	75
Lithium heparin (HEP) plasma		75	81	90
Serum		78	74	73

7.2.2
Isolating glycoproteins using multilectin affinity columns

The multilectin column was prepared by mixing equal amount of agarose-bound Con A, agarose-bound WGA and agarose-bound Jacalin in an empty PD-10 disposable column (GE Healthcare, Piscataway, NJ, USA). The sample of 50 µL serum or plasma (Tab. 1) was diluted with multilectin column equilibrium buffer (20 mM Tris, 0.15 M NaCl, 1 mM Mn^{2+}, and 1 mM Ca^{2+}, pH 7.4) to a volume of 1 mL, and was loaded on a newly packed multilectin affinity column. After a 15 min reaction, the unbound proteins were eluted with 10 mL of equilibrium buffer, and the captured proteins were released with 12 mL of displacer solution (20 mM Tris, 0.5 M NaCl, 0.17 M methyl-α-D-mannopyranoside, 0.17 M N-acetyl-glucosamine, and 0.27 M galactose, pH 7.4). The multilectin affinity column captured fraction was collected and concentrated with 15 mL, 10 kDa Amicon filters (Millipore, Billerica, MA, USA).

7.2.3
Analysis of glycoproteins on LC-LCQ MS

The lectin-captured proteins, 100 µg, were digested with trypsin, using a procedure described previously [11]. Proteins were first denatured with 6 M guanidine chloride in 0.1 M ammonium bicarbonate buffer, pH 8 and reduced by incubating with 5 mM DTT at 75°C for 1 h, and then alkylated for 2 h with 0.02 M iodoacetamide. The samples were solvent exchanged using 0.1 M ammonium bicarbonate buffer (pH 8) with a 10 kDa Amicon filter (0.5 mL capacity; Millipore), and adjusted to the final protein concentration of 0.5 mg/mL. Then, 1 µg trypsin was added to each sample and incubated at ambient temperature overnight. For complete digestion, another aliquot of 1 µg trypsin was added, and the digestion was continued for a total of 24 h. Then, the peptides were separated on a C18 capillary column (in-house packed, 150 × 0.075 mm) using a Surveyor LC pump (Thermo Electron, San Jose, CA, USA). The flow rate was maintained at 300 nL/min. The gradient was started at 5% ACN with 0.1% formic acid and a linear gradient to 40% ACN was achieved in 165 min, and then ramped to 60% ACN in 20 min and to 90% in next 10 min. Ten microliters of each sample containing 2 µg of protein was injected on the column from a Surveyor autosampler (Thermo Electron) using the full-loop injection mode. The resolved peptides were analyzed on an LCQ DECA XP IT mass spectrometer (Thermo Electron) with an ESI ion source. The temperature of the ion transfer tube was controlled at 185°C and the spray voltage was 2.0 kV. The normalized collision energy was set at 35% for MS/MS. Data-dependent ion selection was monitored to select the most abundant five ions from an MS scan for MS/MS analysis. Dynamic exclusion was continued for a duration of 2 min.

7.2.4
Analysis of glycoproteins on LC-LTQ MS

The glycoproteins enriched from Caucasian American serum sample were digested with trypsin, and the digest was separated on a capillary column (Thermo Hypurity, C18, 150 × 0.075 mm) using Ettan MDLC system from GE Healthcare. The separation gradient was similar to that described in Section 2.2, except that the starting point was 0% ACN due to the use of a trap column (Michrom Bioresources, Auburn, CA, USA) in front of the separation column. The resolved peptides were analyzed on an LTQ mass spectrometer (Thermo Electron) with an ESI ion source. The temperature of the ion transfer tube was controlled at 185°C and the spray voltage was 2.0 kV. The normalized collision energy was set at 35% for MS/MS. Data-dependent ion selection was monitored to select the most abundant five ions from an MS scan for MS/MS analysis. Dynamic exclusion was continued for a duration of 2 min.

7.2.5
Protein database search

Peptide sequences were identified using SEQUEST algorithm (version C1) incorporated in BioWorks software (version 3.1) (Thermo Electron) and Swiss-Prot human protein database. The database search was limited to only those peptides that would be generated by tryptic cleavage. The SEQUEST results were filtered by X_{corr} *versus* charge state. X_{corr} was used for a match with 1.5 for singly charged ions, 2.0 for doubly charged ions, and 2.5 for triply charged ions. We set $\Delta C_n \geq 0.085$ and $R_{sp} \leq 5$. The protein identification (ID) was made based on the corresponding peptide IDs.

7.3
Results and discussion

7.3.1
Protein IDs from the plasma and serum samples

The tryptic digests of glycoproteins isolated from the plasma or serum samples were analyzed on an IT mass spectrometer (LCQ-MS), and proteins with two or more peptide IDs from all 12 samples were considered as positive IDs. Tab. 2a lists nine additional glycoproteins, which were identified with only one peptide by using

Tab. 2a Protein IDs confirmed by LTQ analysis[a]

ID[b]	Protein[c]	Rank[d]	Hits[e]
A2AP	Alpha-2-antiplasmin	40	7
C1QB	Complement C1Q subcomponent, B chain	56	4
CO8G	Complement component C8 gamma chain	69	2
FA11	Coagulation factor XI	58	4
KV3F	IG kappa chain V-III region	74	2
LV3B	IG lambda chain V-III region	70	2
CBG	Corticosteroid-binding globulin	84	1
HV3A	IG heavy chain V-III region	92	1
NUEM	NADH-ubiquinone oxidoreductase	182	1

a) Proteins identified with only one peptide ID in LC-LCQ MS were confirmed in the LC-LTQ MS analysis of Caucasian American serum sample
b) Swiss-Prot entry name
c) Protein name
d) SEQUEST rank represents the confidence of the assignment of a protein in the sample
e) Number of peptide IDs for a given protein sequence

Tab. 2b Data comparison between LCQ and LTQ-MS/MS analysis[a]

Mass spectrometer	LCQ[b]	LTQ[c]
Total protein ID	93	185
Two or more peptide IDs	58	81

a) Serum sample from Caucasian American specimen was tested on LC-LCQ MS and LC-LTQ MS
b) LCQ refers to 3-D IT mass spectrometer
c) LTQ refers to the linear IT mass spectrometer

the LCQ system, but for which the IDs were subsequently confirmed by a separate analysis of the Caucasian American serum sample using the more sensitive linear IT LTQ-MS. In this analysis six of these nine proteins also had better quality IDs on LTQ (with two or more peptide IDs). The remaining three protein IDs were made with a single peptide ID on both the LTQ-MS and LCQ-MS studies and were also confirmed by manual inspection of the peptide fragmentation spectra (one example is shown in Fig. 1). In these spectra, the signals were observed with low noise levels and extensive *b* or *y* ion fragments. Therefore, we considered these three proteins as positive IDs, and compilation of all these IDs gave a total of 158 glycoproteins (see Tab. 3 for a listing).

In addition to providing better protein IDs, the number of identified glycoproteins isolated from Caucasian American serum sample was doubled by using the LTQ and half of these proteins were identified with two or more peptides (Tab. 2b).

Tab. 3 Proteins identified from HUPO plasma/serum samples using the platform of multilectin affinity chromatography coupled with LC-MS/MS

#	ID	Reference	AFA rank				AFA Hits				CA rank				CA Hits				AA rank				AA Hits			
			HP	ED	CT	SE	HP	ED	CT	SE	HP	ED	CT	SE	HP	ED	CT	SE	HP	ED	CT	SE	HP	ED	CT	SE
1	A1AG	ALPHA-1-ACID GLYCOPROTEIN 1 *	18	16	17	17	18	15	8	7	16	18	13	14	15	14	15	16	20	18	18	14	12	16	11	15
2	A1AH	ALPHA-1-ACID GLYCOPROTEIN 2 *	25	18		28	12	13		4	22	16	27	18	10	15	6	12	28	23		16	9	11		13
3	A1AT	ALPHA1ANTITRYPSIN *	5	7	6	5	61	45	33	33	3	4	4	5	72	58	57	52	7	3	3	4	52	49	43	42
4	A1AU	ALPHA-1-ANTITRYPSIN-RELATED PROTEIN *			55				1		76			69	1			1			63	69			1	1
5	A1BG	ALPHA-1B-GLYCOPROTEIN *	22	27	50	30	14	5	1	4	28	35	38	28	9	5	3	5	39	30	40	28	5	7	3	7
6	A1S1	CLATHRIN COAT ASSEMBLY PROTEIN AP19											55	58			2	2			71				1	
7	A2AP	ALPHA-2-ANTIPLASMIN **																				60				1
8	A2GL	LEUCINE-RICH ALPHA-2-GLYCOPROTEIN *	74				1													77		64		1		8
9	A2HS	ALPHA-2-HS-GLYCOPROTEIN *	21	20	11	16	14	11	17	9	19	21	18	20	11	12	12	10	19	20	12	22	14	14	15	8
10	A2MG	ALPHA-2-MACROGLOBULIN *	2	1	1	1	95	69	62	54	4	1	3	1	64	72	69	68	1	1	1	1	105	134	69	83
11	AACT	ALPHA-1-ANTICHYMOTRYPSIN *	37	31	22	39	5	5	6	3	34	57	28	68	5	2	5	1	35	50	20	38	6	2	7	4

Tab. 3 Continued

#	ID	Reference	AFA rank				AFA Hits				CA rank				CA Hits				AA rank				AA Hits			
			HP	ED	CT	SE	HP	ED	CT	SE	HP	ED	CT	SE	HP	ED	CT	SE	HP	ED	CT	SE	HP	ED	CT	SE
12	AAKH	5'-AMP-ACTIVATED PROTEIN KINASE, GAMMA-2												57				2								2
13	AC15	ACTIVATOR 1 140 KDA SUBUNIT			66					1				73				1	84				1			1
14	ACDV	ACYL-COA DEHYDROGENASE, VERY-LONG-CHAIN SP											70				1					71				
15	AFAM	AFAMIN *	77	74		47	1	1		2	70	62	69	60	1	2	1	1	43	42		31	4	4		5
16	AFX1	PUTATIVE FORK HEAD DOMAIN TRANSCRIPTION F		65	52			1	1				39				3										
17	ALC1	IG ALPHA-1 CHAIN C REGION *	3	6	7	10	63	46	31	21	5	5	6	6	61	55	43	51	4	6	5	7	55	46	29	30
18	ALC2	IG ALPHA-2 CHAIN C REGION *	56	23	31	33	2	8	3	3	46		29	51	3		5	2	83	39	45	53	1	5	2	2
19	AMBP	AMBP PROTEIN *	41	39		46	4	3		2	35	55	37	44	4	2	3	2	38	47	37	36	5	3	3	4
20	ANGT	ANGIOTENSINOGEN *	46			62	3			1	41	42	45	47	4	4	2	2	37	92		66	5	1		1
21	ANT3	ANTITHROMBIN-III *	32	53	42		7	2	2		36	45	26		4	3	7		41	27	27	41	4	9	5	3
22	APA1	APOLIPOPROTEIN A-I	15	14	15	15	30	20	13	10	14	11	9	10	28	23	33	27	14	11	11	12	28	35	16	24
23	APA2	APOLIPOPROTEIN A-II	27	22	25	20	10	10	4	7	30	32	31	37	9	6	5	3	26	43	19	23	9	4	8	8
24	APB	APOLIPOPROTEIN B-100 *	45	33	53	31	3	4	1	3	33	29	50	34	5	7	2	4	31	26	39		7	9	3	
25	APC3	APOLIPOPROTEIN C-III *	44	49		28	3	2		3	32	26	35	25	6	8	3	6	47	68	31	29	3	1	4	6

Tab. 3 Continued

#	ID	Reference	AFA rank				AFA Hits				CA rank				CA Hits				AA rank				AA Hits			
			HP	ED	CT	SE	HP	ED	CT	SE	HP	ED	CT	SE	HP	ED	CT	SE	HP	ED	CT	SE	HP	ED	CT	SE
26	APD	APOLIPOPROTEIN D *	95			60														73				1		
27	APE	APOLIPOPROTEIN E *																		54				2		
28	APOH	BETA-2-GLYCOPROTEIN I *	38	25	20	21	4	7	7	6	20	37	21	26	11	5	10	6	21	25	24	21	10	10	6	10
29	ASPH	ASPARTYL/ASPARAGINYL BETA-HYDROXYLASE *			70				1										86	91			1	1		
30	BIEA	BILIVERDIN REDUCTASE A																		58				2		
31	C1QB	COMPLEMENT C1Q SUBCOMPONENT, B CHAIN **										67				1										
32	C1QC	COMPLEMENT C1Q SUBCOMPONENT, C CHAIN *											48				2									
33	C1R	COMPLEMENT C1R COMPONENT *	69				1						60	52			1	2	70				1			
34	C4BP	C4B-BINDING PROTEIN ALPHA CHAIN *	23	24		37	14	7		3	37	36	33	50	4	5	4	2	30	40		42	7	5		3
35	CAGC	CMP-N-ACETYLNEURAMINATE-BETA-GALACTOSAMIDE *											54				2		93			44	1			3

Tab. 3 Continued

#	ID	Reference	AFA rank				AFA Hits				CA rank				CA Hits				AA rank				AA Hits			
			HP	ED	CT	SE	HP	ED	CT	SE	HP	ED	CT	SE	HP	ED	CT	SE	HP	ED	CT	SE	HP	ED	CT	SE
36	CAP1	ADENYLYL CYCLASE-ASSOCIATED PROTEIN 1																							1	
37	CBG	CORTICOSTEROID-BINDING GLOBULIN**												56				2	72						1	
38	CERU	CERULOPLASMIN*	13	11	13	14	36	31	15	10	15	13	10	13	27	20	30	20	16	12	13	18	28	30	15	12
39	CFAB	COMPLEMENT FACTOR B*	29	30	19	22	9	5	7	6		41		35		4		4	40	28	23	25	5	9	6	8
40	CFAH	COMPLEMENT FACTOR H*	18	19	21	23	18	12	6	5	17	19	19	22	13	13	12	8	18	16	22	39	14	17	7	3
41	CFTR	CYSTIC FIBROSIS TRANSMEMBRANE CONDUCTANCE*		54		38		2		3	84	81	72	67	1	1	1	1	77			54	1			2
42	CGRR	CALCITONIN GENE-RELATED PEPTIDE TYPE 1 REC*																			61		2			
43	CHK2	SERINE/THREONINE-PROTEIN KINASE CHK2	105				1				83				1				99				1			
44	CLUS	CLUSTERIN*	47	37	24	42	3	3	5	2	49	31	36	59	2	6	3	1	55	31	28		2	7	5	
45	CNRA	ROD CGMP-SPECIFIC 3',5'-CYCLIC PHOSPHODIES	88				1					80	51	53	1	2	2		60				2			

Tab. 3 Continued

#	ID	Reference	AFA rank HP	ED	CT	SE	AFA Hits HP	ED	CT	SE	CA rank HP	ED	CT	SE	CA Hits HP	ED	CT	SE	AA rank HP	ED	CT	SE	AA Hits HP	ED	CT	SE
46	CO3	COMPLEMENT C3 *	14	3	5	3	35	55	34	49	25	17	14	7	10	15	15	45	12	5	32	3	30	47	4	52
47	CO4	COMPLEMENT C4 *	39	40	27	43	4	3	3	2	39	30	47	38	4	6	2	3	29	24	33	15	8	11	4	15
48	CO5	COMPLEMENT C5 *												45				2			82				1	
49	CO8A	COMPLEMENT COMPONENT C8 ALPHA CHAIN *																			46				3	
50	CO8G	COMPLEMENT COMPONENT C8 GAMMA CHAIN *	86				1																			
51	CO9	COMPLEMENT COMPONENT C9 *	70				1				74				1										1	
52	COPB	COATOMER BETA SUBUNIT	55					2			64				2						78				74	
53	CP34	CYTOCHROME P450 3A4																	62	59			2	2		1
54	CRBH	CRUMBS PROTEIN HOMOLOG 1 *	83				1																			
55	CTD1	CATENIN DELTA-1			84					1																
56	DFFA	DNA FRAGMENTATION FACTOR ALPHA SUBUNIT			69					1																
57	DN2L	DNA2-LIKE HOMOLOG	76				1				91				1											
58	DSRA	DOUBLE-STRANDED RNA-SPECIFIC ADENOSINE DEA	82				1																			

Tab. 3 Continued

#	ID	Reference	AFA rank				AFA Hits				CA rank				CA Hits				AA rank				AA Hits			
			HP	ED	CT	SE	HP	ED	CT	SE	HP	ED	CT	SE	HP	ED	CT	SE	HP	ED	CT	SE	HP	ED	CT	SE
59	FA11	COAGULATION FACTOR XI **											73				1									
60	FHR1	COMPLEMENT FACTOR H-RELATED PROTEIN 1 *											46		2											
61	FHR3	COMPLEMENT FACTOR H-RELATED PROTEIN 3 *	66					1													57					1
62	FIBA	FIBRINOGEN ALPHA/ALPHA-E CHAIN *	8				47				11	14			42	18			5		54		53			1
63	FIBB	FIBRINOGEN BETA CHAIN *	9				47				10	15			42	17			11		36		37			3
64	FIBG	FIBROGEN GAMMA CHAIN *	7				55				9	12			45	23			8	45	26		49	3		5
65	FINC	FIBRONECTIN *		63	32				1	3	53	38			2	4			46	63		26	3	1		7
66	GC1	IG GAMMA-1 CHAIN C REGION	19	15	12	13	18	19	16	12	29	23	16	17	9	9	13	13	22	19	16	19	10	15	13	10
67	GC2	IG GAMMA-2 CHAIN C REGION	71				71	1			1		66	62			1	1	65		35		1			5
68	GC3	IG GAMMA-3 CHAIN C REGION *	43	28	30	24	3	5	3	5	51	52	61	29	2	2	1	5	44	36		32	3	5		4
69	GC4	IG GAMMA-4 CHAIN C REGION	36	17	18	19	6	14	7	7	24	25	17	16	10	9	13	15	23	17	15	17	10	17	14	13

Tab. 3 Continued

#	ID	Reference	AFA rank				AFA Hits				CA rank				CA Hits				AA rank				AA Hits			
			HP	ED	CT	SE	HP	ED	CT	SE	HP	ED	CT	SE	HP	ED	CT	SE	HP	ED	CT	SE	HP	ED	CT	SE
70	GPS1	G PROTEIN PATHWAY SUPPRESSOR 1	60	44	44	72	2	3	2	1		47	57	31		3	2	5	51	62	48	57	3	2	2	2
71	GRAA	GRANZYME A *	69			1													54			62	2			1
72	GSHR	GLUTATHIONE REDUCTASE																								
73	HBA	HEMOGLOBIN ALPHA CHAIN	65	38			1	3			52	34			2	5			73	41	44	59	1	4	2	1
74	HBB	HEMOGLOBIN BETA CHAIN *	51	21			2	10			44	20		40	3	13		3	49	22	60	43	3	13	1	3
75	HBG	HEMOGLOBIN GAMMA-A AND GAMMA-G CHAINS									59				2				53				2			
76	HEMO	HEMOPEXIN PRECURSOR (BETA-1B-GLYCOPROTEIN) *	11	10	14	12	41	32	13	16	12	10	11	9	40	28	29	30	9	9	7	10	46	36	21	25
77	HEP2	HEPARIN COFACTOR II *												55				2								
78	HERG	VOLTAGE-GATED POTASSIUM CHANNEL *	108				1																			
79	HPT2	HAPTOGLOBIN-2 PRECURSOR *	10	13	10	8	45	23	21	21	8	8	8	8	49	30	34	44	10	4	9	9	41	48	19	30
80	HPTR	HAPTOGLOBIN-RELATED PROTEIN	4	5	3	4	62	49	48	48	7	6	7	4	58	47	40	54	6	8	6	5	53	45	24	39

Tab. 3 Continued

#	ID	Reference	AFA rank HP	AFA rank ED	AFA rank CT	AFA rank SE	AFA Hits HP	AFA Hits ED	AFA Hits CT	AFA Hits SE	CA rank HP	CA rank ED	CA rank CT	CA rank SE	CA Hits HP	CA Hits ED	CA Hits CT	CA Hits SE	AA rank HP	AA rank ED	AA rank CT	AA rank SE	AA Hits HP	AA Hits ED	AA Hits CT	AA Hits SE
81	HRG	HISTIDINE-RICH GLYCOPROTEIN *	30	48		44	8	2		2	40	69	4					1	36	29	50	35	5	7	1	4
82	HS27	HEAT SHOCK 27 KDA PROTEIN			57				1																	
83	HV1F	IG HEAVY CHAIN V-I REGION	57	68	59	45	2	1	1	2				48				2	58		56	51	2		1	2
84	HV3A	IG HEAVY CHAIN V-III REGION *												64				1								
85	HV3D	IG HEAVY CHAIN V-III REGION			63				1					54				2				73			1	
86	HV3F	IG HEAVY CHAIN V-III REGION									63	49			1											
87	HV3G	IG HEAVY CHAIN V-III REGION			65				1				74		1	2										
88	HV3H	IG HEAVY CHAIN V-III REGION			66				1				75				1									
89	HV3J	IG HEAVY CHAIN V-III REGION																		56				2		
90	HV3P	IG HEAVY CHAIN V-III REGION	63		34	57	1	2	2	1	61	53		36	1	2		3	45		34		3			4
91	HV3T	IG HEAVY CHAIN V-III REGION																			52					1
92	IC1	PLASMA PROTEASE C1 INHIBITOR *	33	64		55	7	1		1	50	44	66	42	2	3	1	2	32	44	43	40	6	3	2	3

Tab. 3 Continued

#	ID	Reference	AFA rank			AFA Hits			CA rank			CA Hits			AA rank			AA Hits		
			HP ED	CT	SE	HP ED	CT	SE	HP ED	CT	SE	HP ED	CT	SE	HP ED	CT	SE	HP ED	CT	SE
93	IGJ	IMMUNOGLOBULIN J CHAIN *	87 43	26	34	1 3	4	3	21 27	22	19	11 8	9	11	75	35	49	1	4	2
94	IMB2	IMPORTIN BETA-2 SUBUNIT	109 56	46	51	1 2	2	2	85 65	74	41	1 2	1	3	81	49	82	1	2	1
95	IRK4	INWARD RECTIFIER POTASSIUM CHANNEL 4							75			1				59			1	
96	ITH1	INTER-ALPHA-TRYPSIN INHIBITOR HEAVY CHAIN *	34 59	33	26	6 1	2	4	42 39	23	27	3 4	8	5	64 34	25	46	1 5	5	2
97	ITH2	INTER-ALPHA-TRYPSIN INHIBITOR HEAVY CHAIN H *	28 29	23	25	9 5	5	5	31 28	20	24	8 7	11	7	24 21	14	20	9 13	14	10
98	ITH4	INTER-ALPHA-TRYPSIN INHIBITOR HEAVY CHAIN *	26 35	36	18	12 4	2	7	27 46	30	23	10 3	5	8	27 33	34	24	9 6	4	8
99	K1CR	KERATIN, TYPE I CYTOSKELETAL 18 *	99 58	63		1 2	1													
100	K1M4	KERATIN, TYPE I CUTICULAR HA4	104			1			70			1								
101	K1M5	KERATIN, TYPE I CUTICULAR HA5	102	35		1		3	76			1					61			
102	K1M6	KERATIN, TYPE I CUTICULAR HA6			48			2												

7.3 Results and discussion

Tab. 3 Continued

#	ID	Reference	AFA rank				AFA Hits				CA rank				CA Hits				AA rank				AA Hits			
			HP	ED	CT	SE	HP	ED	CT	SE	HP	ED	CT	SE	HP	ED	CT	SE	HP	ED	CT	SE	HP	ED	CT	SE
103	K1MB	KERATIN, TYPE I CUTICULAR HA3-II			49					2																
104	KAC	IG KAPPA CHAIN C REGION	12	8	9	7	41	34	21	24	6	3	5	3	59	65	55	56	13	10	8	8	29	35	21	30
105	KCRU	CREATINE KINASE	78		62																					
106	KNG	KININOGEN *	31	32	41	40		1	1		26	24	25	49	10	9	8	2	34	38	38	30	6	5	3	5
107	KV1E	IG KAPPA CHAIN V-I REGION	49	61	49		8	4	2	3	45		43		3		2		53	66	52		2	1	1	
108	KV1F	IG KAPPA CHAIN V-I REGION	53		58		2		1				41				2		69				1			
109	KV1N	IG KAPPA CHAIN V-I REGION		60				1																		
110	KV1Y	IG KAPPA CHAIN V-I REGION	62		48		1		1		57		42		1		2		66	64	41		1	1	2	
111	KV2C	IG KAPPA CHAIN V-II REGION	89	57	40	27	1	2	2	4	47	77	53	65	3	1	2	1	59	57	62	68	2	2	1	1
112	KV2D	IG KAPPA CHAIN V-II REGION	67	50			1	2			60	51			1	2			51				2			
113	KV3F	IG KAPPA CHAIN V-III REGION *																				94				1
114	KV3G	IG KAPPA CHAIN V-III REGION	47	35	41		2	2	2		43	56	44		3	2	2		67	48	42	48	1	2	2	2
115	KV3I	IG KAPPA CHAIN V-III REGION	73				1				65	43	64	46	1	3	1	2								

Tab. 3 Continued

#	ID	Reference	AFA rank HP	ED	CT	SE	AFA Hits HP	ED	CT	SE	CA rank HP	ED	CT	SE	CA Hits HP	ED	CT	SE	AA rank HP	ED	CT	SE	AA Hits HP	ED	CT	SE
116	KV3M	IG KAPPA CHAIN V-III REGION HIC		71				1												74				1			
117	LAC	IG LAMBDA CHAIN C REGIONS	20	12	16	9	17	23	10	21	18	22	15	15	13	11	14	16	17	13	17	11	21	20	12	24
118	LV3B	IG LAMBDA CHAIN V-III REGION *																			76				1		
119	LV4C	IG LAMBDA CHAIN V-IV REGION	66				1				54				2						70				1		
120	LV6C	IG LAMBDA CHAIN V-VI REGION	54				2																				
121	MK07	MITOGEN-ACTIVATED PROTEIN KINASE 7	75	67			1	1					65				1				79				1		
122	MSH3	DNA MISMATCH REPAIR PROTEIN MSH3												89				1									
123	MUC	IG MU CHAIN C REGION *	16	9	8	11	30	33	25	17	13	54	12	12	34	2	21	24	15	14	10	13	28	19	19	20
124	MYH4	MYOSIN HEAVY CHAIN, SKELETAL MUSCLE									55		40	72	2		3	1									
125	MYH9	MYOSIN HEAVY CHAIN, NONMUSCLE TYPE A		72	68	53		1	1	2	82		84	86	1		1	1	87		64		1		1	
126	NIP2	BCL2/ADENOVIRUS E1B 19-KDA PROTEIN-INTERAC											56				2										

Tab. 3 Continued

#	ID	Reference	AFA rank				AFA Hits				CA rank				CA Hits				AA rank				AA Hits			
			HP	ED	CT	SE	HP	ED	CT	SE	HP	ED	CT	SE	HP	ED	CT	SE	HP	ED	CT	SE	HP	ED	CT	SE
127	NUEM	NADH-UBIQUINONE OXIDOREDUCTASE 39 KDA *	93				1																			
128	P85B	PHOSPHATIDYLINOSITOL 3-KINASE REGULATORY B									54				2											
129	PKD2	POLYCYSTIN 2 *	75	60		68		1	1	1	75		75		1		1		74				1			
130	PLE1	PLECTIN 1	98				1							79				1								
131	PLMN	PLASMINOGEN *	40	52	32	56	4	2	3	1	58	50	32		1	2	4		42	52	55		4	2	1	
132	PON1	SERUM PARAOXONASE/ARYLESTERASE 1 *												84				1				45				1
133	PZP	PREGNANCY ZONE PROTEIN *	24	26	29	58	13	6	3	1	23	40	24	21	10	4	8	9	25	15	29	27	9	18	5	7
134	RL32	60S RIBOSOMAL PROTEIN L32				78			1																		
135	RL5	60S RIBOSOMAL PROTEIN L5	84			65		1		1	80		77		1		1				94	80			1	1
136	RYNR	RYANODINE RECEPTOR, SKELETAL MUSCLE *	90			50	1			2		78	78	39		1	1	3	75				1			
137	RYR2	RYANODINE RECEPTOR 2 *	79		54		1		1																	
138	SAMP	SERUM AMYLOID P-COMPONENT *	42	34	39	70	4	4	2	1	66	71	71	82	1	1	1	1	56	55	55	50	2	2		2

Tab. 3 Continued

#	ID	Reference	AFA rank HP	AFA rank ED	AFA rank CT	AFA rank SE	AFA Hits HP	AFA Hits ED	AFA Hits CT	AFA Hits SE	CA rank HP	CA rank ED	CA rank CT	CA rank SE	CA Hits HP	CA Hits ED	CA Hits CT	CA Hits SE	AA rank HP	AA rank ED	AA rank CT	AA rank SE	AA Hits HP	AA Hits ED	AA Hits CT	AA Hits SE
139	SM30	SENESCENCE MARKER PROTEIN-30	101				1												63				2			
140	SMF1	SWI/SNF-RELATED, MATRIX-ASSOCIATED, ACTIN									56	63	52	30	2	2	2	5								
141	SPK	SYMPLEKIN	59	46	45	74	2	3	2	1	49	59		33	3		2	5	50	61	46	56	3	2	2	2
142	SR54	SIGNAL RECOGNITION PARTICLE 54 KDA PROTEIN		81	47			1	2													77				1
143	SYR	ARGINYL-TRNA SYNTHETASE				36				3												86				1
144	THRB	PROTHROMBIN *	48	36	37	29	3	3	2	4	38	33		62	4	6		1	52	37	21	33	2	5	7	4
145	TRFE	SEROTRANSFERRIN *	1	2	4	2	98	64	40	52	1	2	2	2	104	69	75	68	2	2	2	2	105	79	58	53
146	TRFL	LACTOTRANSFERRIN *	58	45	43	73	2	3	2	1	48		58	32	3		2	5	48	60	47	55	3	2	2	2
147	TRHY	TRICHOHYALIN			56	59			1	1																
148	TRI7	THYROID RECEPTOR INTERACTING PROTEIN 7	103				1				82				1											
149	TRT2	TROPONIN T, CARDIAC MUSCLE ISOFORMS		82				1											65				1			
150	TTC3	TETRATRICOPEPTIDE REPEAT PROTEIN 3												71				1	85				1			
151	TTHY	TRANSTHYRETIN			63					1	59		63		1		1		69				1			

Tab. 3 Continued

#	ID	Reference	AFA rank				AFA Hits				CA rank				CA Hits				AA rank				AA Hits			
			HP	ED	CT	SE	HP	ED	CT	SE	HP	ED	CT	SE	HP	ED	CT	SE	HP	ED	CT	SE	HP	ED	CT	SE
152	U2R2	U2 SMALL NUCLEAR RIBONUCLEOPROTEIN AUXILLIA				52				2																
153	VTNC	VITRONECTIN *	39		34		2		4		48	60	34	63	3	2	4	1	33	32	30	37	6	7	5	4
154	WEE1	WEE1-LIKE PROTEIN KINASE	55				2				69	61			1	2			72				1			
155	Y167	HYPOTHETICAL PROTEIN KIAA0167			64				1			72				1										
156	Y188	HYPOTHETICAL PROTEIN KIAA0188			76					1												85				1
157	Z33A	ZINC FINGER PROTEIN 33A	68				1				71				1				57				2			
158	ZA2G	ZINC-ALPHA-2-GLYCOPROTEIN *												43				2	68	71		47	1	1		2

a) Protein ID was confirmed by LC-LTQ
b) Glycosylation site(s) of the protein is(are) assigned in Swiss-Prot database

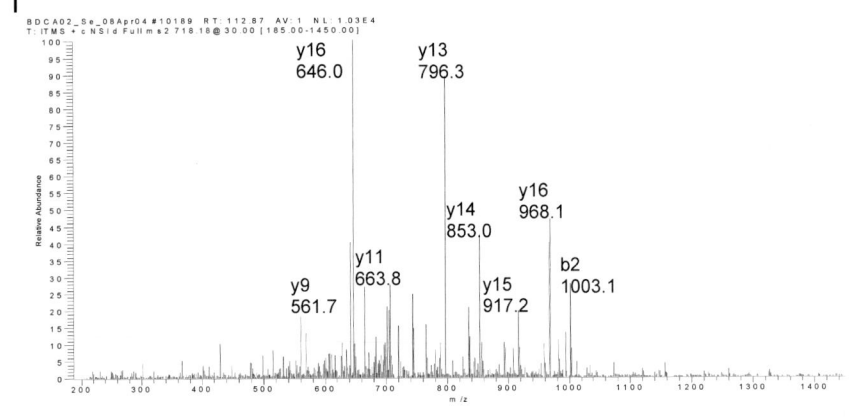

Fig. 1 The peptide, SETEIHQGFQHLHQLFAK, of corticosteroid-binding globulin was detected in serum sample from Caucasian American specimen by LC-LTQ MS/MS. (X_{corr} 3.81 (+ 3); ΔC_n 0.61; and R_{sp} 1). In this spectrum, the signals were observed with low noise level and extensive b or y ion fragments.

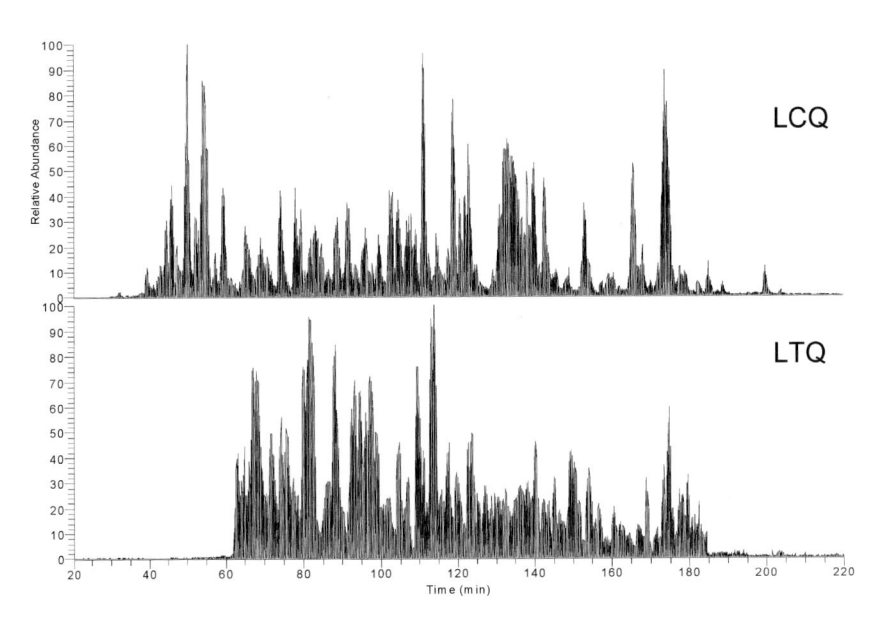

Fig. 2 The chromatograms from the separation of a tryptic digest of glycoproteins isolated from the Caucasian American serum sample. Top profile was from the LC-LCQ MS analysis and the bottom one was from the LC-LTQ MS analysis, which had a solvent delay of 10 min due to a difference in starting conditions of the separation (0% ACN instead of 5% ACN).

Fig. 2 shows a high degree of reproducibility for the HPLC separation and we believe that the improved proteomic analysis was due to both the new mass spectrometer and better chromatographic performance achieved with the MDLC. The observed delay in time for the peptides to elute in the LC-LTQ experiment was due to a difference in starting conditions for the separation (0% ACN instead of 5% ACN).

7.3.2
Comparison between serum and plasma glycoproteomes

We have developed a conservative criteria for protein ID (two or more tryptic peptides) combined with the use of a differential SEQUEST score of >20 in relative ranking for comparative studies of different glycoprotein samples [6]. This approach was validated by the exploration of false positives with independent criteria such as the use of measured pI values [12] and the measurement of peak areas of selected peptide ions. In addition, this ranking requires a high degree of consistency between LC-MS/MS analysis and thus is only applied to samples run in a consecutive series. Using this criterion, we compared the serum sample to plasma samples from each ethnic specimen. The serum proteins ranked 20 lower than in plasma samples were selected. Then, the reproducibility of the selected proteins in three ethnic specimens were evaluated and used to construct a summary protein list (Tab. 4) and assess any observed differences between plasma and serum.

In this manner we found that there were few differences in the glycoproteins that were present in HUPO plasma *vs.* serum samples. The major observed differences were proteins expected to be removed in the coagulation process, namely fibrinogen (Tab. 4), and these results suggested that serum preparation maintained the majority of the plasma glycoproteome. At this stage it is not clear why plasminogen was not identified in the serum samples. In addition, the ID of fibrinogen was improved in heparinized plasma (higher rank and more peptide IDs), which might suggest that heparin is better than EDTA and citrate in stabilizing plasma.

7.3.3
Comparison of the glycoproteins present in the samples collected from three ethnic groups

The serum and plasma samples provided by HUPO consisted of pools of one female and one male subject from three ethnic groups. While it is clear that no overall conclusions on the effect of ethnicity on the expression of the proteome can be made from such a limited sample set, it is of interest to analyze the results of this initial study.

In the preliminary comparison, the proteins that consistently had higher or lower levels (difference in MS rank that was more than 20) in more than one type of plasma/serum samples of one ethnic specimen relative to the other two ethnic specimens were selected. Then, the number of hits of each selected protein was investigated in each sample. By the comparison of the number of hits, the proteins confirmed of having constant higher or lower level in all four types of plasma/serum samples of the ethnic specimen were considered to be either upregulated or downregulated in this particular ethnic specimen, and listed in Tab. 5.

Tab. 4 Comparison of HUPO human serum and plasma samples

Sample type Ethnic group	Rank[a]				Hits[b]			
	HEP	EDTA	CIT	Serum	HEP	EDTA	CIT	Serum
Fibrinogen alpha/alpha-E chain								▼
Caucasian American	11	14			42	18		
Asian American	5		54		53		1	
African American	8				47			
Fibrinogen beta chain								▼
Caucasian American	10	15			42	17		
Asian American	11		36		37		3	
African American	9				47			
Fibrinogen gamma chain								▼
Caucasian American	9	12			45	23		
Asian American	8	45	26		49	3	5	
African American	7				55			
Plasminogen								▼
Caucasian American	58	50	32		1	2	4	
Asian American	42	52	55		4	2	1	
African American	40	52	32	56	4	2	3	1

a) SEQUEST rank represents the confidence of the assignment of a protein in the sample
b) Number of peptide IDs for a given protein sequence
▼: Concentration of the indicated protein was at a much lower level in serum than in plasma

Tab. 5 shows that the Caucasian American sample had increased level of angiotensinogen (AGT) and reduced level of histidine-rich glycoprotein (HRG) relative to the other two samples, and vitronectin (VNT) was present at a lower level in the African American sample. These protein level changes were further evaluated using the peak areas of selected peptides of the proteins in extracted ion chromatograms, and similar regulation changes of these proteins were found. For example, the peak areas of the peptide, SGFPQVSMFTHTFPK, of HRG were 590, 419, and 180 (E + 6 U) in heparinized plasma samples of Asian American, Africa American, and Caucasian American specimens, respectively. This change was confirmed by a second peptide present in this protein, DSPVLIDFFEDTER (peak area: 411, 333, and 199 (E + 6 U)). For another example, the peak areas of peptide, VWELSK, of AGT were 155, 129, and 359 (E + 6 U) for the heparinized plasma samples of Asian American, African American, and Caucasian American specimens, respectively. This observation confirmed the higher level of AGT in Caucasian American specimen, although, in this one case, the SEQUEST rank and the number of hits comparison between the heparinized samples was ambiguous.

Tab. 5 Comparison among HUPO samples from three ethnic specimens

Sample type	Rank[a]			Hits[b]			Change of regulation
	AA	AFA	CA	AA	AFA	CA	
ANGIOTENSINOGEN (Up-regulated in CA)							
Citrated Plasma			45			2	in CA
EDTA plasma	92		42	1		4	▲
Heparin plasma	37	46	41	5	3	4	
Serum	66	62	47	1	1	2	
HISTIDINE-RICH GLYCOPROTEIN (Down-regulated in CA)							
Citrated plasma	50			1			in CA
EDTA plasma	29	48	69	7	2	1	▼
Heparin plasma	36	30	40	5	8	4	
Serum	35	44		4	2		
VITRONECTIN (Down-regulated in AFA)							
Citrated plasma	30	34	34	5	4	4	in AFA
EDTA plasma	32		60	7		2	▼
Heparin plasma	33	39	48	6	2	3	
Serum	37		63	4		1	

a) SEQUEST rank represents the confidence of the assignment of a protein in the sample
b) Number of peptide IDs for a given protein sequence
▲: Protein was found at a higher level in the ethnic specimen
▼: Protein was found at a lower level in the ethnic specimen

Genetic variants, such as M235T polymorphism present in AGT, have a significant association with essential hypertension and associated cardiovascular diseases [13–15], and the AGT gene has shown a marked difference between ethnic groups [16–18]. The presence of this M235T variant is also related to plasma AGT levels where hypertensive subjects have a higher plasma AGT level compared with control subjects [19]. A recent study suggested that M235T was associated with a stepwise increase in AGT levels in white subjects [20]. In our research, the Caucasian American specimen showed higher plasma AGT level than Asian American and African American specimens. However, the AGT genotypes of the three ethnic plasma samples are not clear from this preliminary study and future experiments could be directed at characterizing the complete sequence of AGT in these samples by techniques such as affinity pull-down followed by LC/MS analysis of an enzyme digest [21]. In addition, any conclusion of the relevance of this observation to an elevated risk of hypertension in the Caucasian American ethnic group would require a well-controlled population study.

The physiological function of HRG is not clear, but it has been found to regulate the anticoagulant activity of heparin [22]. The plasma level of HRG is under significant genetic control and includes factors such as blood type and the age [23]. However, the level can also be affected by the environment, for example, women receiving estrogens have a reduced plasma HRG level, ranging from 15 to 26% in a dose-dependent manner [24, 25], and the protein level declines during pregnancy to about 50% of initial values [26, 27]. In addition, the HRG level varies under certain disease situations, such as in a woman with dural arteriovenous fistula where the level of HRG was found to be 50% lower [28]. Therefore, a conclusion that the lower level of HRG in a plasma sample of the Caucasian American group is under genetic control can only be determined after correlating the result with detailed clinical information on individual subjects.

VNT is an essential mediator of adhesion and is a spreading factor found in serum, tissue and many cells *in vitro* [29]. The collagen-binding activity of VNT may be related to the progression of liver disease [30], as many studies have shown that the level of VNT is low in liver disease [31, 32], and could be significantly correlated with alcoholic liver cirrhosis [33]. Our results found a lower level of VNT in the plasma sample of the African American specimen compared with Caucasian American and Asian American specimens. Future studies could be designed to correlate such results with liver disease in different population groups.

7.4
Concluding remarks

In this research, we used a multilectin affinity column to isolate glycoproteins from HUPO plasma and serum samples, and identified glycoproteins using LC-MS/MS analysis of tryptic digests. The multilectin column was useful for both capturing the glycoproteins as well as depleting nonglycosylated proteins, some of which are present in high concentrations, such as albumin. The analysis of the glycoproteins resulted in the characterization of a large number of IDs (approximately 150), which will complement IDs of plasma proteins by other approaches. In addition, the presence of specific glycoproteins will aid efforts on the annotation of the plasma proteome by providing evidence on the tissue of origin for this important subset of the blood proteome.

It is probable that glycosylation results in the biosynthesis of proteins of relatively high stability and solubility, which can be attributed to both the polarity of the carbohydrate group as well as properties such as protease resistance. With this scenario it is perhaps not surprising that, from the glycoprotein perspective, there are not substantial differences between the plasma and the serum. Again there were not substantial differences between the glycoproteome of the samples from three ethnic specimens. There were, however, significant differences in levels of the following proteins: AGT, HRG, and VNT, and these observations will be the subject of follow-up studies. Finally, glycoproteins offer many advantages for potential biomarkers such as increased stability and solubility relative to unmodified proteins so that issues around plasma collection and storage could be minimized.

This work was supported by a grant from HUPO. We wish to thank Thermo Electron and GE Healthcare for the instrumentation support with the LTQ linear IT and Ettan MDLC HPLC system. We would also like to thank Dr. Shiaw-Lin Wu and Dr. Barry Karger for their helpful comments.

7.5
References

[1] Carlson, J., Eriksson, S., Alm, R., Kjellstrom, T., *Hepatology* 1984, *4*, 235–241.

[2] Hanley, J. M., Haugen, T. H., Heath, E. C., *J. Biol. Chem.* 1983, *258*, 7858–7869.

[3] Hahn, T. J., Goochee, C. F., *J. Biol. Chem.* 1992, *267*, 23982–23987.

[4] Ghosh, S., Hevi, S., Chuck, S. L., *Blood* 2004, *103*, 2369–2376.

[5] Anderson, N. L., Anderson, N. G., *Electrophoresis* 1998, *19*, 1853–1861.

[6] Yang, Z., Hancock, W. S., *J. Chromatogr. A.* 2004, *1053*, 79–88.

[7] Seelenmeyer, C., Wegehingel, S., Lechner, J., Nickel, W., *J. Cell Sci.* 2003, *116*, 1305–1318.

[8] Kraus, M. H., Fedi, P., Starks, V., Muraro, R., Aaronson, S. A., *Proc. Natl. Acad. Sci. USA* 1993, *90*, 2900–2904.

[9] Zhang, H., Li, X. J., Martin, D. B., Aebersold, R., *Nat. Biotechnol.* 2003, *21*, 660–666.

[10] Pieper, R., Su, Q., Gatlin, C. L., Huang, S. T., Anderson, N. L., Steiner, S., *Proteomics* 2003, *3*, 422–432.

[11] Wu, S. L., Amato, H., Biringer, R., Choudhary, G., Shieh, P., Hancock, W. S., *J. Proteome Res.* 2002, *1*, 459–465.

[12] Wang, Y., Hancock, W. S., Weber, G., Eckerskorn, C., Palmer-Toy, D., *J. Chromatogr. A.* 2004, *1053*, 269–278.

[13] Frossard, P. M., Hill, S. H., Elshahat, Y. I., Obineche, E. N., Bokhari, A. M., Lestringant, G. G., John, A., Abdulle, A. M., *Clin. Genet.* 1998, *54*, 285–293.

[14] Jeunemaitre, X., Inoue, I., Williams, C., Charru, A., Tichet, J., Powers, M., Sharma, A. M., *et al.*, *Am. J. Hum. Genet.* 1997, *60*, 1448–1460.

[15] Kamitani, A., Rakugi, H., Higaki, J., Yi, Z., Mikami, H., Miki, T., Ogihara, T., *J. Hum. Hypertens.* 1994, *8*, 521–524.

[16] Onipinla, A. K., Barley, J., Carter, N. D., MacGregor, G. A., Sagnella, G. A., *J. Hum. Hypertens.* 1999, *13*, 865–866.

[17] Morise, T., Takeuchi, Y., Takeda, R., *J. Intern. Med.* 1995, *237*, 175–180.

[18] Barley, J., Blackwood, A., Sagnella, G., Markandu, N., MacGregor, G., Carter, N., *J. Hum. Hypertens.* 1994, *8*, 639–640.

[19] Rotimi, C., Cooper, R., Ogunbiyi, O., Morrison, L., Ladipo, M., Tewksbury, D., Ward, R., *Circulation* 1997, *95*, 2348–2350.

[20] Sethi, A. A., Nordestgaard, B. G., Tybjaerg-Hansen, A., *Arterioscler. Thromb. Vasc. Biol.* 2003, *23*, 1269–1275.

[21] Zhu, Z., Becklin, R. R., Desiderio, D. M., Dalton, J. T., *Biochemistry* 2001, *40*, 10756–10763.

[22] Lijnen, H. R., Van Hoef, B., Collen, D., *Thromb. Haemost.* 1984, *51*, 266–268.

[23] Drasin, T., Sahud, M., *Thromb. Res.* 1996, *84*, 179–188.

[24] Hennis, B. C., Boomsma, D. I., Fijnvandraat, K., Gevers Leuven, J. A., Peters, M., Kluft, C., *Thromb. Haemost.* 1995, *73*, 484–487.

[25] Jespersen, J., Petersen, K. R., Skouby, S. O., *Am. J. Obstet. Gynecol.* 1990, *163*, 396–403.

[26] Seki, H., *Nippon Sanka Fujinka Gakkai Zasshi* 1986, *38*, 317–326.

[27] Tatra, G., Nasr, F., Hoyer, E., *Z Geburtshilfe Perinatol.* 1983, *187*, 124–126.

[28] Shigekiyo, T., Yoshida, H., Kanagawa, Y., Satoh, K., Wakabayashi, S., Matsumoto, T., Koide, T., *Thromb. Haemost.* 2000, *84*, 675–679.

[29] Weller, M., Wiedemann, P., Bresgen, M., Heimann, K., *Fortschr. Ophthalmol.* 1990, *87*, 221–225.

[30] Yamada, S., Kobayashi, J., Kawasaki, H., Res. Commun. Mol. Pathol. Pharmacol. 1997, 97, 315–324.

[31] Inuzuka, S., Ueno, T., Torimura, T., Tamaki, S., Sakata, R., Sata, M., Yoshida, H., Tanikawa, K., *Hepatology* 1992, 15, 629–636.

[32] Tomihira, M., *Fukuoka Igaku Zasshi* 1991, 82, 21–30.

[33] Hogasen, K., Homann, C., Mollnes, T. E., Graudal, N., Hogasen, A. K., Hasselqvist, P., Thomsen, A. C., Garred, P., *Liver* 1996, 16, 140–146.

8
Evaluation of prefractionation methods as a preparatory step for multidimensional based chromatography of serum proteins*

Eilon Barnea, Raya Sorkin, Tamar Ziv, Ilan Beer and Arie Admon

Prefractionations of proteins prior to their proteolysis, chromatography, and MS/MS analyses help reduce complexity and increase the yield of protein identifications. A number of methods were evaluated here for prefractionating serum samples distributed to the participating laboratories as part of the human Plasma Proteome Project. These methods include strong cation exchange (SCX) chromatography, slicing of SDS-PAGE gel bands, and liquid-phase IEF of the proteins. The fractionated proteins were trypsinized and the resulting peptides were resolved and analyzed by multidimensional protein identification technology coupled to IT MS/MS. The MS/MS spectra were clustered, combined, and searched against the IPI protein databank using Pep-Miner. The identification results were evaluated for the efficacy of the different prefractionation methodologies to identify larger numbers of proteins at higher confidence and to achieve the best coverage of the proteins with the identified peptides. Prefractionation based on SCX resulted in the largest number of identified proteins, followed by gel slices and then the liquid-phase IEF. An important observation was that each of the methods revealed a set of unique proteins, some identified with high confidence. Therefore, for comprehensive identification of the serum proteins, several different prefractionation approaches should be used in parallel.

8.1
Introduction

8.1.1
The HUPO Plasma Proteome Project (PPP) goals and the serum as a complex sample

The goal of the HUPO PPP is to advance comprehensive analyses of human serum and plasma protein repertoires and to compare these repertoires between healthy and diseased individuals [1, 2]. Such comprehensive analyses of human serum

---

* Originally published in Proteomics 2005, 13, 3367–3375

Exploring the Human Plasma Proteome. Edited by Gilbert S. Omenn
Copyright © 2006 WILEY-VCH Verlag GmbH & Co. KGaA, Weinheim
ISBN: 3-527-31757-0

proteomes are complicated due to their content of thousands of different proteins, which are present in a wide range of concentrations [3–10]. The first phase of the PPP project, described in different articles in this journal, aims at identifying as many proteins as possible from tested samples of human sera or plasma that were distributed by the PPP to participating laboratories.

Most current mass spectrometers have a duty cycle of one or a few seconds for each full MS, followed by one or more MS/MS scans. Therefore, during each chromatography, a few hundred peptides are expected to be resolved, fragmented, and identified. Importantly, in analyses of very complex peptide mixtures, even simple repetitions of the same chromatographies of proteolytic peptides can increase the number of identified peptides just by random selection of different peptides for fragmentations in each chromatography. Such additions of similar chromatographies exhaust themselves after a number of repetitions. In order to obtain new identifications, a different mode of separation needs to be incorporated [6].

By prefractionating the intact serum proteins prior to their proteolysis and analyses by LC-MS/MS, the effective concentration of each protein in different fractions can be increased, while the mixture complexity can be reduced. This would result in less interference between peptides during the LC-MS/MS, thus leading to larger numbers of identified peptides. Very large increases in the number of identified peptides and proteins in complex samples were recently achieved by using multidimensional protein identification technology (MudPIT) combining strong cation exchange (SCX) with long RP gradients [3, 6, 11] (reviewed in [12–14]) and by performing repeated chromatographies with mass segmentation selection of peptides for fragmentation [3, 6]. However, even with the increased number of steps employed for prefractionation, the number of identified proteins eventually reaches a plateau, since the concentration of some rare proteins is still below the needed threshold level of sensitivity for the mass spectrometers. These statements are true for any proteomics project involving complex protein mixtures. Indeed, numerous attempts were made to develop effective, simple, reproducible, and inexpensive protein prefractionation schemes [15, 16] with an outstanding example described in [9].

The most common prefractionation techniques for proteins are electrophoresis, IEF, ion exchange, RP, size exclusion, and affinity based approaches. The focus of this study is to evaluate some of these prefractionation schemes after depletion of the abundant serum proteins to identify the largest number of proteins, with the highest level of confidence.

The most abundant serum proteins should be first depleted to increase the relative amounts of the less abundant proteins. Depletion of these most abundant proteins should be approached carefully, since a significant number of serum proteins are usually bound to them, especially to albumin and immunoglobulin G (IgG) [7]. The remaining proteins can be further separated into subfractions, either by chromatography, electrophoresis, or both. The protein preproteolysis fractionation scheme selected for high throughput should be robust, effective, and reproducible. It should also be compatible with the subsequent steps, which include proteolysis, followed by capillary chromatography in one or more dimensions, and identification of the peptides by MS/MS.

The effects of depletion of the abundant proteins, preproteolysis fractionation, multidimensional chromatographies of the proteolytic peptides, and selection of peptides for fragmentation by mass segmentation, can be additive and extremely beneficial in improving the number of peptides selected for fragmentation.

8.1.2
The scope of this manuscript

In this study, different schemes of preproteolysis and postproteolysis fractionation were compared for their effectiveness in identifying the largest number of peptides and proteins, and in raising the confidence level for the individual identifications. The serum proteins were depleted of HSA and IgG and prefractionated by SDS-PAGE followed by cutting of gel slicing, liquid-phase IEF, or SCX chromatography. Furthermore, these were followed by proteolysis and analyses by 1-D or 2-D capillary LCs of the peptides.

8.2
Materials and methods

8.2.1
Depletion from serum albumin and antibodies

One of the distributed human serum samples, labeled BDFA01 by the PPP organization, was used for the entire comparative project described here. HSA and IgG were depleted using the Aurum serum protein minikit (Bio-Rad) based on Affi-Gel Blue and Affi-Gel protein-A resins, according to the manufacturer. After the depletion, aliquots containing 100 μg of the remaining proteins were stored at −80°C. These were thawed only once immediately before use.

8.2.2
MudPIT and mass segmentation

Aliquots containing 100 μg of depleted serum proteins were diluted in 8 M urea and 100 mM ammonium bicarbonate, reduced with 10 mM DTT, incubated at 60°C for 30 min, carboxyamidomethylated with 10 mM iodoacetamide at room temperature for 30 min, and proteolyzed with 2 μg modified trypsin (Promega) overnight at 37°C. The resulting peptides were desalted using disposable Silica-C18 tip (Harvard) and fractionated using SCX MicroTip Column (Harvard) with 0.1% acetic acid at pH 3 and 5% ACN. This was followed by collection of the flow-through and increasing concentrations of ammonium acetate in steps of 10, 20, 40, 60, 80, 100, 150, 250 to 500 mM. The collected fractions were dried by vacuum centrifugation, dissolved in 0.1% formic acid, and analyzed by μRP-LC-MS/MS. Three of the salt elutions containing the most complex mixture of peptides (10, 20, 40 mM) were rerun a total of three times by μRP-LC-MS/MS, limiting the m/z boundaries of the precursor ions in each run to 400–800, 800–1200, and 1200–2000 m/z (mass segmentation).

8.2.3
Protein separation by SDS-PAGE

One hundred fifty micrograms of proteins after depletion were resolved by running them in three lanes of a 10% mini-SDS-PAGE. After staining with CBB, one lane was cut into 20 slices for in-gel proteolysis with 0.2–0.5 μg trypsin adjusted according to the staining level of the proteins in the cut slices. The resulting peptides were recovered from the gel slice and analyzed by one dimension μRP-LC-MS/MS. The slices from the other two gel lanes were combined, cut into six slices for proteolysis, and the peptides were subsequently analyzed by MudPIT. Trypsin (0.5–1 μg) was used for in-gel proteolysis of each slice. Each fraction was brought to 50% ACN, 0.1% acetic acid, and analyzed by offline MudPIT on SCX MicroTip Column (Harvard) with an increasing salt concentration of ammonium acetate and 5% ACN at pH 3 as described in Section 2.2. The fractions were lyophilized and dissolved in 0.1% formic acid and resolved by μRP-LC-MS/MS without repeated chromatographies for mass segmentations.

8.2.4
SCX separation of intact proteins followed by MudPIT

SCX separation of the intact proteins was performed on a disposable minicolumn (Bio-Rad) filled with 200 μL of S-Sepharose beads (Sigma), prewashed with 8 M urea, 5% ACN, and 1 M NaCl adjusted to pH 3 with 0.1% acetic acid (column regeneration solution). The column was equilibrated with 8 M urea, 5% ACN, 0.1% acetic acid, pH 3 (equilibration solution). Four hundred micrograms of proteins after depletion were diluted ten times into equilibration solution, loaded on the column, followed by collection of the flow-through and 12 increasing salt steps of 50, 100, 150, 200, 250, 300, 350, 400, 450, 500, 750 mM, and 1 M of NaCl in 8 M urea, 0.1% acetic acid, 5% ACN at pH 3. Tris-base was added to each one of the 13 fractions to raise the pH level to approximately 8.5. The fractions were reduced and carboxyamidomethylated and proteolyzed with 0.4–0.8 μg trypsin each, according to the protein concentration in each sample. The fractions were desalted using Silica-C18 tip (Harvard). Twenty percent of each fraction was dried by vacuum centrifugation, dissolved in 0.1% formic acid, and analyzed by μRP-LC-MS/MS.

The remaining parts of each of the 13 fractions were combined into five fractions (two to three fractions in each). Each combined fraction was brought to 50% ACN and 0.1% acetic acid, pH 3, and loaded on an SCX tip. Its peptides were separated using ten step MudPIT with ammonium acetate elutions as described in Section 2.2. Each one of the 50 fractions was analyzed by μRP-LC-MS/MS, as described in Section 2.3.

8.2.5
Liquid-phase IEF followed by MudPIT

Four hundred micrograms of proteins after depletion were mixed with 1 mL carrier ampholytes (pH range 3–10; Bio-Rad) and 17 mL of 8 M urea focused with the Mini-Rotofor for 4 h at 10 W. Sixteen fractions were collected, and the protein concentration in each fraction was determined using Bradford reagent. Tris-base was added to each fraction to raise the pH to about 8.5. The proteins were reduced, carboxyamidomethylated, and proteolyzed with 0.4–0.8 µg trypsin each, according to the protein concentration in each sample. The peptide pools were desalted using Silica-C18 tip (Harvard) and 20% of each desalted peptide mixture was dried by vacuum centrifugation, dissolved in 0.1% formic acid, and analyzed by µRP-LC-MS/MS.

The remaining parts of each of the 16 fractions were combined into five pools according to the protein complexity in the fractions used for each of these peptide pools (2–3 fractions in each pool). Each pooled fraction was brought to 0.1% acetic acid, 50% ACN, pH 3, and analyzed by MudPIT with an SCX MicroTip as described in Section 2.3. The peptides were eluted with ten increasing ammonium acetate steps as described in Section 2.2 (resulting in a total of 50 fractions) and analyzed by µRP-LC-MS/MS.

8.2.6
Capillary RP-LC-MS/MS

All the final steps of the peptide analyses were performed by RP capillary chromatography on 30 cm fused silica capillaries (J&W, 100 µm id) self-packed with POROS R2, 10 µm (Applied Biosystems). The peptides were eluted using a 50 min gradient from 5 to 50% ACN, containing 0.1% formic acid followed by a wash step of 95% ACN for 15 min. The flow rate was about 0.4 µL/min and the peptides were electrosprayed into an LCQ-DecaXP mass spectrometer (Thermo Electron, San Jose, CA, USA). The MS was performed in the positive ion mode using full MS scans that were followed by CID of the three most abundant ions detected in the full MS scans.

8.2.7
MS data processing and peptide/protein identifications

The MS data was clustered and analyzed by Pep-Miner [17], and identified against the IPI human database with semitryptic settings. A peptide was considered as "high quality" if its Pep-Miner identification score was greater than 80/100 (roughly corresponding to SEQUEST X_{corr} of 1.4 for singly-charged peptides, 2.2 for doubly-charged peptides, and 3 for triply-charged peptides [18]), and if the scores of competing sequences were lower by at least 3/100. Only proteins having at least one high quality peptide were examined, and those identified with less than five peptides were pronounced as positively identified only after visual inspection by a trained mass spectrometrist.

8.3
Results

The objective of this study was to evaluate the added value of protein prefractionation with respect to the yield of protein identifications. Serum (0.6 mL) of the (HUPO PPP) BDFA01 donor was first depleted of HSA and IgG and aliquots from

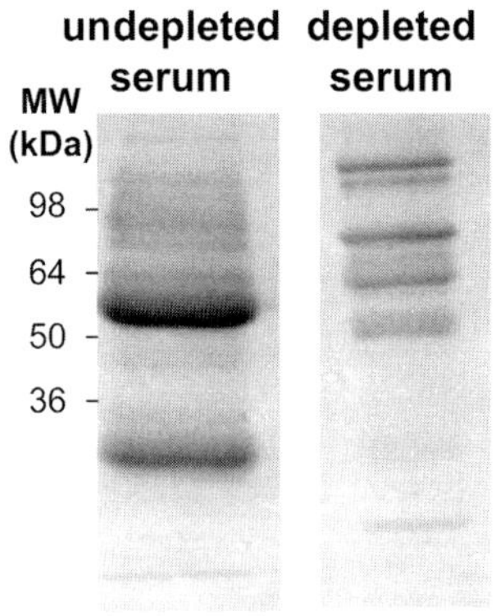

Fig. 1 SDS-PAGE of serum before and after depletion of the HSA and of the IgG.

the depleted serum were used for each of the analytical methods evaluated here. The effect of depletion of the serum from the HSA and from the IgG is demonstrated in Fig. 1, showing that the vast majority of these proteins were successfully removed. However, some HSA and IgG molecules remain even after the depletion indicated by the detection of their tryptic peptides (supplementary material). The analysis of the fraction of proteins and peptides bound to the HSA and IgG is the subject of another study conducted as part of this project.

The serum proteins in the HSA and IgG depleted serum were prefractionated by the three different approaches depicted in Fig. 2: (A) SDS-PAGE followed by cutting of gel slices; (B) liquid-phase IEF based on the Rotofor instrument (Bio-Rad); (C) SCX chromatography at low pH. The proteins in the different fractions or slices were trypsinized and the resulting peptides were analyzed by 1-D or 2-D chromatography. Each of the approaches described above was performed in two different ways. One way was to fractionate the serum into large numbers of protein fractions, followed by one LC/MS/MS for each fraction. The other alternative was to use smaller numbers of protein fractions and then perform a more comprehensive MudPIT analysis of each fraction. A MudPIT analysis was also performed on the entire albumin and antibody depleted serum, without any prefractionation of the remaining proteins, thus serving as a control for the prefractionation approaches. The MS/MS spectra of the entire project were clustered, arranged, and searched against the IPI human databank using Pep-Miner [17]. Pep-Miner was also used to compare the protein lists that resulted from the various prefractionation methods.

8.3.1
Comparisons between the prefractionation methods

The different schemes of preproteolysis and postproteolysis fractionation (Fig. 2) resulted in a large diversity of protein identifications. The clustered MS/MS data of all the different LC-MS/MS runs identified a total of 470 proteins, based on the 2210 identified peptides. A list of the proteins identified by each of the methods is supplied in the electronic supplementary material. The list includes the names of the proteins, the number of identified peptides, coverage, identification confidence

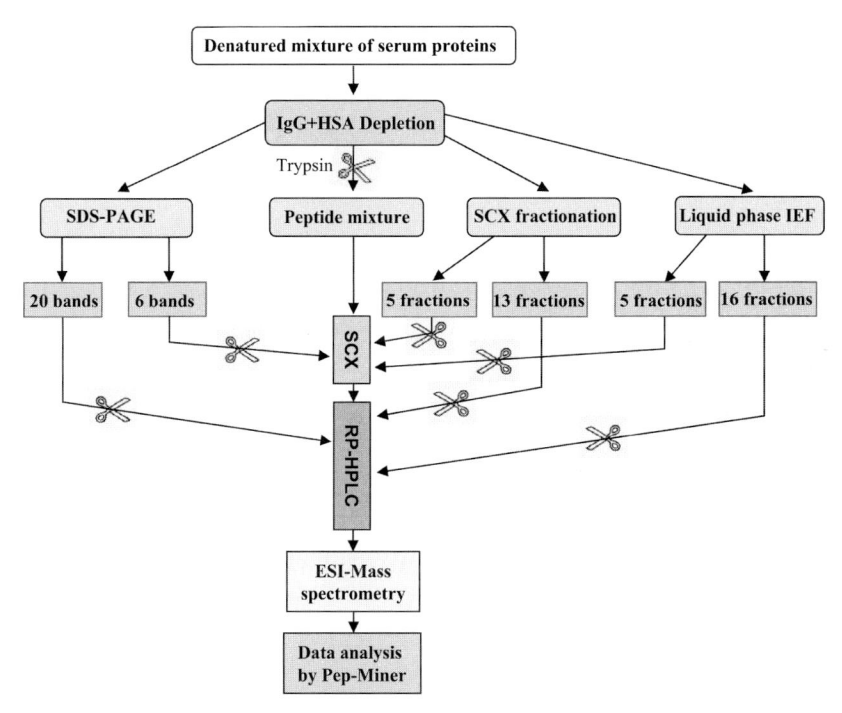

Fig. 2 A schematic view of the fractionation and analytical process.

level, and the pointers to the methodology used for their identification. About half (47%) of the proteins were identified only by a single peptide, 37% by two or three peptides, 7% by four to nine peptides, and 9% with more than ten identified peptides. Similar percentages were observed for the different approaches. The number of proteins identified by each of the procedures is listed in Tab. 1.

It seems that the SCX prefractionation of the proteins, when followed by MudPIT analyses on the different protein fractions, resulted in both the identification of the largest number of different proteins and in the best coverage of each protein. The SDS-PAGE slicing method resulted in a bit lower protein identification, while the liquid-phase IEF prefractionation seemed to be less effective in producing protein identifications.

As expected, the more abundant serum proteins (other than HSA and IgG) were identified with many peptides and appeared in most of the fractionation schemes. Of the 44 proteins that were identified by more than ten peptides, 28 proteins (64%) were detected by all seven schemes and 42 (96%) appeared in at least four of them.

8.3.2
Identification of different protein subsets

Some of the prefractionation methods resulted in identifying rather similar sets of proteins, while others resulted in identifying different sets of proteins. The lists of proteins that were identified using the different prefractionation procedures were

Tab. 1 Comparison between the prefractionation methods

Protein prefractionation	Peptide analysis	No. of runs	Total proteins	Total peptides	Proteins identified with >10 peptides	%	Proteins identified with 4–9 peptides	%	Proteins identified with 2–3 peptides	%	Proteins identified with 1 peptide	%
–	MudPIT + mass segment	15	101	675	21	21	14	14	12	12	54	53
SDS-PAGE	LC-MS/MS	20	110	595	17	15	16	15	26	24	51	46
SDS-PAGE	MudPIT	60	156	868	25	16	17	11	30	19	85	54
SCX	LC-MS/MS	13	68	396	10	15	16	24	9	13	33	49
SCX	MudPIT	50	182	849	25	14	13	7	34	19	110	60
Rotofor	LC-MS/MS	16	68	359	9	13	9	13	8	12	42	62
Rotofor	MudPIT	50	108	499	12	11	10	9	29	27	57	53

Tab. 2 Unique proteins in the different prefractionation methods

Method	Total proteins	Total peptides	Proteins identified with >4 peptides	%	Proteins identified with 2–3 peptides	%	Proteins identified with 1 peptide	%
MudPIT + mass segment	22	22			2		18	90
SDS-PAGE[a]	57	69			9		48	84
SCX[a]	80	91			11		69	86
Rotofor[a]	58	91	2	4	16	28	40	68

a) Combined data of LC-MS/MS and by MudPIT from each of the prefractionation approaches

compared to MudPIT with mass segmentation without prefractionation. Fig. 3A displays the numbers of proteins uniquely identified in the MudPIT with mass segmentation relative to the compared prefractionation method and the numbers of identified proteins shared by both methods. Relatively large numbers of identified proteins that were not identified using the MudPIT with mass segmentation were identified by other methods. It should be emphasized that some proteins were observed only in MudPIT without prefractionation, possibly due to their loss during the prefractionation steps. In all the three approaches, the addition of MudPIT instead of a single LC-MS/MS following the prefractionation phase dramatically increased the number of uniquely identified proteins. This effect was further emphasized by comparing the numbers of proteins uniquely identified in each prefractionation method between LC-MS/MS alone and MudPIT (Fig. 3B).

8.3.3
Proteins identified by only one prefractionation method

To further investigate the subsets of uniquely identified proteins we combined the data gathered by LC-MS/MS and by MudPIT from each of the prefractionation approaches. As can be seen in Tab. 2, prefractionation approaches that resulted in larger numbers of identified proteins also brought about the identification of more unique proteins by each of the methods. Most of the unique proteins were identified by one to three peptides. Interestingly, the Rotofor, which was not as efficient as the other methods in separating the proteins, resulted in the separation of a very small, yet significant numbers of unique proteins, each identified with a relatively large number of peptides (two forms of cytochrome c; IPI00099564.2 and IPI00258043.1). These proteins were not identified using the other approaches; therefore, if only one of the methods had been used, these relatively abundant proteins would have been overlooked. It can be concluded that at this early stage of the project, when comprehensive lists of proteins are sought, it is advisable to use different prefractionation and postfractionation methods to ascertain that the maximum number of proteins present in the serum are detected and accounted for.

Tab. 3 presents a complementary point of view. It displays the number of proteins missed by each prefractionation method. Inverse correlations were observed between the numbers of identified and missing proteins in the different methods. The Rotofor prefractionation resulted in a higher percentage of missing abundant proteins. The MudPIT with mass segmentation that did not include any prefractionation still resulted in the identification of most of the abundant proteins.

8.3.4
Different methods resulted in diverse peptide coverage

The added value of the usage of multiple approaches is demonstrated in Fig. 4, displaying the peptide coverage of four abundant proteins identified in all the approaches by large numbers of peptides (α2-macroglobulin, haptoglobin-2, antitrypsin, and serotransferrin). Each pie chart shows the percentage of unique pep-

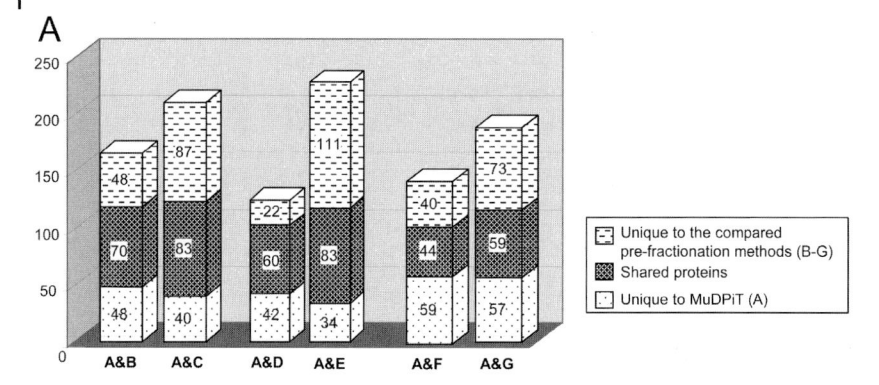

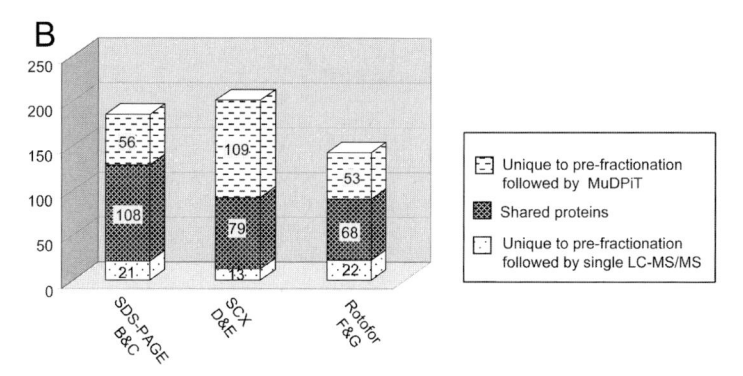

Fig. 3 Determination of the number of proteins identified uniquely in one prefractionation method and or not in the other. A, Comparison between MudPIT with mass segmentations (A) to each of the prefractionation methods: SDS-PAGE slices followed by single LC-MS/MS (B); SDS-PAGE followed by MudPIT (C); SCX proteins followed by single LC-MS/MS (D); SCX proteins followed by MudPIT (E); Rotofor followed by single LC-MS/MS (F); Rotofor followed by MudPIT (G). Indicated inside the histograms are the numbers of proteins identified uniquely by the MudPIT (bottom), shared proteins identified by both the MudPIT and the compared method (middle), and proteins identified only by the compared method (top). B, Numbers of proteins identified uniquely when each of the prefractionations was analyzed by LC-MS/MS (bottom), when it was followed by a MudPIT (top) and the number of proteins shared by both (middle). SDS-PAGE histogram compares analyses of 20 SDS-PAGE gel slices followed by single LC-MS/MS (B) to analyses of six SDS-PAGE slices followed by MudPIT on each (C); SCX histogram compares analyses of 13 SCX protein fractions followed by single LC-MS/MS (D) to analyses of five SCX proteins fractions followed by MudPIT on each fraction (E); Rotofor histogram compares analyses by LC-MS/MS of 16 Rotofor fractions (F) to analyses of five Rotofor fractions followed by MudPIT on each (G).

tides identified in each of the seven methods described, with the percentages of the shared peptides. This analysis resulted in very diverse peptide coverage for these proteins when the different prefractionation methods were used. α2-macroglobulin, for example, was detected with a relatively large number of peptides (43%) using SDS-PAGE slices followed by single LC-MS/MS (method C), while its coverage was only 2% when the Rotofor followed by single LC-MS/MS was used for prefractionation (method F). In contrast, haptoglobin-2 was detected with only 7%

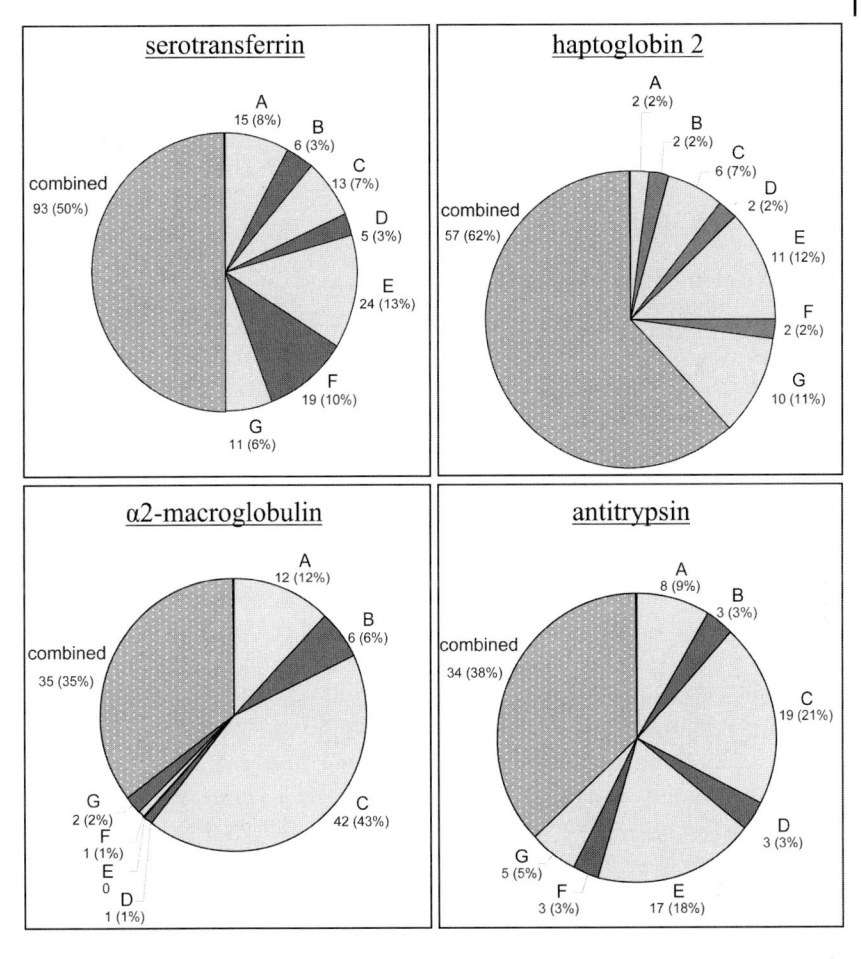

Fig. 4 Peptide coverage of individual relatively abundant proteins following different prefractionations. Pie charts display the effect of the different prefractionations on the peptide coverage of four abundant proteins (α2-macroglobulin, haptoglobin-2, antitrypsin and serotransferrin). Both total numbers and percentages of the peptides are given. Section (combined) indicates the number of peptides identified in at least two methods. Other sections indicate the numbers of peptides identified only in one of the seven methods described: MudPIT and mass segmentations (A); SDS-PAGE slices followed by single LC-MS/MS (B); SDS-PAGE followed by MudPIT (C); SCX proteins followed by single LC-MS/MS (D); SCX proteins followed by MudPIT (E); Rotofor followed by single LC-MS/MS (F); Rotofor followed by MudPIT (G).

coverage in SDS-PAGE slices and with 11% using the Rotofor. These results emphasize the advantage of using multiple separation methods to obtain better coverage and identifications.

Tab. 3 Missing proteins in the different prefractionation methods

Method	Total proteins	Total peptides	Proteins identified with >10 peptides	%	Proteins identified with 4–9 peptides	%	Proteins identified with 2–3 peptides	%	Proteins identified with 1 peptide	%
MudPIT + mass segment	323	527	2	1	9	2	126	39	186	58
SDS-PAGE[a]	222	351	1	1	9	4	69	31	143	64
SCX[a]	184	312	2	1	8	5	61	33	112	61
Rotofor[a]	247	462	6	3	10	4	70	28	161	65

a) Combined data of LC-MS/MS and by MudPIT from each of the prefractionation approaches

8.4
Discussion

8.4.1
Giving every peptide a chance

The different fractionation schemes were chosen for their merit in increasing the concentrations of individual proteins and peptides in each fraction. This way more proteins reached the threshold needed for detection. As expected, adding more dimensions to both the protein and peptide fractionation steps resulted in better identification of more proteins. Combining single protein preproteolysis fractionation with postproteolysis peptide fractionation turned out to be more effective relative to the extensive fractionation of peptides in a single dimension. However, methods (such as liquid-phase IEF) that were less effective in identifying a large number of proteins, but resulted in identification of unique proteins, should not be overlooked as long as the goal is to obtain identification of the largest possible protein repertoire from the samples.

Each one of the different prefractionation and postfractionation methods resulted in the identification of a very small but significant number of unique proteins that were identified with a very large number of peptides. These proteins were not detected using the other approaches. When one fractionation method had been used, those proteins that were clearly present in the serum in relatively large amounts would have been overlooked and the information on them would have been lost. Therefore, we can conclude that at this early stage of attempting to create the most comprehensive list of proteins in human serum, it is advisable to combine all the available prefractionation and postfractionation methods to ensure that the highest number of proteins present in the serum are detected and accounted for.

8.4.2
How to identify more of the marginal proteins

When very complex mixture of proteins and their proteolytic peptides are analyzed using a limited number of chromatographies, many proteins are identified only by one peptide. The level of confidence for the identification of some of the proteins is thus often not very high. Combining different approaches and increasing the repetitions of the same methodologies raises the number of identified peptides (up to a certain limit) possibly just by increasing the number of collected MS/MS spectra of peptides. Because significant numbers of the more abundant proteins were identified in all the different methods, the question arises as how to further increase the number of positive identifications of the rare ones. Better preproteolysis fractionation of the proteins was clearly beneficial and prefractionation SCX resulting in the largest numbers of proteins identified at a higher level of confidence when compared to liquid-phase IEF and SDS-PAGE. The SCX prefractionation results in proteins free of the acrylamide gel, which possibly interferes with the free diffusion of the trypsin and of the proteolyzed peptides. The liquid-phase IEF requires large amount of proteins to function properly due to the relatively large surface area in contact with the sample within the instrument. The use of diluted protein samples for the Rotofor resulted in loss of proteins due to nonspecific binding. We have used the liquid-phase IEF without adding detergent to prevent contamination of the proteolytic peptides during the next step with the detergent. The amount of protein loaded on the SCX is not limited. In this case we loaded the minimal amount required for the next step, namely, MudPIT. The amount of protein loaded on the SDS gel was the maximum loadable amount on one gel lane without reducing the resolution due to overloading.

8.4.3
Clustering and comparing raw data

The use of the Pep-Miner software tool to cluster the raw MS/MS data of the entire projects was of tremendous value. Pep-Miner organized the data, reduced its volume, improved the spectra quality, and increased the confidence in identification results. MS/MS data from hundreds of different chromatographies and different fractionation schemes were treated in a unified manner. Moreover, by clustering at the raw data level, Pep-Miner enabled us to perform a precise comparison between different fractionation methods, both at the raw data level and at the identified peptide level.

8.4.4
High throughput and ruggedness *versus* high sensitivity

The number of identified proteins in this study is relatively small, since a rugged approach was used for the LC-MS/MS step. We selected 100 µm POROS analytical capillary and relatively large flow rates (between 0.4 and 1 µL/min) in order to increase ruggedness and facilitate a comparative study of different approaches. In order to increase the total number of proteins in a limited time frame, more rapid scanning (possible on the new LTQ mass spectrometer), more rapid chromatographies (possible on monolithic columns), and higher sensitivity chromatographies (possible with thinner internal diameter columns) can all lead to larger numbers of identified proteins *per* instrument time. The number of identified peptides and proteins is limited by the known phenomenon of peptide coelution, which is a prevalent problem in analyses of very complex mixtures of proteins. To overcome the problem associated with coeluting peptides, the solution is the use of more rapid scanning mass spectrometers, the use of longer gradients, and more fractionation of both the proteins and the tryptic peptides. We have used one full MS followed by three data dependent MS/MS scans, a common practice that aims at maximizing the peptides fragmented at the tip of their peak. In our data, the large majority of the peptides appeared in only one MS/MS scan *per* run. The use of larger numbers of dependent MS/MS spectra after each full MS is also commonly performed, and is beneficial mostly for very complex mixtures in which the fragmentation of the minor peptides still gives meaningful data. A comparison of the optimal number of dependent CIDs after each full MS was not performed here.

8.4.5
The cost effectiveness of the different methods

MudPIT seems to be the most effective method for returning the largest number of proteins *per* working hour of the mass spectrometer. Collecting ten SCX peptide fractions followed by regular 1 h gradient, LC-MS/MS resulted in the largest number of proteins identified in the shortest time period among all the different experiments. To further increase the number of identified proteins, additions of more preproteolytic and postproteolytic fractionations would be beneficial, but the cost effectiveness of added fractionations diminishes with added fractionation steps.

High-resolution MudPIT analyses would require at least ten chromatographies, therefore comparisons between the protein repertoires of hundreds or even thousands of people's sera, each occupying the HPLC mass spectrometer for days, is clearly beyond the capabilities of most research laboratories. Therefore a totally different analytical approach is needed for the future large-scale phase of the HUPO-PPP study.

8.5
Concluding remarks

A relatively large comparison between different methodologies for preproteolysis and postproteolysis analyses of serum proteins using one serum sample for the entire study is described. It can be concluded that when the aim of the analyses is to detect as many proteins as possible, the use of different protein prefractionation methods coupled with MudPIT was the most effective in identifying many proteins and in a better coverage of individual proteins. Since a significant set of unique proteins were identified following the use of each of the prefractionation approaches, the use of only one method runs the risk of missing significant serum proteins. On the other hand, in order to detect as many proteins as possible *per* mass spectrometer working hour, MudPIT without prefractionation is still the method of choice.

The authors wish to thank HUPO PPP and the Israel Ministry of Science for funding the experimental part of the research. The experimental part was performed in the Smoler Proteomics Center at the Technion.

8.6
References

[1] Omenn, G. S., *Proteomics* 2004, *4*, 1235–1240.
[2] Hanash, S., *Mol. Cell. Proteomics* 2004, *3*, 298–301.
[3] Adkins, J. N., Varnum, S. M., Auberry, K. J., Moore, R. J., et al., *Mol. Cell. Proteomics* 2002, *1*, 947–955.
[4] Pieper, R., Gatlin, C. L., Makusky, A. J., Russo, P. S. et al., *Proteomics* 2003, *3*, 1345–1364.
[5] Xiao, Z., Conrads, T. P., Lucas, D. A., Janini, G. M. et al., *Electrophoresis* 2004, *25*, 128–133.
[6] Shen, Y., Jacobs, J. M., Camp, D. G., 2nd, Fang, R. et al., *Anal. Chem.* 2004, *76*, 1134–1144.
[7] Zhou, M., Lucas, D. A., Chan, K. C., Issaq, H. J. et al., *Electrophoresis* 2004, *25*, 1289–1298.
[8] Anderson, N. L., Polanski, M., Pieper, R., Gatlin, T. et al., *Mol. Cell. Proteomics* 2004, *3*, 311–326.
[9] Rose, K., Bougueleret, L., Baussant, T., Bohm, G. et al., *Proteomics* 2004, *4*, 2125–2150.
[10] Harper, R. G., Workman, S. R., Schuetzner, S., Timperman, A. T. et al., *Electrophoresis* 2004, *25*, 1299–1306.
[11] Link, A. J., Eng, J., Schieltz, D. M., Carmack, E. et al., *Nat. Biotechnol.* 1999, *17*, 676–682.
[12] Link, A. J., *Trends Biotechnol.* 2002, *20*, S8–S13.
[13] Wang, H., Hanash, S., *J. Chromatogr. B Analyt. Technol. Biomed. Life Sci.* 2003, *787*, 11–18.
[14] Evans, C. R., Jorgenson, J. W., *Anal. Bioanal. Chem.* 2004, *378*, 1952–1961.
[15] Wienkoop, S., Glinski, M., Tanaka, N., Tolstikov, V. et al., *Rapid Commun. Mass Spectrom.* 2004, *18*, 643–650.
[16] Lecchi, P., Gupte, A. R., Perez, R. E., Stockert, L. V. et al., *J. Biochem. Biophys. Methods* 2003, *56*, 141–152.
[17] Beer, I., Barnea, E., Ziv, T., Admon, A., *Proteomics* 2004, *4*, 950–960.
[18] Yates, J. R., 3rd, Morgan, S. F., Gatlin, C. L., Griffin, P. R. et al., *Anal. Chem.* 1998, *70*, 3557–3565.

9
Efficient prefractionation of low-abundance proteins in human plasma and construction of a two-dimensional map*

Sang Yun Cho, Eun-Young Lee, Joon Seok Lee, Hye-Young Kim, Jae Myun Park, Min-Seok Kwon, Young-Kew Park, Hyoung-Joo Lee,, Min-Jung Kang, Jin Young Kim, Jong Shin Yoo, Sung Jin Park, Jin Won Cho, Hyon-Suk Kim and Young-Ki Paik

Human plasma is the most clinically valuable specimen, containing not only a dynamic concentration range of protein components, but also several groups of high-abundance proteins that seriously interfere with the detection of low-abundance potential biomarker proteins. To establish a high-throughput method for efficient depletion of high-abundance proteins and subsequent fractionation, prior to molecular analysis of proteins, we explored how coupled immunoaffinity columns, commercially available as multiple affinity removal columns (MARC) and free flow electrophoresis (FFE), could apply to the HUPO plasma proteome project. Here we report identification of proteins and construction of a human plasma 2-DE map devoid of six major abundance proteins (albumin, transferrin, IgG, IgA, haptoglobin, and antitrypsin) using MARC. The proteins were identified by PMF, matching with various internal 2-DE maps, resulting in a total of 144 nonredundant proteins that were identified from 398 spots. Tissue plasminogen activator, usually present at 10–60 ng/mL plasma, was also identified, indicative of a potentially low-abundance biomarker. Comparison of representative 2-D gel images of three ethnic groups (Caucasian, Asian-American, African-American) plasma exhibited minor differences in certain proteins between races and sample pretreatment. To establish a throughput fractionation of plasma samples by FFE, either MARC flow-through fractions or untreated samples of Korean serum were subjected to FFE. After separation of samples on FFE, an aliquot of each fraction was analyzed by 1-D gel, in which MARC separation was a prerequisite for FFE work. Thus, a working scheme of MARC → FFE 1-D PAGE → 2-D-nanoLC-MS/MS may be considered as a widely applicable standard platform technology for fractionation of complex samples like plasma.

* Originally published in Proteomics 2005, 13, 3386–3396

Exploring the Human Plasma Proteome. Edited by Gilbert S. Omenn
Copyright © 2006 WILEY-VCH Verlag GmbH & Co. KGaA, Weinheim
ISBN: 3-527-31757-0

9.1
Introduction

Human plasma is the most obtainable and clinically valuable specimen [1]. Plasma components, derived from tissues and organs, vary in concentration at least nine to ten orders of magnitude [2]. Human plasma contains several groups of high-abundance protein that seriously interfere with the detection of low-abundance protein components [3]. For example, albumin constitutes approximately 51–71% of the total protein present in human serum and immunoglobulin G constitutes 8–26% [4]. To date, compiled reports from various groups indicate less than 3000 nonredundant plasma proteins. These have been identified using combinations of 2-DE-based separation and LC-based separation techniques. Because plasma normally perfuses tissues, plasma proteins are very attractive quantitative biomarkers, yet analysis of the plasma proteome may be very challenging. Therefore, this project requires highly advanced separation techniques as well as bioinformatics tools. Of many challenges, the most important issue appears to be improvement in detecting subnanomolar concentrations of proteins (*e.g.*, cytokines and tissue leakage proteins) [2]. To detect these lower abundance proteins in plasma at least two problems must be settled: (1) efficient throughput depletion strategy of high-abundance proteins and (2) post-depletion fractionation. Depletion of albumin or IgG has been demonstrated to enable greater sensitivity for the remaining proteins in the complex mixture of blood fluids [5, 6]. Recently, Steel *et al.* [7] reported a method of albumin removal using immunoaffinity resins. However, this method is limited to albumin, which may be not enough to detect low-abundance proteins, which are usually masked by other high-abundance proteins (*e.g.*, IgG, IgA, transferrin, haptoglobin). Pieper *et al.* [8] also reported fractionation of serum proteins by immunoaffinity chromatography that results in removal of eight highly abundant proteins, followed by sequential anion-exchange and SEC. However, this method is not commercially available and not easily assessed. They resolved about 3700 protein spots and identified 1800 by MS, which were recognized as 325 distinct proteins after sequence homology and similarity searches to eliminate redundancies. With these issues resolved, it should be feasible to establish a reference protein 2-D gel map.

Since one of the major scientific objectives of HUPO Plasma Proteome Project (HPPP) in its pilot phase was to compare the advantages and limitations of many technology platforms using different reference specimens of human plasma, we were particularly interested in establishing the depletion and prefractionation platform for the analysis of human plasma. As one of the participating teams in the pilot phase of HPPP, we have explored two different strategies for analyzing human plasma proteome. For the first strategy, we employed multiple affinity removal column (MARC), a throughput-potential immunoaffinity column to remove major abundance proteins, followed by a 2-DE technique to analyze plasma proteins and assess differences present between the plasma of ethnic groups and the differently pretreated plasma. From this experiment, we anticipated that 2-DE would provide better resolving power for protein variants [9] or their proteolytic cleavage products present in each ethnic group or differently pretreated plasma/serum samples. For the second strategy, we attempted to use free flow electrophoresis (FFE) for prefractionation of the MARC-depleted plasma proteins and have made a cross comparison between this and untreated samples for their resolution patterns. This is because FFE has been found to have advantages in both improved sample recovery, probably due to absence of gel media or membranous material, and higher sample loading capacity with continuous sample feeding. Because of these advantages, FFE has been widely used for prefractionation of samples [10–12]. Thus, the main purpose of this study is to establish a platform technology system for analyzing human plasma proteome. To achieve this goal, we attempted to optimize a working scheme of the prefractionation of human plasma samples prior to running the whole project (profiling of human plasma proteome). Here we present supporting data that our prefractionation system including MARC and FFE are efficient for both removal of high-abundance proteins and resolution of low-abundance proteins.

9.2
Materials and methods

9.2.1
Plasma sample preparation

Unless otherwise described, samples used for this experiment, Caucasian-American (CA), African-American (AF), and Asian-American (AA), are HUPO reference specimens provided by Becton Dickinson Diagnostics (Franklin Lakes, NJ, USA). Each pool is 1 U of blood each from one male and one postmenopausal female healthy, fasting donor, collected in a standard donor set-up after informed consent, immediately pooled, then divided into four equal volumes in bags with appropriate concentrations of K-EDTA, lithium heparin, or sodium citrate for plasma and without prevention of clotting for serum. This procedure required 2 h at room temperature. Each pool was then aliquoted into numerous 250 µL portions, then frozen and stored at −70°C. The aliquots were tested for HIV, HBV, and HCV. We also used Korean serum (K1) for FFE/1-DE/2-D-nanoLC-MS/MS, which was prepared as described above.

9.2.2
Depletion of major abundance proteins with an immunoaffinity column

Depletion of the six most abundant proteins (*i.e.*, albumin, transferrin, IgG, IgA, haptoglobin, and antitrypsin) in serum or plasma was carried out using a MARC (Agilent, Wilmington, DE, USA). A 4.6 × 100 mm MARC with binding capacity for 20 µL of human plasma was used. Chromatographic separation of the abundance proteins by MARC was carried out using a mobile phase reagent kit according to a standard LC protocol provided by the manufacturer. Briefly, crude human serum and plasma samples were diluted five times with Buffer A containing protease inhibitors (COMPLETE™, Roche) and filtered through 0.22 µm spin filters by spinning at 16 000 × g at room temperature for 1–2 min. The sample was injected and flow-through fractions were collected and stored at −20°C until use. To resolve depleted plasma proteins on 2-D gels, flow-through fractions from MARC were pooled and precipitated with precooled solution of 10% TCA for 1 h at −20°C. After washing with ice-cold acetone, the pellets were resolublized in the sample buffers of 2-DE and FFE.

9.2.3
2-DE

2-DE was carried out essentially as described [13]. Aliquots in sample buffer (7 M urea, 2 M thiourea, 4.5% CHAPS, 100 mM DTE, 40 mM Tris, pH 8.8) were applied to immobilized pH 3–10 nonlinear gradient strips (Amersham Biosciences, Uppsala, Sweden). IEF was performed at 80 000 Vh. The second dimension was analyzed on 9–16% linear gradient polyacrylamide gels (18 cm × 20 cm × 1.5 mm) at constant 40 mA *per* gel for approximately 5 h. After protein fixation in 40% methanol and 5% phosphoric acid for 1 h, the gels were stained with CBB G-250 for 12 h. The gels were destained with H_2O, scanned in a Bio-Rad (Richmond, CA) GS710 densitometer and converted into electronic files, which were then analyzed with Image Master Platinum 5.0 image analysis program (Amersham Biosciences).

9.2.4
Identification of proteins by MS

For 2-D gel mapping of the plasma proteome, proteins were identified by mass fingerprinting or matching with various internal 2-DE maps. Protein spots excised from 2-D gels were destained, reduced, and alkylated and then digested with trypsin (Promega, Madison, WI, USA) as previously described [14]. Tryptic peptides were desalted and purified as described [15]. Recovered peptides were prepared for MALDI-TOF MS by mixing with CHCA, 1% formic acid in 50% ACN, and droplets were allowed to dry on the MALDI sample plate. PMF was performed using a Voyager DE-PRO MALDI-TOF mass spectrometer (Applied Biosystems, Foster City, CA), operating in delayed reflector mode. Proteins were identified from the peptide mass maps using MS-Fit (http://prospector.ucsf.edu), MASCOT (http://www.matrixscience.com/search_form_select.html), and ProFound (http://129.85.19.192/profound_bin/WebProFound.exe) to search for the protein databases, Swiss-Prot (version 44.1) and GenBank.

9.2.5
Fractionation of the plasma samples by FFE

We perfomed FFE using ProTeam™ FFE instrument (Tecan, Munich, Germany) [12]. Human serum or TCA precipitated flow-through fractions of MARC were diluted 1:10 with separation media 3 (see below). Traces of the red, acidic dye 2-(4-sulfophenylazo)-1,8-dihydroxy-3,6-naphthalenedisulfonic acid (SPADNS, Aldrich) were added to ease the optical control of the migration of the sample within the separation chamber. Final protein concentration was approximately 6 mg/mL. FFE separations were conducted at 10°C using the following media. Anodic stabilization medium (I_1): 14.5% w/w glycerol, 8 M urea, 0.03% w/w, hydroxy propyl methyl cellulose (HPMC), 100 mM H_2SO_4; separation medium (I_2): 14.5% w/w glycerol, 8 M urea, 0.03% w/w HPMC, 14.5% w/w Prolyte™ 1; separation medium (I_{3-5}): 14.5% w/w glycerol, 8 M urea, 0.03% w/w HPMC, 14.5% w/w Prolyte™ 2; separation medium (I_6): 14.5% w/w glycerol, 8 M urea, 0.03% w/w HPMC, 14.5% w/w Prolyte™ 3; cathodic stabilization medium (I_7): 14.5% w/w glycerol, 8 M urea, 0.03% w/w HPMC, 100 mM NaOH; counterflow medium (I_8): 14.5% w/w glycerol, 8 M urea; anodic circuit electrolyte: 100 mM H_2SO_4; cathodic circuit electrolyte: 100 mM NaOH. The experiments were run in a horizontal separation using a 0.4 mm spacer. A flow rate of ~60 mL/h (inlet I_{1-7}) was used in combination with a voltage of 1500 V, which resulted in a current of 24 mA. Samples were perfused into the separation chamber using the cathodal inlet at ~0.7 mL/h [12]. Residence time in the separation chamber was ~33 min. Fractions were collected in polypropylene minititer plates, numbered 1 (anode) through 44 (cathode). The protein fractions were analyzed by SDS-PAGE using an XCell SureLock™ Mini-Cell (Invitrogen Carlsbad, USA) in combination with precast NuPAGE® Novex 4–12% Bis-Tris gels. CBB-staining of the proteins was carried out using the SimplyBlue™ SafeStain kit (Invitrogen).

9.2.6
LC-MS/MS

The 2-D and 1-D separations of tryptic peptides were performed on a Nano Proteomics Solutions system (Agilent), comprising an autosampler, one capillary pump, one nanoflow pump, a micro 6-port column-switching valve, and an MSD XCT IT mass spectrometer with a nanoelectrospray interface. For the chromatographic separation in the first- and second dimension it is necessary to set up two different methods, one for SCX chromatography and another for RP separation. The salt solution gradient for elution of the peptides from the SCX column was delivered from the capillary pump and the gradient for RP separation was delivered from the nanoflow pump. The nanoflow gradient started with 5% ACN and increased to 65% ACN with a slope of 0.5%/min for each RP analysis. The salt gradient was pumped in steps as 1.5, 5, 15, 50, and 100% of a 1 M ammonium acetate solution. The salt solution gradient was developed for 15 min in each step and then, prior to the washing step, the SCX column was switched to bypass with the micro 6-ports valve in the autosampler to retain current conditions. Therefore, each step contributes to salt gradient steps on the SCX column. The nanoflow pump supplied these solvents: A = water + 0.1% formic acid and B = ACN + 0.1% formic acid. The column flow rate was 300 nL/min. Stop time was 155 min *per* step with a post running time of 5 min. The capillary pump supplied these solvents: A = water + 3% − ACN + 0.1% formic acid and B = 500 mM NaCl + 3% ACN + 0.1% formic acid. Column flow rate was 15 µL/min. The ionization mode was positive nanoelectrospray with an Agilent orthogonal source. Drying gas flowed at 5 L/min and drying gas temperature was 325°C. Vcap was typically 2000–2200 V, with skim 1 at 30 V and capillary exit offset at 75 V. The trap drive was set at 85 V with averages of 1 or 2. ICC (ion charge control) was on with maximum accumulation time of 150 ms, smart target was 125 000, and MS scan range was 300–2200. Automatic MS/MS was in ultra scan mode, with the number of parents 2, averages of 2, fragmentation amplitude of 1.15 V, SmartFrag on (30–200%), active exclusion on (after 2 spectra for 1 min), prefer +2 on, MS/MS scan range of 100–1800, and ultra scan on. Columns used for the 2-D-LC/MS/MS were as follows: nanocolumn (C18): Zorbax 300SB-C18 (3.5 μm, 150 × 0.075 mm), Enrichment column (C18): Zorbax 300SB-C18 (5 μm, 5 × 0.3 mm), SCX column: Zorbax Bio-SCX Series 2 (0.8 × 50 mm).

9.2.7
Bioinformatics

Each acquired MS/MS spectrum was searched against the nonredundant protein sequence database using the Spectrum Mill software tool. Sequences of uninterpreted CID spectra were identified by correlation with the peptide sequences present in the protein sequence database (NCBInr 2004.09) using the Spectrum Mill MS Proteomics Workbench (Rev A.03.00.015, Agilent). The SpectrumMill search results were initially assessed by "score" and "SPI" (Scored Peak Intensity). The software creates theoretical peptides for all, or a limited group of, database proteins; calculates corresponding MS/MS spectra; and compares them to an experimental spectrum (submitted for the database search) to find the match. Score means points to the matched (Bonus) or unmatched (Penalty) peaks. Bonus points are awarded for each matched peak, at one point *per* peak regardless of peak height. Penalty points for unmatched peaks are based on peak height/height of tallest peak. Scored peak intensities are calculated as follows: from peaks remaining after peak detection, this is the percentage of total intensity in the query set spectrum, which is matched to peaks in the library spectrum. Scored peak intensities lower than 50% suggest a poor match, or presence of noncorresponding fragment ion types in the query set spectrum. Adjusting the value of minimum matched peak intensity to less than 50% (default value) will enable reporting of poorer quality matches. As a general rule, for declaring a protein hit, protein score > 13, peptide score > 10, and SPI (%) > 70 were applied throughout the data analysis procedures as suggested by the manufacturer. All the proteins identified in this paper are based on assignment of at least two peptides.

9.3
Results and discussion

9.3.1
2-DE map of human plasma devoid of high-abundance proteins

Seeking which proteomic platform might be the best approach for identifying low-abundance proteins in plasma, we first employed a strategy (Fig. 1) by which the depletion or fractionation of plasma sample was achieved. In an attempt to exa-

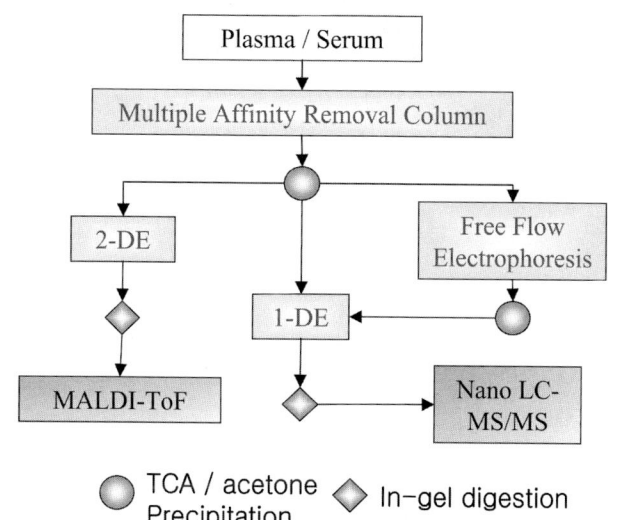

Fig. 1 Strategy for identification of human plasma proteins using 2-DE, LC-MS/MS, and FFE.

mine the efficiency of MARC in depletion of major six abundance proteins, plasma CA-Heparin (20 µL) was loaded onto MARC and the flow-through fractions of plasma proteins that had been pooled three times (*i.e.*, 60 µL of plasma corresponding to 300 µg of depleted proteins) were subjected to 2-DE analysis which resulted in 778 spot images. For whole plasma, about 1 mg proteins were loaded onto a 2-D gel, which showed about 740 spot images. As expected, proteins from the control sample showed typical overloaded spots on these six major protein positions (Fig. 2A), while the flow-through fractions showed very clean areas that usually were occupied by these six abundant proteins (Fig. 2B). When compared with proteins not loaded onto MARC, proteins of the flow-through fractions displayed a very different pattern without detectable amounts of the six major abundance proteins (Fig. 2B). For example, proteins that comigrate with the albumin and IgG enrichment region (heavy and light) are now visible and can also be identified (Fig. 2B and C and Tab. 1). Very similar results were also obtained from other samples such as CA-EDTA (CA-E), CA-citrate (CA-C), and CA-serum (CA-S). Of 778 protein spots, 389 spots were successfully identified by MALDI-MS where 144 proteins were found to be nonredundant when filtered from those having <90% homology with protein DB search. An expanded map of Fig. 2B with spot numbers is presented in Fig. 2C where 79 numbered spots out of 398 identified spots have low percent volume (0.003–0.1%) with respect to total plasma proteins as listed on Tab. 1. These percent volumes were almost equal to 0.01–4.8 µg/mL plasma (55–60 mg protein). From the 2-DE results, when the sum of 778 spots were set as 100 percent volume, the percent volume of tissue plasminogen activator (TPA) was approximately 0.0176 which is usually estimated to be present in 10–60 ng/mL [2] and may be indicative of a potential low-abundance biomarker. In general, MARC depletion appears to be simpler than methods previously reported by Pieper *et al.* [8] although there are differences in the number of identified pro-

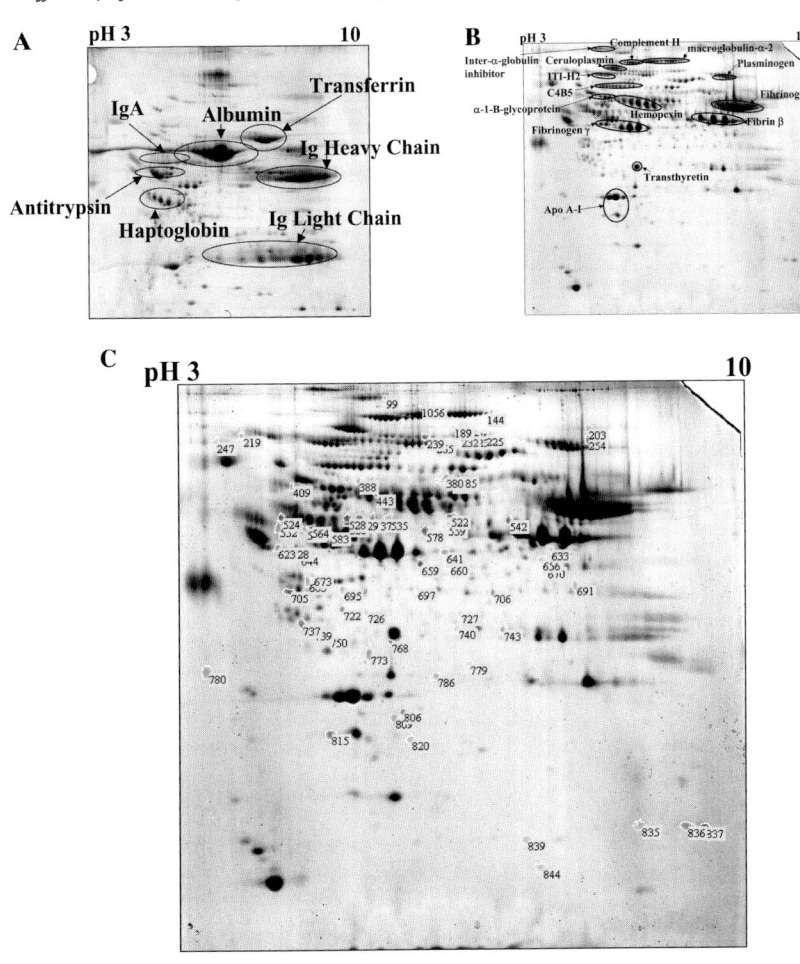

Fig. 2 2-DE of total human plasma proteins and of proteins fractionated on MARC. Proteins were resolved by 2-DE with separation by IEF in pH 3–10NL IPG strips in the first dimension and 9–16% SDS-PAGE in the second dimension. Acidic end of the first dimension is on the left. (A) Intact Caucasian serum (CA-S). (B) Flow-through from MARC (CA-H). A clickable map of flow-through from MARC (CA-H) is provided at YPRC-PDS (http://yprcpds.proteomix.org/) [15]. (A)–(C) are the same gel. Seventy-nine numbered spots with a low percent volume to total plasma proteins (0.17–4.8 µg/60 mg plasma protein) were identified and are listed in Tab. 1.

teins and detection limits between these two approaches. Nevertheless, both results indicate that depletion methods are crucial for identification of biomarker candidates. The question is which system(s) offer superior throughput potential and reproducibility, such as with MARC, where an automatic HPLC system can be routinely coupled with a commercially available immunoaffinity column. Results presented here suggest that MARC successfully removed these abundant proteins

Tab. 1 List of identified proteins in flow-through of MARC on 2-DE

Spot ID	Accession no.	Protein name	Score	Major decision DB	Matched peptides no.	Sequence coverage (%)	Theoretical value M_r	pI
99	gi|30722344	Hypothetical protein	101	MASCOT	20/72 (28%)	16	248 918	5.92
144	Q15742	NGFI-A binding protein 2 (EGR-1 binding protein 2) (Melanoma-associated delayed early response protein)	1.108e+04	MS-Fit	7/95 (7%)	18	56 595	6.5
189	gi|47777671	RAD54-like	48	MASCOT	5/26 (19%)	15	84 299	8.85
203	Q96FN5	Kinesin-like protein KIF12	6.685e+05	MS-Fit	9/94 (10%)	26	60 407	9.3
219	gi|37748641	Kininogen 1	86	MASCOT	10/41 (24%)	24	47 871	6.29
225	gi|2120082	Retrovirus-related hypothetical protein II -human retrotransposon L1NE-1	79	MASCOT	13/39 (33%)	15	83 588	9.74
232	gi|20306882	F-box and WD-40 domain protein 10	65	MASCOT	10/56 (18%)	17	120 599	9.45
234	Gi|5052951	Unknown	72	MASCOT	8/10 (80%)	6	148 921	9.69
239	gi|22035674	Pericentrin B; kendrin; pericentrin-2	59	MASCOT	33/78 (42%)	9	377 754	5.39
247	gi|3277542	ATP synthase, H$^+$ transporting, mitochondrial F0 complex, subunit f, isoform 2	48	MASCOT	3/37 (8%)	54	10 967	9.70
254	gi|13994374	Haspin	2.328e+04	MS-Fit	7/67 (10%)	13.0	88 461	9.2
265	3745750	X-ray crystal structure of C3d: A C3 fragment and ligand for complement receptor 2	55	Mascot	7/65 (11%)	27	32 845	6.34
272	gi|729884	Mitogen-activated protein kinase kinase kinase 8 (COT proto-oncogene serine/ threonineprotein kinase)	1.593e+06	MS-Fit	8/96 (8%)	27.0	52 898	5.5
274	P 11802	Cell division prolein kinase 4 (EC 2.7.1.37)	53	MASCOT	6/59 (10%)	32	33 655	6.32
305	gi|51466818	PREDICTED: hypothetical protein XP_499233	57	MASCOT	5/54 (9%)	43	21 131	8.70

Tab. 1 Continued

Spot ID	Accession no.	Protein name	Score	Major decision DB	Matched peptides no.	Sequence coverage (%)	Theoretical value M_r	pI
309	gi|1362789	DNA-activated protein kinase, catalytic subunit-human	1.58	ProFound	14/74 (19%)	7	465300	6.8
311	P11940	Polyadenylate-binding protein 1 (Poly(A)-binding protein 1)	54	MASCOT	8/67 (12%)	22	61142	9.12
360	gi|7437388	Protein disulfide-isomerase (EC 5.3.4.1) ER60 precursor	131	MASCOT	17/80 (21%)	37	56761	5.98
365	gi|5834584	Hypothetical protein	63	MASCOT	6/31 (19%)	30	35948	8.05
380	gi|182309	Factor XIII a subunit	122	MASCOT	17/68 (25%)	29	83231	5.75
385	gi|27805091	Surfactant, pulmonary-associated protein D	50	MASCOT	6/59 (10%)	28	37675	6.25
388	gi|4826772	Insulin-like growth factor binding protein, acid labile subunit	111	MASCOT	9/17 (53%)	23	65994	6.33
392	gi|14602658	IGHM protein	1.013e + 06	MS-Fit	13/53 (25%)	31	67922	6.0
409	gi|4504489	Histidine-rich glycoprotein precursor; histidine-proline rich glycoprotein	4.707e + 07	MS-Fit	14/65 (22%)	29	59579	7.1
443	P05546	Heparin cofactor II precursor (HC-II) (Protease inhibitor leuserpin 2) (HLS2)	1.875e + 07	MS-Fit	13/77 (17%)	24	57071	6.4
522	gi|1335098	Unnamed protein product	56	MASCOT	8/52 (15%)	17	49264	6.43
524	gi|42659813	Similar to RIKEN CdNA 3830422K02	56	MASCOT	5/52 (10%)	40	16298	7.53
528	gi|4504965	l-plastin; plastin 2; lymphocyte cytosolic protein-1 (plasmin)	55	MASCOT	9/57 (16%)	19	70245	5.20
529	gi|90030	l-plastin polypeptide	111	MASCOT	14/65 (22%)	34	63839	5.41
535	gi|34529119	Unnamed protein product	3.488e + 06	MS-Fit	12/59 (20%)	38	53322	6.1
537	gi|547198	LPS-binding protein, LBP	71	MASCOT	12/58 (21%)	28	53350	6.2
542	gi|7384823	TU12B1-TY	50	MASCOT	7/51 (14%)	14	54630	6.61

Tab. 1 Continued

Spot ID	Accession no.	Protein name	Score	Major decision DB	Matched peptides no.	Sequence coverage (%)	Theoretical value M_r	pI
552	P01042	Kininogen precursor (alpha-2-thiol proteinase inhibitor)	81	MASCOT	10/59 (17%)	31	47 883	6.29
553	gi|48425163	Chain 1, crystal structure of P13 alanine variant of antithrombin	112	MASCOT	12/59 (20%)	35	48 934	6.12
554	gi|51475988	PREDICTED: hypothetical protein XP_373953	48	MASCOT	4/65 (6%)	44	15 173	7.57
559	gi|21361302	Serine (or cysteine) proteinase inhibitor, clade A (alpha-l antiproteinase, antitrypsin)	93	MASCOT	11/62 (18%)	37	48 511	7.33
564	Gi|4557287	Angiotensinogen precursor; angiotensin II precursor	2.22	ProFound	6/24 (25%)	18	53 140	5.9
568	P05543	Thyroxine-binding globulin precursor (T4-binding globulin)	92	MASCOT	10/54 (19%)	31	46 295	5.87
578	Gi|4768514	T-cell receptor beta chain	60	MASCOT	4/59 (7%)	93	4675	9.30
583	Gi|999514	Chain B, antithrombin III	154	MASCOT	18/84 (21%)	42	48 916	5.95
623	P42566	Epidermal growth factor receptor substrate 15 (protein Epsl 5) (AF-lp protein)	3.707e + 04	MS-Fit	10/73 (14%)	16	98 675	4.5
628	gi|4945688l	EPHXI	61	MASCOT	6/49 (12%)	21	52 927	6.77
633	Q14964	(R39A_HUMAN) Ras-related protein Rab-39A (Rab-39)	62	MASCOT	6/50 (12%)	39	24 991	7.57
641	gi|39644990 1	BCL2L12 protein	61	MASCOT	7/60 (12%)	34	29 736	8.31
644	gi|11321640	Basic beta 1 syntrophin; syntrophin, beta 1 (dystrophin-associated protein A1, 59 kDa, basic component 1)	51	MASCOT	8/67 (12%)	22	58 025	8.81

Tab. 1 Continued

Spot ID	Accession no.	Protein name	Score	Major decision DB	Matched peptides no.	Sequence coverage (%)	Theoretical value M_r	pI
656	P61421	(VA0D_HUMAN) vacuolar ATP synthase subunit d (EC 3.6.3.14)	47	MASCOT	6/60 (10%)	20	40 303	4.89
659	gi|423038	Pigment epithelial-differentiating factor precursor	1.97	ProFound	9/69 (13%)	29	46 310	5.8
660	gi|509 79288	DKFZp451A211 protein	50	C MASCOT	8/58 (14%)	18	78 522	7.82
670	gi|4106984	R30923_1	54	MASCOT	4/7 (57%)	10	62 430	6.73
673	gi|3859855	Intersectin long form	53	MASCOT	11/52 (21%)	10	195 438	7.57
683	gi|21753163	Unnamed protein product	56	MASCOT	4/22 (18%)	46	13 995	5.20
691	gi|106874	Finger protein MTF34	61	MASCOT	5/57 (9%)	49	22 375	9.24
695	P53420	Collagen alpha 4 (IV) chain precursor	59	MASCOT	10/41 (24%)	9	16 3500	8.90
697	gi|11544455	(AL158091) dJ824A14.1 (tissue plasminogen activator protein)	57	MASCOT	5/29 (17%)	28	17 105	10.05
705	gi|39725942	Twinkle; gp4-like protein with intra-mitochondrial nucleoid localization	8.366e + 04	MS-Fit	7/34 (21%)	11	77 106	9.13
706	gi|28317367	TPA: mitochondrial inner membrane translocase	55	MASCOT	5/61 (8%)	41	17 926	6.56
722	Q9H254	Spectrin beta chain, brain 3 (spectrin, nonerythroid beta chain 3)	51	MASCOT	10/31 (32%)	8	24 3274	5.47
726	gi|28839607	BRDT protein	56	MASCOT	7/64 (11%)	22	52 893	9.34
727	gi|107920	Transcription factor NF-kappa-B 50K chain precursor	48	MASCOT	9/64 (14%)	16	105 304	5.20
737	gi|1386193	Zinc finger protein 197; zinc finger protein 20; VHL-associated KRAB-A domain-containing protein; zinc finger protein 166	63	MASCOT	8/18 (44%)	10	118 771	8.91

Tab. 1 Continued

Spot ID	Accession no.	Protein name	Score	Major decision DB	Matched peptides no.	Sequence coverage (%)	Theoretical value M_r	pI
739	gi|5921999	Dual-specificity tyrosine-(Y)-phosphorylation regulated kinase 1B isoform b	46	MASCOT	6/55 (11%)	21	64 893	9.21
740	gi|3124654	106 kDa O-GlcNAc transferase-interacting protein	51	MASCOT	8/75 (11%)	13	105 973	5.59
743	P42696	(K117_HUMAN) Probable RNA-binding protein KIAA0117 (HAL845)	56	MASCOT	6/32 (19%)	20	48 535	10.11
750	gi|47059125	G protein-coupled receptor 158	49	MASCOT	9/54 (17%)	13	135 128	8.57
768	gi|5706444	dj 1071N3.1 (endothelin converting enzyme 1)	48	MASCOT	3/60 (5%)	100	4948	4.14
773	gi|7705933	Chemokine-like factor isoform e; transmembrane proteolipid; chemokine-like factor 1	50	MASCOT	4/58 (7%)	61	7647	9.82
779	P33897	Adrenoleukodystrophy protein (ALDP)	60	MASCOT	8/57 (14%)	20	82 757	9.03
780	gi|7770217	PRO2675	63	MASCOT	8/65 (12%)	26	32 553	6.14
786	P00966	Argininosuccinate synthase (EC 6.3.4.5)	50	MASCOT	6/59 (10%)	27	46 500	8.54
806	gi|6006001	Plasma glutathione peroxidase 3 precursor	3.083e + 05	MS-Fit	12/69 (17%)	45	25 403	8.3
809	gi|42661119	PREDICTED: KIAA0565 gene product	50	MASCOT	11/71 (15%)	14	165 016	5 27
815	gi|31565469	RAGE protein	50	MASCOT	6/54 (11%)	38	27 102	9.10
820	gi|11359934	Hypothetical protein DKFZp434C0927.1	67	MASCOT	7/58 (12%)	38	27 268	9.05
835	gi|404722	Guanine nucleotide regulatory protein	49	MASCOT	7/63 (11%)	29	44 036	8.11
836	gi|22760822	Unnamed protein product	50	MASCOT	5/47 (11%)	31	38 595	6.34
837	gi|10835095	Serum amyloid A4, constitutive; C-SAA	69	MASCOT	5/21 (24%)	42	14 797	9.27
839	gi|16306550	Selenium binding protein 1	50	MASCOT	7/64 (11%)	25	52 358	5.93
844	gi|31075807	LYK5 protein	58	MASCOT	6/58 (10%)	37	38 572	6.30
1056	gi|24797065	Zinc finger protein 212; Zinc linger protein C2H2-150	62	MASCOT	5/8 (62%)	10	55 413	6.97

regardless of anticoagulant pretreatment (*e.g.*, EDTA, heparin, sodium citrate plasma, and serum). After the removal of six major abundant proteins from serum, resulting in enrichment of lower abundance proteins, we were able to map those present in lower concentrations (Fig. 2C). With regard to the potential drawback of depletion such as loss of some low-abundance proteins bound to these six high-abundance proteins, we tested this possibility by analyzing the bound proteins by 2-DE and found no protein other than the six abundant proteins, suggesting that at the limit of silver stain 2-D gel detection, low-abundance proteins are not lost during this initial depletion. This remains to be confirmed as more sensitive techniques are available. According to the information provided by the manufacturer (Agilent), the only nontargeted proteins detected in the bound fraction are apolipoprotein A-1, complement C3, and complement C4 when analyzed by LC/MS/MS (source: http://www.chem.agilent.com/cfusion/faq/faq2.cfm?subsection=45&section=7&faq=808). Data from this experiment have been deposited in the human plasma reference database linked 2-DE image used by MARC-treated plasma on YPRC-PDS (http://yprcpds.proteomix.org/) [15]. This reference database also includes clickable spots on the 2-DE image, linked to the information of identification flow (MALDI-TOF spectrum, monoisotopic peak list, and resultant html files).

9.3.2
Expression of different anticoagulant-treated plasma

Since HUPO provided various reference specimens, we were interested in examining differences in expression profiles of low-abundance protein groups present in the plasma and serum of three ethnic groups, and samples treated with different anticoagulants. Comparison of representative 2-D gel images of these samples allowed detection of variations between races and sample types (Fig. 3). As a whole, 2-DE images obtained from different specimens showed very similar display patterns in each plasma sample but there are differences in some proteins between them. For example, fibrinogen and its related proteins (*e.g.*, fibrinogen α, fibrinogen γ, and fibrin β) were not seen on 2-DE images of serum (Fig. 3C *vs.* 2B). When samples were prepared differently before 2-DE, the number of spots on 2-DE were similar, but not identical. For example, the average spot numbers on 2-D gels was 894 ± 55.15 ($n = 12$) while the average spot number of citrate-treated plasma, EDTA-treated, heparin-treated, and serum were 948 ± 31, 902 ± 23, 872 ± 81, and 824 ± 33, respectively. However, these numbers may not indicate the best preparation method for 2-DE analysis. Depending on the experiment, each preparation method may have advantages and disadvantages. For comparison of ethnic groups with respect to differential patterns of protein spots, we assumed that our normal sample represents each race.

Differences in intensity of spots among the CA (readily detected), AA, and AF (in apparent) specimens were noted for two proteins (kininogen I (gi|37748641) marked "1" in Fig. 3B) and apolipoprotein A-IV precursor (gi|178779, marked "2") in all four preparations. This observation could indicate either individual or ethnic differences, which cannot be evaluated with the total of three pooled donor specimens for this pilot phase of the HPPP. Other proteins appeared very similar across the three specimens.

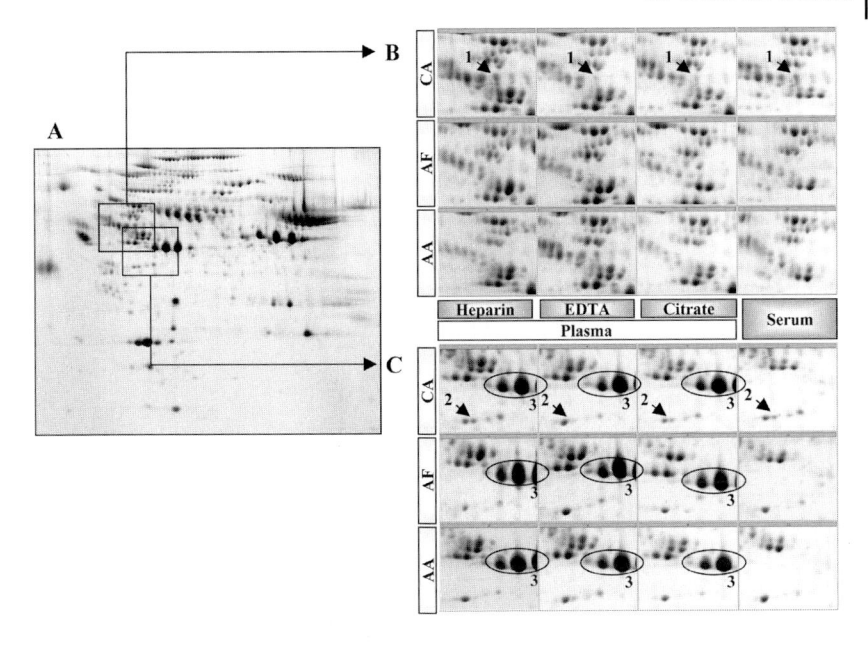

Fig. 3 2-DE images for comparison between samples. A and B sections were zoomed, respectively. 1, kininogen I (gi|37748641). 2, apolipoprotein A-IV precursor (gi|178779). 3, fibrinogen gamma (gi|223170).

9.3.3
FFE/1-DE/nanoLC-MS/MS and 2-DE/MALDI-TOF

To establish a high-throughput fractionation of plasma samples, either MARC flow-through fractions or untreated samples, the serum sample (K1) was subjected to FFE. We used Korean serum (K1) for this pilot experiment due to a shortage of HUPO reference specimen (*i.e.*, CA-S). After separation on FFE, an aliquot of each fraction was analyzed by SDS-PAGE. There are two reasons behind this approach (using 1-DE after FFE): one is that the fractions obtained from FFE contain stabilizing agents (*e.g.*, glycerol and HPMC) that are usually found to interfere with column operation in LC-MS/MS. We used a 1-D gel to remove HPMC and excessive glycerol each time. The other reason is that 1-DE shows a visual pattern of bands present in each column, which indicates how many proteins could be identified out of each band during the pilot phase of this work. In an attempt to optimize FFE conditions, three different samples (AA-H, K1-serum, and intact MARC-depleted) were tested. This confirmed that MARC is a prerequisite for running FFE because sample depleted through MARC exhibited better resolution (Fig. 4B) compared to control (Fig. 4A), where the former displayed much lower abundance proteins than the latter. The fact that many protein bands resolved in just one lane indicates that FFE is a very efficient method for prefractionation of complex samples like plasma (Fig. 4B). In MARC-untreated samples, the relatively broad distribution of the albumin band may represent several isoforms or their complexes (Fig. 4A). Since

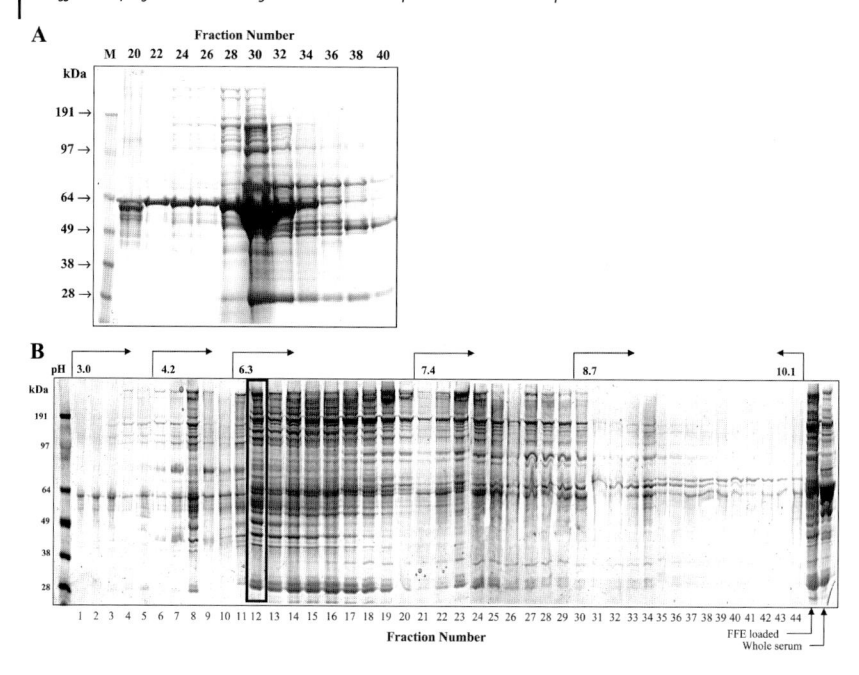

Fig. 4 Analytical SDS-PAGE analysis of FFE fractions. Whole precipitated proteins from FFE fractions are shown. Proteins were visualized by CBB. (A) An intact nondepleted Korean serum. (B) Depleted Korean serum by MARC was fractionated with FFE to 44 fractions according to the p*I* values of proteins. Each fraction was precipitated and loaded to 1-D SDS-PAGE. Selected lane (boxed) was fully identified by 1-D-LC-MS/MS as listed in Tab. 2.

albumin is well known to interact with several other proteins, this could result in suboptimal resolution. We treated serum with MARC and analyzed one lane (1 out of 44 fractions) from FFE and subsequently analyzed them by 1-D/LC-MS/MS, which takes normally 70 min *per* CBB-stained protein band, respectively. As a first step, we took protein bands in a whole lane, which corresponds to one pH range and analyzed those proteins by 1-D LC-MS/MS, resulting in 39 proteins identified. The number of peptides identified from each single lane varied from 2 to 56 (Tab. 2). Combined use of FFE and 1-D/LC-MS/MS yielded more proteins identified as compared to 1-D/LC-MS/MS alone. For example, although two proteins (complement C3 precursor and alpha-2-macroglobulin precursor) were identified by two methods, approximately 1.5–2-fold more peptides were produced by the combined method (26 *vs.* 18 for complement C3 precursor; 42 *vs.* 22 for alpha-2-macroglobulin precursor). Even more, the score values of each protein (493.35 for complement C3 precursor; 663.44 for alpha-2-macroglobulin precursor) obtained from this combined method were much higher than those (281.53 for complement C3 precursor; 305.05 for alpha-2-macroglobulin precursor) from a single method (*i.e.*, 1-D/LC-MS/MS). This may enable not only enrichment of low-abundance proteins but also eliminate the possibility of low-abundance proteins mixing with major abundance proteins.

Tab. 2 Identified proteins by FFE, 1-D SDS-PAGE and nano LC-MS/MS. From 1 out of a total of 44 fractions shown as in bands (Fig. 4B), 39 proteins identified and listed

No.	Peptides	Score	Intensity	M_r	pI	Accession number	Protein name
1	56	845.4	1.48E + 07	515565.2	6.61	4502153	Apolipoprotein B-100 precursor
2	5	72.3	7.29E + 06	268746.4	5.75	31874109	Hypothetical protein
3	7	88.7	1.89E + 07	259227.3	5.49	16933542	Fibronectin 1 isoform 3 preproprotein
4	9	128.1	8.89E + 06	249075.6	5.92	30722344	Hypothetical protein
5	10	158.0	1.49E + 07	192862.6	6.66	2144577	Complement C4A precursor [validated]
6	3	48.7	3.59E + 07	192337.0	6.80	14577919	Complement component 4A preproprotein
7	17	240.4	5.67E + 07	187149.1	6.02	40786791	Complement component 3
8	11	144.2	1.17E + 07	163278.8	6.00	4557225	Alpha-2-macroglobulin precursor
9	2	26.3	6.39E + 06	139126.3	6.28	4504375	Complement factor H
10	3	38.6	8.12E + 06	122652.3	5.42	1070458	Ferroxidase (EC 1.16.3.1) precursor [validated]
11	6	83.1	2.63E + 07	106437.0	6.40	125000	Inter-alpha-trypsin inhibitor heavy chain H2 precursor
12	6	81.6	3.11E + 07	101389.7	6.31	2851501	Inter-alpha-trypsin inhibitor heavy chain H1 precursor
13	2	32.2	2.45E + 07	76684.9	4.86	41393602	Complement component 1, s subcomponent
14	4	49.3	1.43E + 07	69069.6	5.64	4501987	Afamin precursor
15	3	28.2	1.85E + 07	67033.6	7.15	4502503	Complement component 4 binding protein, alpha
16	3	49.4	1.17E + 07	59578.6	7.09	4504489	Histidine-rich glycoprotein precursor
17	2	24.8	9.68E + 06	58649.4	5.84	106929	Lysine carboxypeptidase (EC 3.4.17.3) 83K chain
18	3	53.8	6.22E + 07	54305.9	5.55	14326449	Vitronectin
19	5	64.7	1.08E + 08	54272.8	5.58	46577680	Alpha-1B-glycoprotein precursor
20	2	35.3	2.31E + 07	53114.4	5.87	30582541	Angiotensinogen
21	2	33.5	2.35E + 07	52951.0	5.34	455970	Vitamin D-binding protein/group specific component
22	4	50.4	4.11E + 07	52691.8	6.12	576554	Antithrombin III variant
23	4	58.7	5.59E + 07	52236.7	6.06	32891795	Clusterin
24	9	117.5	5.53E + 07	51676.7	6.55	11321561	Hemopexin

Tab. 2 Identified proteins by FFE, 1-D SDS-PAGE and nano LC-MS/MS. From 1 out of a total of 44 fractions shown as in bands (Fig. 4B), 39 proteins identified and listed

No.	Peptides	Score	Intensity	M_r	pI	Accession number	Protein name
25	8	119.1	1.20E + 08	49846.6	5.50	1340142	Alpha1-antichymotrypsin
26	2	22.3	1.48E + 07	48556.2	7.34	1708609	Kallistatin precursor (Kallikrein inhibitor)
27	2	26.9	1.82E + 07	47901.5	6.29	37748641	Kininogen 1
28	5	85.9	2.38E + 07	47360.8	5.97	27692693	Similar to alpha-fetoprotein
29	2	19.8	1.62E + 07	46342.5	5.97	15217079	Pigment epithelium-derived factor
30	2	22.2	5.16E + 07	39777.6	5.15	298532	Paraoxonase/arylesterase
31	4	54.3	1.25E + 07	38999.7	5.95	4502067	Alpha-1-microglobulin/bikunin precursor
32	2	29.1	1.70E + 07	37646.8	6.08	12054070	Immunoglobulin heavy chain constant region alpha 1
33	6	73.6	1.34E + 07	36481.4	8.70	1340170	C1 esterase inhibitor
34	10	153.9	3.92E + 08	30778.0	5.56	37499465	Apolipoprotein A-I
35	2	24.4	1.58E + 07	26148.5	8.58	1196442	Galactose-specific lectin
36	2	18.8	3.70E + 07	25650.5	9.78	41147276	PREDICTED: similar to hypothetical protein (L1H 3 region)
37	7	133.5	4.55E + 07	15887.1	5.52	4507725	Transthyretin precursor
38	2	28.6	4.83E + 07	15649.4	9.24	32880065	FK506 binding protein 2, 13 kDa
39	2	38.6	1.91E + 07	11175.1	6.27	4502149	Apolipoprotein A-II precursor

9.4
Concluding remarks

We report here the construction of a human plasma 2-DE map, aided by depletion of six major abundance proteins (albumin, transferrin, IgG, IgA, haptoglobin, and antitrypsin) after immunoaffinity column separation using MARC. We also fractionated plasma or serum proteins by FFE prior to either display on 2-D gels or runs on 1-D (RP) nanoLC-MS/MS, allowing more efficient analysis of low-abundance proteins, such as TPA. Thus, the combined procedures of MARC and FFE are an efficient platform technology for profiling the complex human plasma proteome.

This study was supported by a grant of the Korea Health 21 R&D Project, Ministry of Health & Welfare, Republic of Korea (03-PJ10-PG6-GP01-0002 to YKP) and HUPO PPP.

9.5
References

[1] Ardekani, A. M., Liotta, L. A., Petricoin, E. F. 3rd, *Expert Rev. Mol. Diagn.* 2002, *2*, 312–320.
[2] Anderson, N. L., Anderson, N. G., *Mol. Cell. Proteomics* 2002, *1*, 845–867.
[3] Marshall, J., Jankowski, A. *et al.*, *J. Proteome Res.* 2004, *3*, 364–382.
[4] Putnam, R. W., *The Plasma Proteins*, Academic Press, New York 1975.
[5] Lollo, B. A., Harvey, S., Liao, J., Stevens, A. C. *et al.*, *Electrophoresis* 1999, *20*, 854–859.
[6] Adkins, J. N., Varnum, S. M., Auberry, K. J. *et al.*, *Mol. Cell. Proteomics* 2002, *1*, 947–955.
[7] Steel, L. F., Trotter, M. G., Nakajima, P. B., Mattu, T. S., Gonye, G., Block, T., *Mol. Cell. Proteomics* 2003, *2*, 262–270.
[8] Pieper, R., Gatlin, C. L., Makusky, A. J., Russo, P. S. *et al.*, *Proteomics* 2003, *3*, 1345–1364.
[9] Park, K. S., Cho, S. Y., Kim, H, Paik, Y.-K., *Int. J. Cancer* 2002, *97*, 261–265.
[10] Hoffmann, P., Ji, H., Moritz, R. L., Connolly, L. M., Frecklington, D. F. *et al.*, *Proteomics* 2001, *1*, 807–818.
[11] Zischka, H., Weber, G., Weber, P. J., Posch, A., Braun, R. J., Buhringer, D. *et al.*, *Proteomics* 2003, *3*, 906–916.
[12] Weber, G., Islinger, M., Weber, P., Eckerskorn, C., Volkl, A. *et al.*, *Electrophoresis* 2004, *25*, 1735–1747.
[13] Park, K. S., Kim, H., Kim, N. G., Cho, S. Y. *et al.*, *Hepatology* 2002, *35*, 1459–1466.
[14] Shevchenko, A., Wilm, M., Vorm, O., Mann, M., *Anal. Chem.* 1996, *68*, 850–858.
[15] Cho, S. Y., Park, K. S., Shim, J. E., Kwon, M. S. *et al.*, *Proteomics* 2002, *2*, 1104–1113.

10
Comparison of alternative analytical techniques for the characterisation of the human serum proteome in HUPO Plasma Proteome Project*

Xiaohai Li, Yan Gong, Ying Wang, Songfeng Wu, Yun Cai, Ping He, Zhuang Lu, Wantao Ying, Yangjun Zhang, Liyan Jiao, Hongzhi He, Zisen Zhang, Fuchu He, Xiaohang Zhao and Xiaohong Qian

Based on the same HUPO reference specimen (C1-serum) with the six proteins of highest abundance depleted by immunoaffinity chromatography, we have compared five proteomics approaches, which were (1) intact protein fractionation by anion-exchange chromatography followed by 2-DE-MALDI-TOF-MS/MS for protein identification (2-DE strategy); (2) intact protein fractionation by 2-D HPLC followed by tryptic digestion of each fraction and microcapillary RP-HPLC/microESI-MS/MS identification (protein 2-D HPLC fractionation strategy); (3) protein digestion followed by automated online microcapillary 2-D HPLC (strong cation-exchange chromatography (SCX)-RPC) with IT microESI-MS/MS; (online shotgun strategy); (4) same as (3) with the SCX step performed offline (offline shotgun strategy) and (5) same as (4) with the SCX fractions reanalysed by optimised nanoRP-HPLC-nanoESI-MS/MS (offline shotgun-nanospray strategy). All five approaches yielded complementary sets of protein identifications. The total number of unique proteins identified by each of these five approaches was (1) 78, (2) 179, (3) 131, (4) 224 and (5) 330 respectively. In all, 560 unique proteins were identified. One hundred and sixty-five proteins were identified through two or more peptides, which could be considered a high-confidence identification. Only 37 proteins were identified by all five approaches. The 2-DE approach yielded more information on the p*I*-altered isoforms of some serum proteins and the relative abundance of identified proteins. The protein prefractionation strategy slightly improved the capacity to detect proteins of lower abundance. Optimising the separation at the peptide level and improving the detection sensitivity of ESI-MS/MS were more effective than fractionation of intact proteins in increasing the total number of proteins identified. Overall, electrophoresis and chromatography, coupled respectively with MALDI-TOF/TOF-MS and ESI-MS/MS, identified complementary sets of serum proteins.

* Originally published in Proteomics 2005, 13, 3423–3441

Exploring the Human Plasma Proteome. Edited by Gilbert S. Omenn
Copyright © 2006 WILEY-VCH Verlag GmbH & Co. KGaA, Weinheim
ISBN: 3-527-31757-0

10.1
Introduction

A worldwide effort is underway to characterize all human proteins and establish their structural and functional relationships [1]. Of all the proteomes, the proteins of the blood are perhaps of the greatest biological, medicinal and economic importance. Serum or plasma is of unusual biological complexity, reflecting its communication with all cells, tissues and organs. Thus, the task of the HUPO Plasma Proteome Project (PPP) has been distributed among collaborating labs around the world [2]. Among the scientific goals of the pilot phase research of this project, the most important is to compare the results from different types of technologies, and establish robust and reproducible techniques for sample preparation, protein fractionation and identification with extended dynamic range. The fractionation and identification of serum/plasma proteins has a long history [3], and a limited nonredundant protein list has been published [4].

Serum contains a small group of proteins of unusually high abundance, including albumin, immunoglobulins, transferrin and macroglobulin, which constitutes about 85% of the total serum protein and which severely interfere with the identification of proteins of lower abundance. Thus, the characterisation of the serum proteome necessitates extensive fractionation prior to mass spectrometric analyses. The strategy used to remove proteins of high abundance and enrich low-abundance ones is crucial. Recently, a variety of techniques have been developed for depletion of highly abundant proteins from serum or plasma [5–9]. In particular, specific affinity removal of the most abundant proteins and subsequent chromatographic separation of peptides or intact proteins has been widely used in the comprehensive characterisation of the human plasma or serum proteome [10, 11]. With removal of the highly abundant proteins, the remaining proteins can be identified over a relatively high dynamic range. However, no single analytical approach is likely to identify all the major proteins in any proteome [12]. Therefore, the collaborating labs in the HUPO PPP are comparing a variety of technological strategies. At the workshop of HUPO PPP held in Montreal in 2003, the preliminary data sets submitted from different labs had little overlap in proteins identified, which was also confirmed in the review by Anderson *et al.* [4]. The lack of overlap can be attributed in part to the different technological strategies or methods of sample preparation used. Recently it was demonstrated that optimisation of 2-D chromatographic separation of peptides in the offline mode [13], performing intact protein fractionation prior to MS identification [14] and optimisation of both the separation of peptides and the sensitivity of ESI MS identification [15] would markedly extend protein identifications for complex proteomic samples, especially human serum. In a systematic approach to this issue, we here compare the results from five different techniques using the same HUPO PPP reference serum specimen with the six proteins of highest abundance depleted by affinity chromatography. The approaches compared were (1) intact protein fractionation by anion-exchange chromatography (WAX) followed by 2-DE and MALDI-TOF-MS/MS for protein identification (2-DE strategy); (2) intact protein fractionation by 2-D HPLC and then coupled with solution digestion of each fraction and microcapillary RP-HPLC microESI-MS/MS identification (protein prefractionation strategy); (3) digestion of mixed proteins by trypsin followed by automated online microcapillary 2-D HPLC with

IT microESI-MS/MS (online shotgun strategy); (4) same as (3) with the strong cation-exchange chromatography (SCX) step performed offline (offline shotgun strategy) and (5) same as (4) with the SCX fractions reanalysed by optimised nanoRP-HPLC-nanoESI-MS/MS. (offline shotgun-nanospray strategy). The protein identification results of each strategy were compared and their particular features were summarised.

10.2
Materials and methods

10.2.1
Materials

Sequencing grade porcine trypsin was purchased from Promega (Madison, WI, USA); DL-DTT and iodoacetamide were obtained from Pierce (Rockford, IL, USA); Tris(2-carboxyethyl)phosphine hydrochloride (TCEP) and ammonium bicarbonate were purchased from Sigma-Aldrich (St. Louis, MO, USA); HPLC grade water was supplied by Millipore (Billerica, MA, USA). UV grade ACN was purchased from Merck (Whitehouse Station, NJ, USA). All other chemical reagents were obtained from Sigma.

10.2.2
Human serum samples

All sera (the HUPO reference specimen, C1-serum), provided by the Chinese Academy of Medicinal Science, were prepared according to the BD protocol with minor modifications [2]. Twenty healthy donors were carefully selected (ten men and ten women) who tested negative to human immunodeficiency virus (HIV), hepatitis B virus (HBV), hepatitis C virus (HCV) and syphilis. Samples were obtained after approved by the Institutional Review Board (IRB) and informed consent by donors. Donors were required to fast and to avoid medicine and alcohol within 12 h before sampling. Human blood was obtained by venipuncture from each donor into evacuated blood collection tubes that contained no anticoagulant. The specimens were centrifuged at 4000 rpm for 15 min at 4°C. The resultant sera were transferred to a second set of centrifuge tubes, which were then centrifuged at 2400 rpm for 5 min to remove any residual cells. Volumes of the sera equal to the lowest volume obtained from any donor were pooled in a new container and mixed gently. Two hundred and fifty microlitres of aliquots were then dispensed into labelled cryovials, which were frozen with dry ice within 2 h. All the tubes were stored at $-80°C$ and transported on dry ice.

10.2.3
Integrated strategy for characterising analytical approaches

Fig. 1 outlines the five different strategies for protein separation and identification that were assessed in this project. 2-DE is suited for getting information on pI-altered protein isoforms but encounters difficulties in the identification of proteins representing

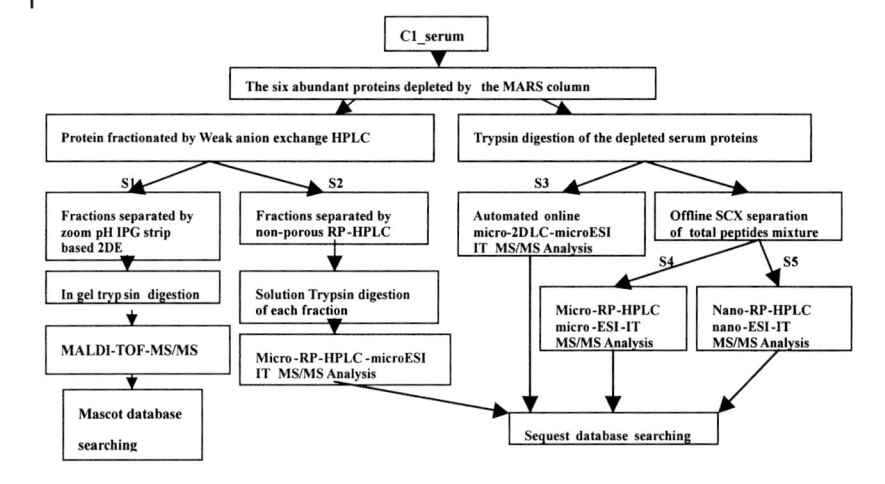

Fig. 1 Schematic overview of integrated strategies for characterising the serum proteome. Details of the five approaches are described in Section 2. S1, S2, S3, S4 and S5 represent the 2-DE, 2-D HPLC protein fractionation, online SCX shotgun, offline SCX shotgun and offline SCX shotgun-nanospray strategies respectively.

extremes of molecular weight, p*I* value or hydrophobicity [16]. On the other hand, HPLC-based strategies involving sample fractionation at the peptide or intact protein level can be suitable for almost all kinds of proteins, and have been applied to comprehensive proteome analysis [17]. Accordingly, the five strategies were combined as alternatives and compared in this research.

10.2.4
Depletion of the highly abundant serum proteins by MARS

Serum samples were thawed, diluted with 5 vol of buffer A (Product No. 5185–5987, pH 7.4) and centrifuged through a 0.45 µm filter membrane at 12 000 rpm. The prepared samples were maintained at 4°C and immediately used for further treatment. On an Elite 230 LC system (Dalian, China), each aliquot of the sample (equal to 30 µL original serum) was injected on a Multiple Affinity Removal System® (MARS) HPLC column (Agilent Technologies, Palo Alto, CA, USA) for the depletion of the six serum proteins of highest abundance according to the manufacturer. The flow-through fractions and the retained fractions were collected manually. All these fractions were stored at −80°C if not used immediately for further treatment.

10.2.5
Desalting and concentrating the flow-through fractions by centrifugal ultrafiltration

The Centriplus centrifugal concentrators (YM-3, MWCO 3kDa; Millipore) were rinsed and used according to the manufacturer's protocol. A 10 mL aliquot of the flow-through fraction was loaded each time. The sample was centrifuged (Biofuge

Stratos, Heraeus Instruments, Germany) at 4500 rpm, 4°C until the volume had decreased to about 200 μL. Finally, the sample solution was buffer-changed gradually into 20 mM ammonium bicarbonate solution (pH 8.5, containing 2 mM EDTA). The concentrated solution was transferred into an Eppendorff vial. Desalted sample solutions that had been prepared separately were combined, mixed gently, divided into 0.8 mL aliquots and stored at −20°C prior to further separation. The protein content was colorimetrically assayed using a revised Bio-Rad RCDC method.

10.2.6
Fractionation of depleted serum samples by anion-exchange HPLC

Anion-exchange chromatography was performed for further fractionation of the serum proteins on the Elite LC system. An aliquot containing 3 mg of total protein was injected on a WAX column (PolyWAX LP, 200 × 4.6 mm, 5 μm, 1000 Å; PolyLC, Columbia, MD, USA) preceded by a PolyWAX LP guard column. The column was eluted with buffer A (15 mM Tris-HCl buffer with 25% ACN, pH 7.4) at a rate of 0.30 mL/min for 15 min, then eluted with a 50 min linear gradient to buffer B (15 mM Tris-HCl buffer with 25% ACN, pH 7.4, containing 0.8 M sodium perchlorate) from 0 to 60% at a rate of 0.60 mL/min, followed by another gradient from 60 to 100% buffer B in 20 min. After elution with 100% buffer B for 40 min, the column was equilibrated with buffer A at 0.60 mL/min for 24 min. Absorbance was monitored at 280 nm. Fractions were collected manually in accordance with the peaks observed in the absorption profile. In total 12 mg proteins were separated in four runs for 2-DE strategy and about 24 mg proteins in eight runs for protein 2-D HPLC fractionation strategy. The same fractions were combined, desalted and buffer-exchanged to 20 mM ammonium bicarbonate solution (pH 8.5) by centrifugal ultrafiltration. Each fraction was reduced in volume to about 2.0 mL and frozen at −20°C until further analysis. The protein content of each fraction was colorimetrically assayed using a revised Bio-Rad RC DC method. With one batch of the sample separated by WAX, six fractions were collected for further 2-DE fractionation, while with another batch, nine fractions were prepared for further protein fractionation by 2-D HPLC.

10.2.7
Protein fractionation by 2-D HPLC with nonporous RP-HPLC

The nonporous RP-HPLC column of the Beckman ProteomeLab PF 2-D system (Beckman Coulter, Fullerton, CA, USA) was used for the RP-HPLC separation of the proteins from the WAX fractions. This was an analytical column (33 × 4.6 mm) containing 1.5 μm ODS1 nonporous silica beads (Eprogen, Darien, IL, USA). Solvent A was 0.1% TFA with 5% ACN in water and solvent B was 0.1% TFA with 5% H_2O in ACN. A gradient was run from 0 to 65% B in 60 min and 65 to 90% in 2 min, then held at 90% for another 10 min. The column was then reequilibrated with a decreasing gradient to 0% B in 2 min and held at 0% for another 26 min. Flow rate was 0.75 mL/min throughout. Effluents were monitored at

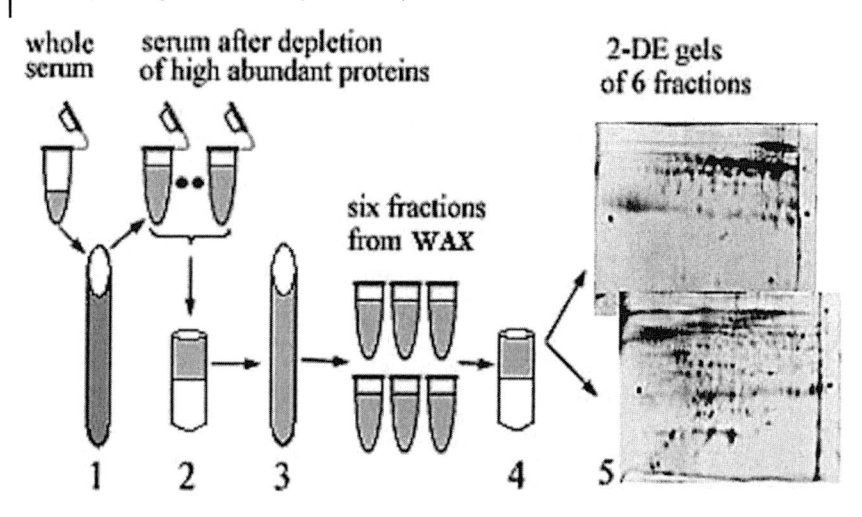

Fig. 2 Scheme of 2-D HPLC/2-DE approach for the separation of serum proteins. 1: MARS column; 2 and 4: centrifugal filter device for the concentration and desalting of samples; 3: WAX fractionation for depleted serum samples (six fractions obtained); 5: zoom pH (4–7 and 6–9) IPG strip based 2-DE separation of the six fractions.

214 nm and fractions were collected using an automated Gilson 215 Liquid handler (Gilson, Middleton, WI, USA). The system was controlled by 32 Karat chromatography software (Beckman Coulter) and UniProt software (Gilson) for the collector. The corresponding RP-HPLC fractions (guided by comparing absorbance profiles) from subsequent runs of the same sample were pooled in the same Eppendorff vials. The ACN in each fraction was evaporated with a stream of nitrogen and the remainder dried with a SpeedVac (Thermo Savant). All the fractions were stored at −20°C pending further sample processing.

10.2.8
The 2-DE strategy for the analysis of serum proteins

10.2.8.1 2-DE
The sample preparation scheme is shown in Fig. 2. After depletion with the MARS, the remaining low- and medium-abundance serum proteins from about 2.0 mL sera were fractionated by WAX. The six resulting protein fractions were dried by Speed-Vac until the volume of each was reduced to about 50 µL. Then 150 µL of rehydration buffer (9 M urea, 2% w/v CHAPS, a trace of bromophenol) was added to each fraction and vortexed to dissolve the proteins completely. 2-DE equipment (IPGphor system, Ettan DALT Six Electrophoresis Unit and related software) and consumables were all from Amersham Biosciences. Prior to IEF, each fraction was mixed with DTT and IPG buffer; the final concentrations were 65 mM and 0.5% respectively. The mixed solutions were loaded onto immobilised DryStrip gels with a pH range of 4–7 or 6–9. In the IEF process, proteins were focussed for 46 000 Vh (pH 4–7) or 58 000 Vh

(pH 6–9) respectively. After completion, the focussed IEF gels were equilibrated according to the manufacturer. The equilibrated IPG strips were then placed on the top of Ettan DALT precast gels and held in place with 0.5% w/v agarose and run on the Ettan DALT Six at 3 W/gel for 45 min for the initial migration and then at 17 W/gel for separation, until the dye front reached the bottom of the gels. Finally, all the gels were stained with mass spectrometric-compatible silver stain [18].

10.2.8.2 In-gel digestion *via* automated workstation
According to the manufacturer's instructions, the images of silver-stained precast gels were scanned using ImageMaster Labscan software in transmission mode. By using ImageMaster software, all visible protein spots were selected. Based on the input spots position information, Ettan Spot Handling Workstation performed spot cutting, gel destaining, drying, digestion and peptide extraction automatically. When all steps were over, the extracted peptides were dried on the bottom of bar-coded 96-well microplates.

10.2.8.3 Protein spot identification by MALDI-TOF-MS/MS
Saturated matrix (2.5 µL) (CHCA dissolved in a solution of 50% v/v ACN, 0.5% v/v TFA) was added to each well of the microplates. After thoroughly mixing the contents of the well, 0.6 µL of the mixture was spotted manually onto an ABI MALDI target plate. The spots were allowed to dry and then put into a 4700 Proteomics Analyzer (Applied Biosystems, Framingham, MA) equipped with a 200 Hz frequency-tripled Nd:YAG laser operating at a wavelength of 355 nm and 200 Hz repetition rate in both MS and MS/MS modes. When acquiring MS spectra, laser intensity was set at 2700 and ions were collected between 700 and 4000 Da. All the acquired MS spectra represented signal averaging of 2000 laser shots. When carrying out the MS/MS acquisition, the six most intensive peptides with S/N exceeding 100 of each spot were selected and subjected to subsequent MS/MS analysis. Each MS/MS spectrum was complied from 3500 shots with laser intensity set to 4500. The collision energy was set at 1 kV and the collision gas was air. Before acquiring data, the instrument was calibrated in plate mode with tryptic peptides of myoglobin as an internal standard. All mass data were searched by GPS Explorer Software v2.0 (with MASCOT as the database search engine) with peptide and fragment ion mass tolerance of 0.25 Da. The database used here is the IPI human fasta database (v2.27). Peptide differential modifications allowed during the search were carbamidomethylation of cysteines and oxidation of methionines. The maximum number of missed cleavages was set to 1 with trypsin as the protease. Protein identification results were manually evaluated. Identified proteins for which the mass tolerance of matched peptides was randomly distributed and almost at the limit, as well as most sequences of matched peptides with one missed cleavage and obvious interference by self-cleavage fragments of trypsin, were rejected as possible false-positive hits. Only identification results with significant MOWSE scores exceeding 58 (confidence level more than 95%) were confirmed as positive hits.

10.2.9
Shotgun strategy for the analysis of serum proteins

10.2.9.1 Trypsin digestion of serum proteins
The concentrated sample (obtained as described in Section 2.5 containing about 1.6 mg/mL proteins) was divided into two aliquots (each equal to 250 µL crude serum). Solid urea was added to the samples, then 40 mM TCEP and 200 mM DTT solutions were added to yield concentrations of 8 M urea, 10 mM DTT and 1 mM TCEP in 100 mM NH_4HCO_3, 10% ACN v/v solution. The samples were mixed and reduced at 37°C for 4 h and then concentrated iodoacetamide in 100 mM NH_4HCO_3 solution was added to give a concentration of 60 mM. The mixture was incubated for an additional 60 min in darkness. After the second incubation, the samples were spun through a Centricon YM-3 filter to a low volume and buffer-exchanged with 100 mM NH_4HCO_3. Finally, the sample solution (about 1 mL) was transferred to an Eppendorff vial, trypsin solution (in 100 mM NH_4HCO_3) was added (100:1, substrate to enzyme) and the resulting mixture was gently vortexed and incubated at 37°C. To ensure complete digestion, a second dose of trypsin was added (50:1, substrate to enzyme) after 4 h and the reaction mixture was continuously incubated for a total of 24 h [19]. After incubation, the sample was acidified by formic acid and dried by SpeedVac. The resulting peptides were resuspended in a solution of 5% ACN + 0.1% formic acid before analysis.

10.2.9.2 Protein identification by micro2-D LC-ESI-MS/MS
The 2-D LC separation was performed on a ProteomeX system (Thermo Finnigan, San Jose, CA, USA), the configuration of which has been described elsewhere [19]. The flow rate was maintained at 1.6 µL/min after the flow splitter. The solution of tryptic peptides was injected to an SCX capillary column (BioBasic-SCX, Thermo Keystone-Hypersil; 320 µm id × 10 cm), which was connected to a 10-port column-switching valve and programmed to sequentially load salt-eluted fractions onto a C18 capillary column (BioBasic C18, 300 Å, 5 µm silica, Thermo Keystone-Hypersil; 180 µm id × 10 cm). The RP capillary column was connected to an ion source chamber with a sheath gas flow at 12 U for MS analysis. The temperature of the ion transfer capillary was set at 160°C and the spray voltage was set at 3.2 kV. Peptide ions were detected in a survey scan in a specified mass range (two microscans) followed by three data-dependent MS/MS scans on the three most intense ions (three microscans each, isolation width 3 Da, 35% normalised collision energy, dynamic exclusion for 5 min) in a completely automated mode. In addition, sample was divided into two equal aliquots, each of which was respectively analysed by segmenting the survey scan mass range into 400–1300 and 1000–2000. For the capillary RP-HPLC separation, the gradient was started at 5% ACN, ramped to 55% ACN with 0.1% formic acid in 120 min and finally ramped to 90% ACN in 0.1% formic acid for another 60 min gradient. A multistep salt gradient was developed to elute peptides from the SCX capillary column onto the RP column, accomplished by mixing different ratios of solvent C (5% ACN, 0.1% formic acid) and solvent D (5% ACN, 0.1% formic acid, 1 M ammonium chloride). The 14 salt steps were performed with 20, 50, 75, 100, 150, 200, 250, 300, 400, 500, 600, 700, 800

and 1000 mM ammonium chloride. The flow rate was 2.0 µL/min throughout. The 10-port valve allows loading of the successive salt elution fraction onto the second capillary RP column while the first one is performing LC-MS/MS analysis.

10.2.9.3 Data processing
All MS/MS spectra were searched using the SEQUEST algorithm incorporated into the Thermo Finnigan Bioworks software (v3.1) [20]. The database is the IPI human fasta database (v2.27). Database searching was performed allowing for differential modification on cysteine residues (carbamidomethylation, +57 Da) and methionine residues (oxidation, +16 Da), with peptide mass tolerance of 1.5 Da. MS/MS spectra were searched against the database using the following criteria: (1) As *per* a similar rule according to Washburn *et al.* [21], positive protein identification was accepted for a peptide with X_{corr} value of greater than or equal to 3.75 for triply-, 2.2 for doubly- and 1.9 for singly-charged ions, ΔC_n equal to or greater than 0.1 and R_{sp} values equal to or lower than 4. (2) Additionally, proteins with less than three peptide hits were manually evaluated to confirm the identification based on the following criteria. First, all the MS/MS spectra must have good quality of fragment ions having acceptable S/N; second, the fragment ions should consist of at least three consecutive fragmented series of b or y ions; third, some identified more intense fragment ions either correspond to +2 fragment ions or the dehydrated or deammoniated ions; last, if the same MS/MS spectrum from the same scan was matched to different peptides with different sequences, identification was assigned to only one protein, favouring the protein supported by the largest number of peptides which had the best quality of MS/MS spectra matched with high X_{corr} value.

10.2.10
Protein fractionation strategy for the analysis of serum proteins

10.2.10.1 2-D LC fractionation of serum proteins
Depleted serum protein samples (22 mg protein) obtained as described in Section 2.5 were fractionated by WAX in successive eight runs (Section 2.6, nine fractions obtained *per* run and matching fractions pooled) followed by nonporous RP-HPLC (Section 2.7; 130 fractions obtained in all). The final fractions were lyophilised and stored at −20°C pending further sample processing.

10.2.10.2 Digestion of the 2-D LC separated fractions
All fractions were dissolved in 100 µL solution containing 8 M urea and 100 mM ammonium bicarbonate. After vortexing and complete dissolution, the pH of each fraction was adjusted with dilute ammonia solution to about pH 8.5, then samples were reduced and alkylated by the same method as in Section 2.9.1. Ammonium bicarbonate solution (100 mM) was then added to reduce the concentration of urea to 1 M. Finally, trypsin was added and the mixtures incubated at 37°C for 4 h, then a

second dose of trypsin was added and the reaction mixtures incubated for a total of 24 h. After digestion, samples were lyophilised to dryness and stored at $-20°C$ before analysis.

10.2.10.3 1-D microRP-HPLC-ESI-MS/MS identification of digested serum proteins

1-D separation was performed on a ProteomeX system. The system was configured in high-throughput mode. Two RP C18 trap columns (100 µm id × 5 mm) were connected with the 10-port column-switching valve. While one trap column was being loaded with sample by one pump, the other could be eluted to the C18 RP separation column (Vydac; 0.15 mm id, 15 cm) by the other pump. Tryptic peptides were resolved in 0.1% formic acid solution with 5% ACN and directly analysed by the system with different sample volumes loaded according to the intensity of the UV absorbance of each fraction. The column was eluted with a 60 min gradient from 2 to 55% ACN at a flow rate of 1.6 µL/min followed by another gradient from 55 to 90% ACN in 20 min. Elution was continued for 10 min at 90% ACN. The column was then reequilibrated with 2% ACN, 0.1% formic acid before loading another sample. The MS parameters were the same as in Section 2.9.2. The survey scan ms range was set at 400–2000. MS/MS spectra database searching was based on the same criteria as in 2.9.3.

10.2.11
Offline shotgun strategy for the analysis of serum proteins

10.2.11.1 Offline SCX for first-dimension chromatographic separation of peptides

The concentrated sample (described in Section 2.5; containing about 1.6 mg/mL proteins, equivalent to 500 µL crude serum) was digested with trypsin as *per* Section 2.9.1. After digestion, the resulting peptide mixture was concentrated by SpeedVac and desalted by RP-HPLC. The desalted solution was concentrated by SpeedVac to a lesser volume and stored at $-20°C$ prior to analysis.

Offline SCX separation of peptides from a desalted sample was performed on the LC system configured as in Section 2.7. An analytical Hypersil SCX column (Thermo-Keystone; 4.6 mm id × 25 cm) was used for the first-dimension separation. The following conditions were used: Solution A was 5 mM NH_4Cl solution (adjusted to pH 3.0 with formic acid) containing 25% ACN and solution B was 800 mM NH_4Cl solution (adjusted to pH 3.0 with formic acid) containing 25% ACN. A gradient was run from 0 to 60% B in 60 min, maintained for 5 min and ramped to 100% B in 20 min, then held for another 30 min. The column was then eluted with a decreasing gradient of solvent B back to 0% in 1 min, and reequilibrated for another 30 min. The flow rate was 0.75 mL/min and the detection wavelength was 214 nm. The sample collection method and the subsequent preparation method was the same as described in Section 2.7. All effluent fractions were combined into 46 fractions after comparison of chromatograms and stored at $-20°C$ prior to analysis.

10.2.11.2 1-D capillary RP-HPLC/microESI-IT-MS/MS analysis for the SCX-separated peptide fractions

The experiment was performed on a Finnigan ProteomeX system. The system was configured in a high-throughput mode as in Section 2.10.3. A microcapillary C18 RP column (Vydac; Cat. No. 238EVS. 1525, 0.15 mm id, 25 cm) was used for the separation of peptide fractions. The flow rate after splitting was kept between 1.6 and 1.8 µL/min. Peptide fractions were resolved in 0.1% formic acid solution with 2% ACN and directly analysed by the system with different sample loading volume according to the intensity of the UV absorbance of each fraction. The column was eluted over a 110 min gradient from 5 to 50% ACN followed by another gradient from 50 to 90% ACN in 15 min. After maintaining the elution for 20 min at 90% ACN, the column was reequilibrated by 2% ACN, 0.1% formic acid solution for 30 min before the next sample loading. The MS parameters were the same as in Section 2.9.2. The survey scan ms range was set at 400–2000. The MS/MS spectra database searching criteria were the same as in Section 2.9.3.

10.2.12
Optimised nanoRP-HPLC-nanoESI IT-MS/MS for the reanalysis of offline SCX-separated peptides (offline-nanospray strategy)

The Finnigan ProteomeX system was configured similar to that in Section 2.10.3. The orthogonal microspray source was substituted for the Finnigan nanospray ion source. A microcapillary C18 RP column (Vydac; Cat. No. 238EVS. 1515, 0.15 mm id, 15 cm), coupled in tandem with a PicoFrit™ column (BioBasic® C18, 5 µm, 75 µm id × 10 cm, with integrated 15 µm id spray tip, purchased from New Objective, Woburn, MA, USA) was used for the separation of peptides. Elution conditions were the same as specified in Section 2.11.2 but the flow rate through the nanocolumn was reduced to about 400 nL/min. The MS parameters were almost the same as in Section 2.9.2. The spray voltage was set at 1.8 kV with no sheath gas applied. The temperature of the ion transfer capillary was set at 175°C. The survey scan ms range was set at 400–2000. The MS/MS spectra database searching criteria was the same as in Section 2.9.3.

10.3
Integrated analysis of the whole data sets

10.3.1
Protein grouping analysis

Occasionally the positively identified peptide sequence can be assigned to a group of proteins rather than a single protein, and it may be impossible to determine which one in the group is present in the actual biological sample without any additional evidence. So, a model named as group model is used for the data processing. However, in more cases, the identified peptide would match only one protein in the database, which is

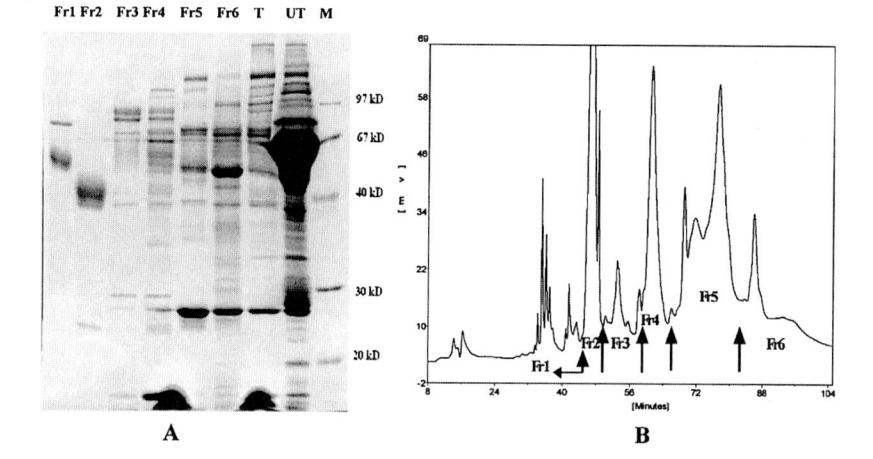

Fig. 3 (A) SDS-PAGE gels of intact serum protein fractions (Fr1-Fr6) obtained by anion-exchange chromatography combined with affinity depletion of high-abundant proteins by MARS column as compared to unfractionated sample (T, depleted sample) and crude sera (UT). Proteins were resolved on 12% Tricine gels and stained by CBBR. (B) Chromatogram of the separation of depleted serum sample by WAX, arrows indicate the starting and ending points of the six fractions collected. Fractions between arrows were sequentially combined as Fr1-Fr6.

also considered as a group in our model for the convenience of further integration. For the protein identification results by the strategies except for 2-DE, proteins with shared peptides are collapsed into a group and reported as a single identification, with the highest-scoring protein entry as the anchor. The protein groups can also be obtained for the PMF data (from 2-DE strategy), ensuring all the grouping proteins containing all of those identified peptides that were clustered by MASCOT to deduce a protein. Finally, the group lists were obtained for all the identified proteins. Furthermore, set-operation was carried out for the data integration analysis, which included two steps. First, all the groups that shared one or more proteins in common with others were clustered. Second, the intersection was deduced from the clustering of groups. If null set was obtained by the intersection operation, the clusters were split into the least subclusters covering all the groups, confirming each intersection contained at least one protein. Finally, all the intersections are taken as the unique identified groups. Similarly, in order to compare our data sets with the high-confident data list ($N = 3020$, high-confident protein identification result of HUPO PPP), we reanalysed the data sets against the IPI2.21 database and got the unique identified protein groups for each strategies.

10.3.2
Sequence clustering

In order to compare our data set with the nonredundant serum proteins list reported by Anderson et al. [4], we applied the same sequence clustering strategy [4] to find out which proteins in our data set were also concluded in the reported list.

10.4
Results and discussion

10.4.1
Depletion of the highly abundant serum proteins

MARS was an easy-to-use method for simultaneously removing albumin, IgG, IgA, haptoglobin, transferrin and antitrypsin in serum with relatively high specificity as shown in Fig. 3A. There was no obvious loss of nontargeted proteins in the depleted serum sample (labelled with "T") and the six fractions (Fr1-Fr6) separated by WAX.

10.4.2
The 2-DE strategy for the analysis of serum proteins

As shown in Fig. 3A, the sample complexity of the six fractions obtained from WAX separation was significantly reduced. The fractionated serum proteins by WAX resulted in markedly different profiles viewed by SDS-PAGE and CBBR-staining (Fig. 3A). All these fractions yielded different 2-DE separation patterns as illustrated in Fig. 4. About 1128 spots were visualised and the masking effect of highly abundant proteins was obviously reduced. On the other hand, some medium-abundance proteins such as glycoproteins and apolipoproteins were significantly increased in the concentration with a resultant masking effect on the detection of proteins of low abundance.

In all, 1128 separated protein spots were excised entirely, digested and spotted on targets, of which 318 spots yielded positive protein identifications with high-confidence MASCOT scores. These turned out to represent 78 unique proteins, reflecting distribution of the same protein across different fractions and the existence of isoforms in some cases. It should be noted that with effective depletion of highly abundant proteins and extensive fractionation of intact serum proteins, more isoforms of serum proteins were evident in 2-DE separation. As shown in Fig. 5, at least three or more pI-altered isoforms of kinnigen precursor, α-2-HS-glycoprotein, leucine-rich α-2-glycoprotein precursor, histidine-rich glycoprotein etc. were clearly separated and identified. The number of identified isoforms of some serum proteins is listed in Supplementary Tab. 1. If known, the detailed glycan analysis of serum glycoproteins of patients could reveal a specific defective glycan processing step and provide useful information about the pathogenesis of disease [22]. The strategy used here demonstrates that it is possible to perform detailed structure analysis and a comparison of relative abundance of different isoforms of the same serum protein. Relative abundance information on some serum proteins could be acquired based on normalised spot abundance value (data not shown).

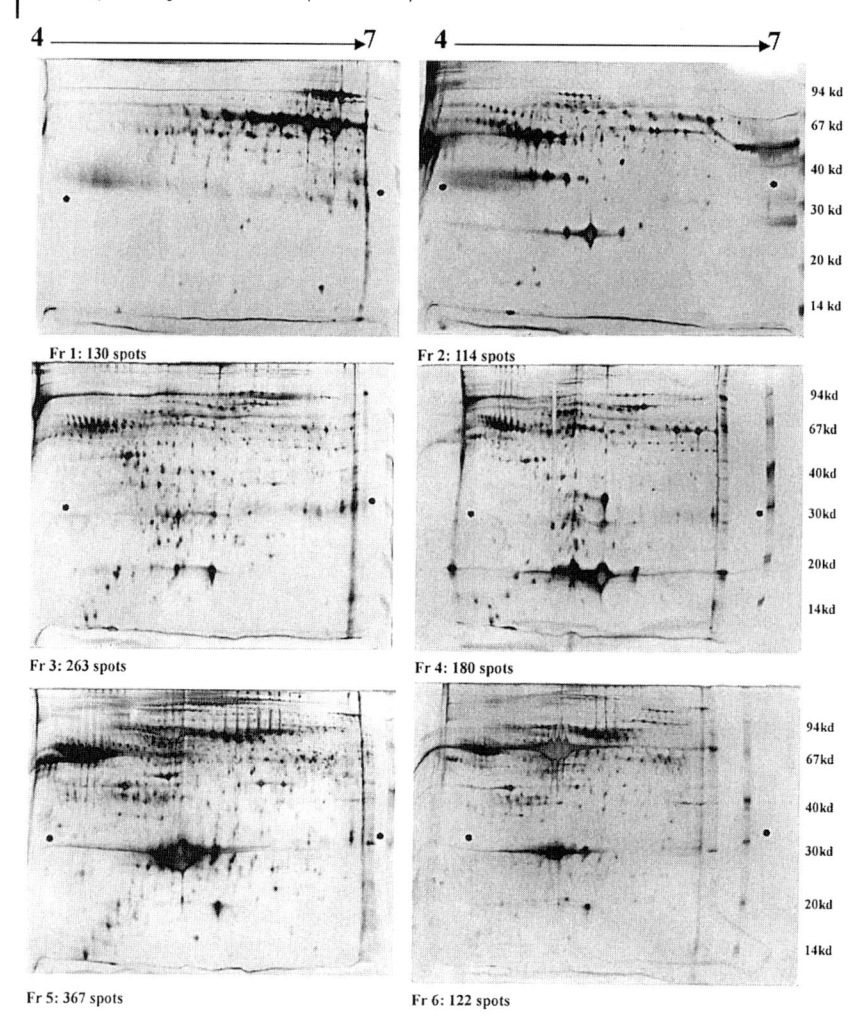

Fig. 4 Narrow pH IPG (4–7) based 2-DE separation pattern of six fractions of human serum proteins after removal of the six highest-abundance proteins by MARS and separation by WAX. 2-DE gel run conditions are described in the text. Gels were silver-stained. Fr1-Fr6 corresponds to the order of fractions eluted from the WAX.

10.4.3
2-D HPLC fractionation for the analysis of serum proteins

Chromatographic protein prefractionation has been applied for comprehensively characterising complex sample proteomes [14]. A 2-D HPLC fractionation strategy was assessed here. Nine fractions were obtained and evaluated by SDS-PAGE (Fig. 6). The fractions obtained by WAX separation were further separated by nonporous RP-HPLC as a second dimension. As shown (in Supplementary

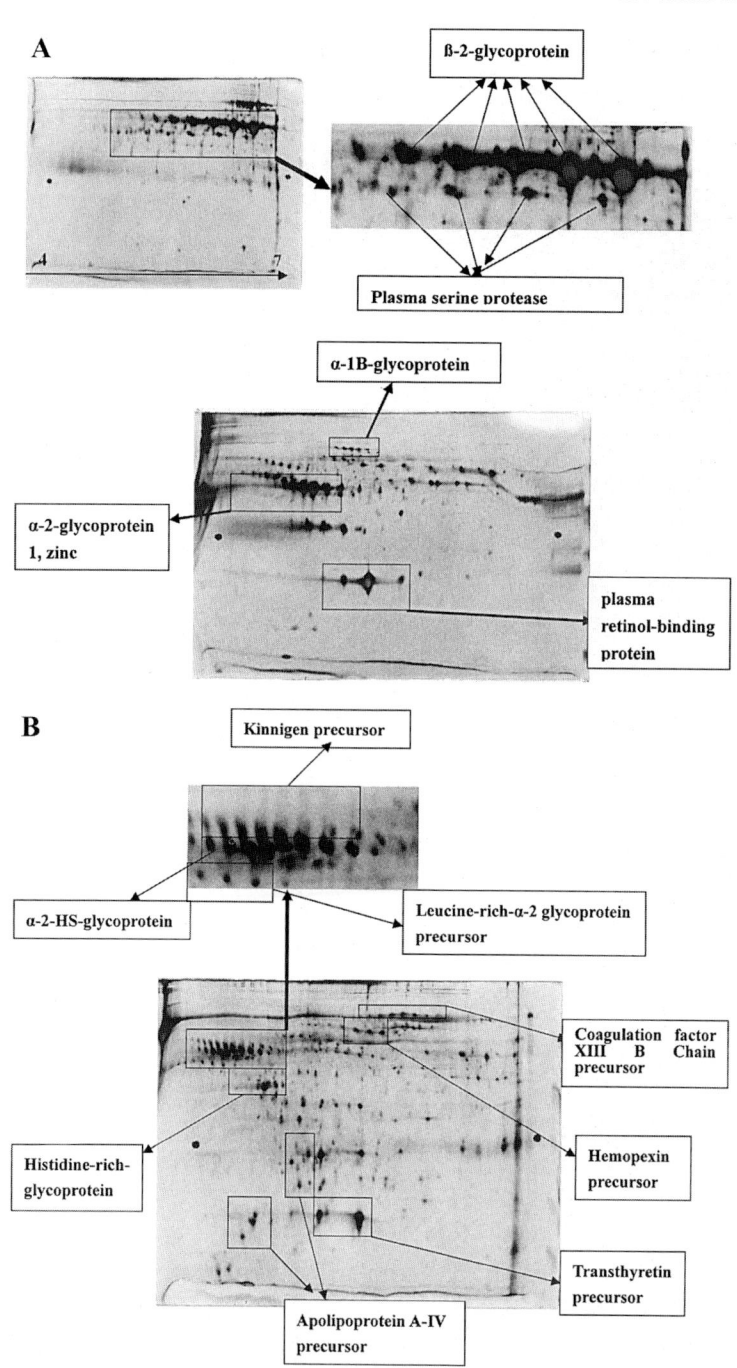

Fig. 5 (A and B) Isoforms of some serum proteins with different p*I* values separated by 2-DE.

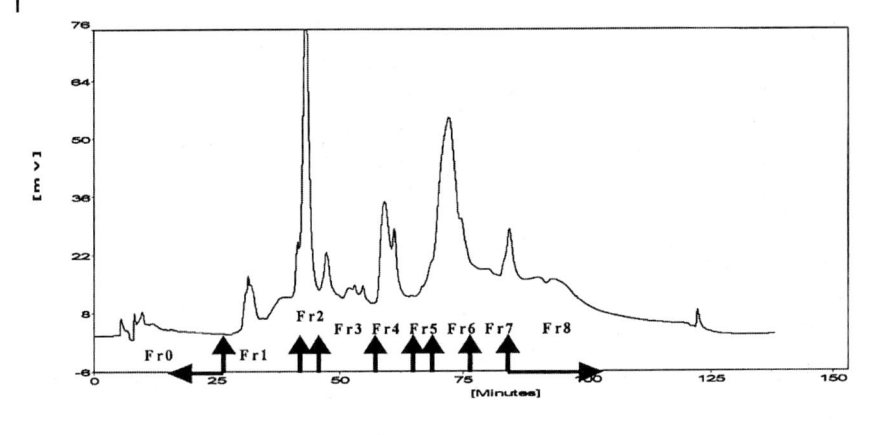

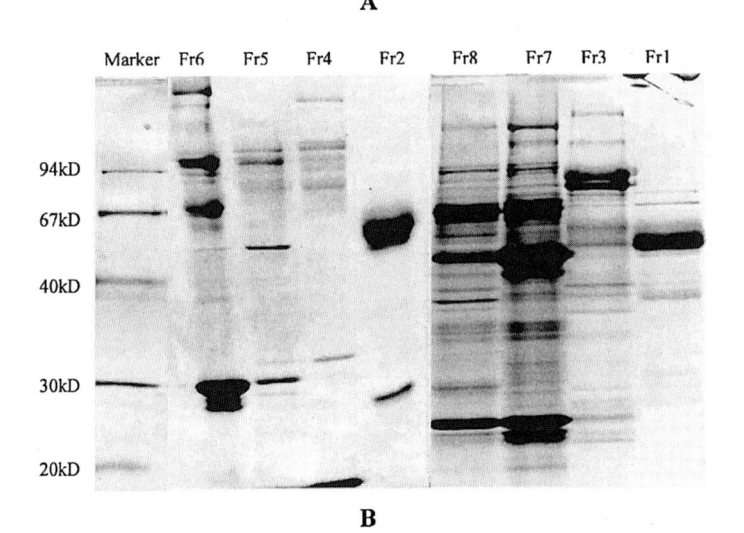

Fig. 6 (A) Chromatogram of the separation of depleted serum sample proteins by WAX. Arrows indicate the starting and ending points of the nine fractions obtained. Fractions between arrows were sequentially combined as Fr0-Fr8. (B) SDS-PAGE gels of WAX separated fractions of intact serum protein (Fr0-Fr8). Proteins were resolved on 12% tricine gels and stained by CBBR.

Tab. 2), different fractions from WAX afforded RP-HPLC chromatograms of different appearance except for Fr7 and Fr8. The results demonstrate the good selectivity of WAX for proteins. The obvious carry-over effect between Fr7 and Fr8 indicates that proteins in these two fractions adsorb to the anion-exchange sites strongly and are coeluted by rapidly ramping to high salt concentration without enough separation. Nevertheless, the 2-D HPLC approach effectively confined the serum proteins of medium abundance, such as β-2-glycoprotein I precursor, α-2-glycoprotein 1 zinc, plasma protease C1 inhibitor precursor and apolipoprotein A-I precursor into several highly enriched fractions and separated them from the less-abundant protein fractions. It should be noted that apolipoprotein A-I precursor was identified in two different nonadjacent fractions, Fr0 and Fr4 (shown in the Supplementary Tab. 2; the No. 3 peak in Fr0 with t_R 37.75 min and the No. 4 peak in Fr4 with t_R 45.82 min). We infer that this reflects variants of apolipoprotein A-I precursor with different pIs and hydrophobic properties. The results also demonstrate the potential of multidimensional chromatography to facilitate the identification of protein variants by MS [23]. Neither mode alone would have sufficed; note that the most abundant proteins in RP-HPLC fractions 4, 5 and 8 would have coeluted in RP-HPLC if not separated first by WAX. We could also get information this way about proteins of medium abundance and their relative abundance in serum. A total of 179 proteins were identified with confidence with this strategy, which was more than from shotgun or 2-DE strategy. The identification results were improved by the extensive fractionation of intact serum proteins. The results show that substantial removal of serum proteins of high and medium abundance is beneficial for the relatively comprehensive characterisation of the serum proteome.

10.4.4
Shotgun strategy for the analysis of serum proteins with online SCX

Recently, the multidimensional shotgun (bottom-up) approach has been shown to be able to identify hundreds of proteins from a single complex protein sample [12, 24, 25] and to be a valid complementary alternative to 2-DE analysis. In the present study, we applied this technique, coupled with MS, for the analysis of the human serum proteome and compared it with the other two strategies. With this approach, about 132 proteins were definitely identified with a total of 428 unique peptides *via* a SEQUEST search of the human fasta database (IPI v2.27). When manually evaluating the protein identifications, we applied more stringent criteria. As shown in Fig. 7, when the MS/MS spectrum obtained from the same MS/MS scan was matched with two different peptides with entirely different sequences, from different proteins, we selected the proteins supported by the peptide that had the higher X_{corr} values and better fragment ions matched with the result. This comprehensive assessment of the data makes the protein IDs of higher confidence.

10.4.5
Shotgun strategy for the analysis of serum proteins with offline SCX

In the shotgun approach to proteomics, data-dependent acquisition is always used to automatically acquire MS/MS spectra of peptides eluting into the mass spectrometer. Occasionally, when analysing more complicated mixtures, data-dependent acquisition incompletely samples among the peptide ions present rather than acquiring tandem mass spectra for all ions available [26]. It then becomes a prerequisite to optimise the separation of the peptides in order to decrease the number in any one fraction being analysed. This allows the mass spectrometer enough time

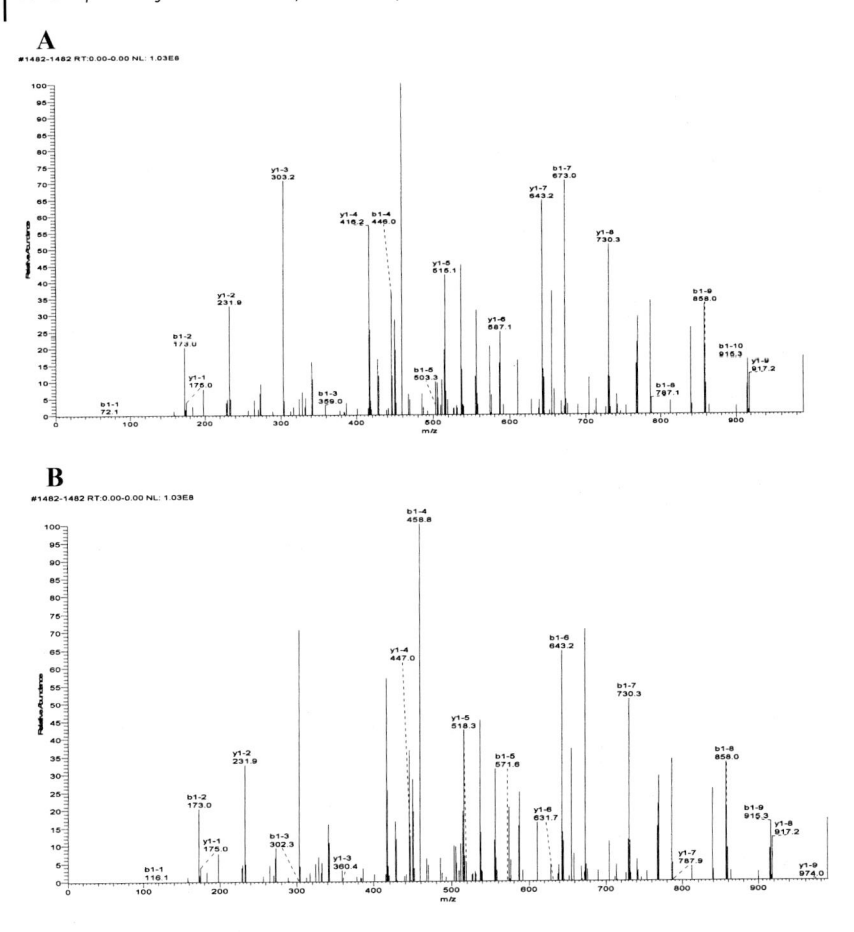

Fig. 7 Same MS/MS spectrum matched to different peptides with different sequences, which resulted in different protein identifications. (A) Spectrum matched to the peptide ATWSGAVLAGR with 3 717 (X_{corr}) and 0.358 (ΔC_n); the protein identified as alpha 1B-glycoprotein (IPI00216722.1). (B) Same spectrum matched to the peptide sequence DGERLASQGR with 2 327 (X_{corr}) and 0.376 (ΔC_n) of hypothetical protein KIAA0657 (IPI00005216.3).

to acquire the qualified MS/MS spectra of all eluting peptides. In order to implement this principle here, we used offline SCX for the first dimension of peptide separation. The advantages of the offline approach are (1) the ability to elute with a linear salt gradient instead of steps, making it less likely that a particular peptide will be split between adjacent fractions; (2) to use ACN in the mobile phases, thereby sharpening the peaks and improving their separation (data not shown) and (3) ease in collection of more numerous fractions, since larger columns with faster flow rates can be used offline than in true online. A mixture of tryptic peptides could be successfully and reproducibly separated this way, as shown in Fig. 8. The effluents collected from

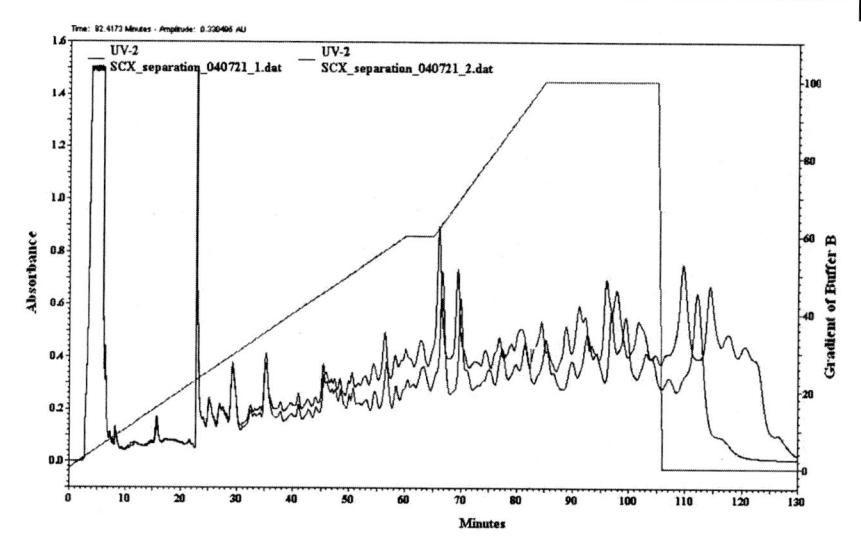

Fig. 8 Offline SCX separation chromatogram of tryptic peptides mixture of serum proteins. Black line indicates the first run of the separation and the brown line indicates the second run of the separation. All the effluent was combined into 46 fractions according to the peak resolution and intensity.

sequential runs were combined into 46 fractions based on peak resolution and intensity, yielding more samples for microcapillary RP-HPLC-microESI-MS/MS analysis. In this manner, the peptide fractions entering the MS instrument were reduced dramatically in complexity and the overall peak capacity of a 2-D LC system could be increased significantly. Ultimately, 516 unique peptides identified led to 224 unique proteins, 73 of which were from two or more peptides. This result was significantly better than that from the online SCX shotgun approach, from which only 131 unique proteins were identified, of which only 53 were through two or more peptides.

10.4.6
Offline SCX shotgun-nanospray strategy for the analysis of serum proteins

Recent research has dramatically demonstrated that ultrahigh sensitivity can be obtained by high-resolution capillary LC separations that provide very low flow rates to an ESI interface [27]. Specifically, ionisation efficiencies can be greatly improved by using longer and narrower capillary columns for the separation of peptides and reducing the electrospray flow rate to the level of low nanolitres *per* minute. Shen *et al.* [15] have successfully applied ultrahigh efficient SCX-RPLC-MS/MS to the characterisation of the human plasma proteome over a wide dynamic range. Accordingly, in order to optimise the separation of peptides and improve sensitivity in ESI, we reconfigured our ProteomeX system as shown in Fig. 9, using a narrower column with an integrated spray tip for both the separation of peptides and as a nanospray tip for the ESI interface. With the improvement of

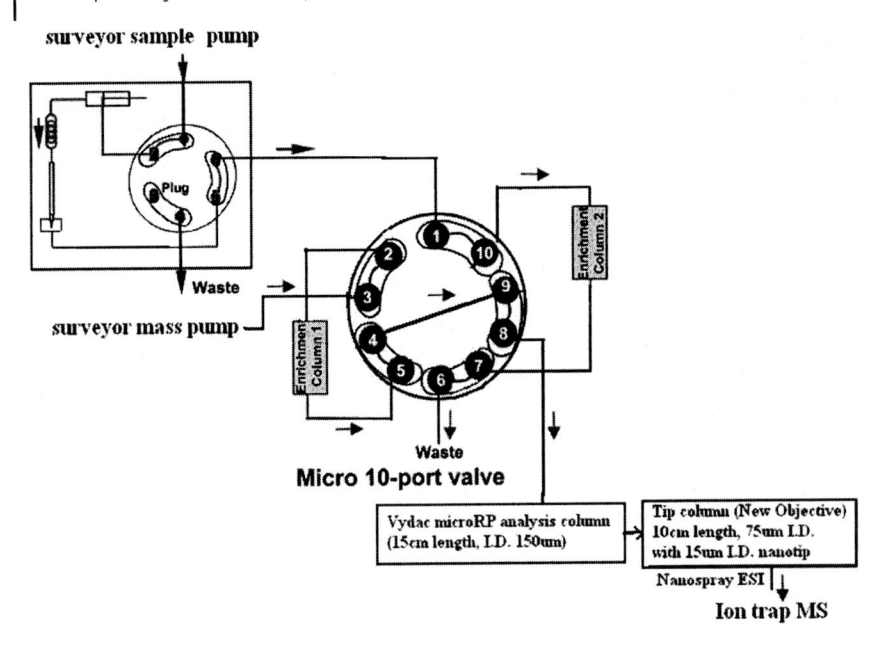

Fig. 9 Flow diagram of the high-throughput configuration of the ProteomeX system. Two enrichment microRP trap columns were used for sequential loading and desalting of peptide samples. A Vydac microRP column coupled in tandem with a PicoFrit column was used for the separation of peptides. ESI interface was the Finnigan nanospray source.

the system configuration, the separation of the peptides and sensitivity of the ESI IT MS/MS was remarkably improved, as shown in Fig. 10. The result also confirmed that the sensitivity of LC-MS coupling via ESI is inversely related to the LC flow rate and thus to the inner diameter of the LC column [28]. The peak capacity was improved by use of the longer microcapillary RP column with integrated spray tip, which has very little postcolumn dead volume. This optimised system resulted in the identification of 778 unique peptides corresponding to 330 unique proteins. There were 106 proteins identified by two or more peptides. The result was obviously better than that of the offline SCX shotgun strategy. A total of 398 proteins were identified in the cumulative data set of both offline shotgun and offline shotgun/nanospray strategies. Of these, 157 proteins (39%) were found by both approaches while 241 (61%) were found by only one. We conclude that optimising both the peptide separation and the sensitivity of the mass spectrometer significantly expands the number of proteins identified, making this combination indispensable for shotgun proteomics. We did note that even when the same samples and the similar analytical strategies were used, the reproducibility of different LC-MS/MS runs was still not very satisfactory. This could be attributed in part to the random sampling of all peptide ions available by the mass spectrometer [26].

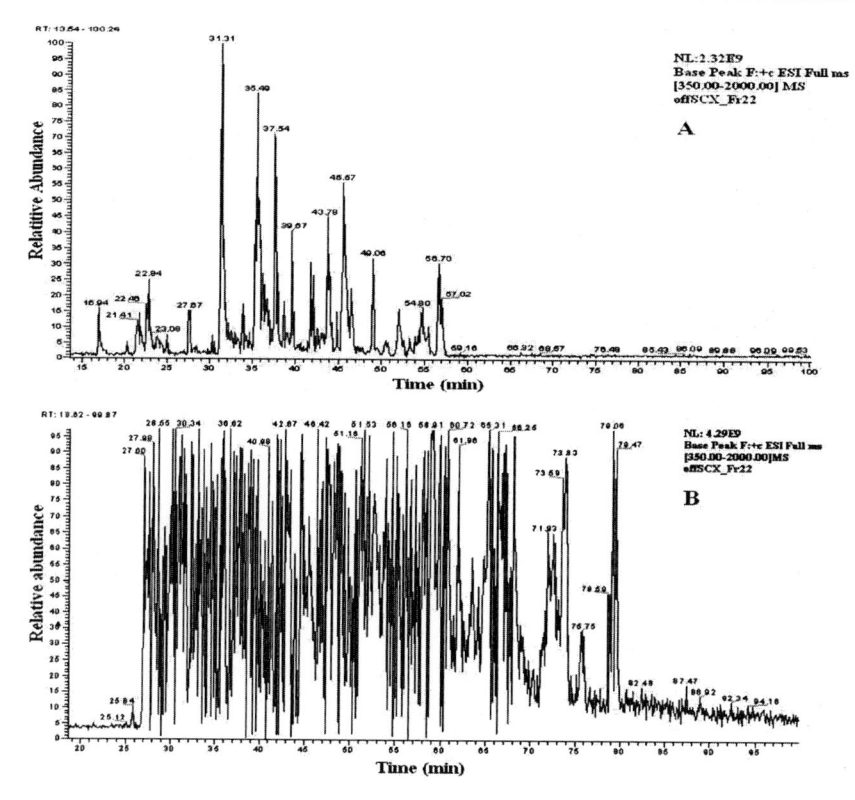

Fig. 10. Comparative base peak chromatograms of offline SCX-separated Fraction 22 analysed by (A) conventional ESI setup in offline SCX shotgun strategy; (B) nanospray ESI setup in offline SCX shotgun-nanospray strategy.

10.4.7
Comparison of the five strategies for the analysis of the human serum proteome

It is now well accepted that no single proteomics technique will accomplish the identification of all the proteins in a really complex sample. MALDI and ESI, though competing methods, may yield results that reflect only the most abundant sample components. Fractionation of proteins and peptides prior to ionisation will remain a prerequisite for proteomic research, especially for the serum proteome. Here, in an effort to determine the best way to characterise the serum proteome, we have compared five alternative strategies for characterising the proteome in the same serum sample with the same preparation method. The number of unique proteins identified by all five approaches was just 32 out of the total of 560. The overlap was greater between any two methods. For example, the overlap between the offline SCX shotgun approach (with and without the nanospray modification) and online SCX shotgun is 89 proteins; between offline SCX and 2-DE, 45 proteins; between offline SCX and 2-D HPLC of intact proteins, 81 proteins (perhaps reflecting their similar nongel-based sample preparation process). It might be supposed that the three data sets of offline SCX shotgun, online SCX shotgun, and intact protein 2-D HPLC, obtained *via* liquid-based separations, the same kind of mass spectrometer and the same database search algorithm would have a higher percentage of overlap of protein identifications, but the number of overlapping proteins identified by all three approaches was only 52, as shown in Fig. 11. The result could be attributed to variations in the sample preparation process, differences in peptide separation efficiency and random sampling of the peptide ions by the mass spectrometer. Even so, the degree of overlap between the three strategies was greater than that of a comparable comparison already reported [4].

Similarly, the distribution of proteins identified by the shotgun, protein 2-D HPLC and 2-DE approaches are summarised in Fig. 12. Only 37 proteins were common to all three. Presumably, the three strategies identified different proteins mainly due to variations in sample preparation, the different separations and the kind of MS identification system applied [29]. For all five strategies, protein identifications were generally in agreement at the level of high- and medium-abundance proteins but were complementary for lower-abundant proteins. In practice, all the techniques have their advantages and limitations, and all can provide complementary information and identify unique proteins. The 2-DE strategy provides much more information on isoforms of serum proteins and relatively quantitative data for proteins in general. 2-D HPLC protein fractionation facilitated the identification of lower-abundant proteins to some extent (data not shown). The offline SCX shotgun strategy afforded better chromatographic separation of peptides than online SCX shotgun, resulting in more unique peptides and proteins identified, especially with the optimised sensitivity of the nanospray setup. It has been well-documented that the dynamic range and coverage obtained from MS/MS proteome analysis is a function of both the quality of the separation(s) applied and the MS apparatus used [15]. Thus, greater sensitivity should broaden the dynamic range of identifications in serum proteomic studies. This is most effectively accomplished by reducing the LC column inner diameter and implementing nanoESI. Here, the identification of less-abundant proteins often depended on MS/MS data of good quality from just one or two peptides, which were randomly acquired during the LC-MS/MS analysis [26]. Our results indicate that by increasing the number of well-separated fractions collected in offline SCX and optimisation of RP-HPLC-ESI-MS/MS, more serum proteins of low abundance can be identified (Fig. 13), such as coagulation factor IX precursor (μg/mL), L-selectin precursor and hepatocyte growth factor-like protein precursor, identified through high-confidence MS/MS spectra of their peptides.

As far as the cost (time, labour and consumables) and convenience were concerned, the 2-D HPLC protein fractionation strategy required much more time and labour for manual sample handling and lent itself less to automation than did the other strategies. The longer the time taken for sample fractionation, the greater the chance that some low-abundance protein might be lost in the process. On the other hand, HPLC fractionation at the intact protein level can conveniently be performed with a large quantity of serum. This would increase the absolute amount of proteins of low abundance in the resulting fractions, increasing the chances of their

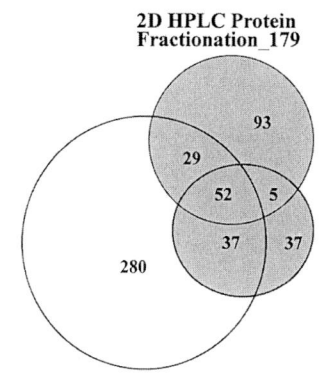

Fig. 11. Venn diagram showing overlapping protein identifications among the three liquid-based strategies (2-D HPLC protein fractionation, online SCX shotgun and offline SCX shotgun) and the number of individual proteins identified with each strategy.

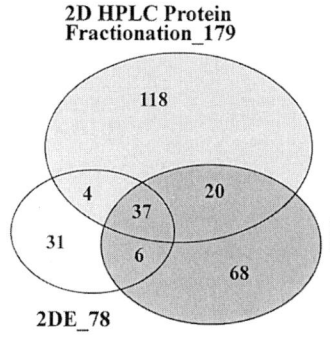

Fig. 12. Venn diagram of overlapping protein identifications among the three strategies (2-D HPLC protein fractionation, online shotgun and 2-DE) and the number of individual proteins identified with each strategy.

being identified. Unfortunately, the same consideration applies to contaminants; epidermal keratins were identified in many fractions in high abundance and through multiple peptide hits (data not shown) which inevitably interfered with the identification of low-abundance proteins, but these contaminants were seldom identified by the shotgun approach. As for the 2-DE strategy, although the six highest-abundance proteins had been removed, when the quantity of loading of the resulting sample was increased, some medium-abundance proteins such as apolipoprotein A-I, hemopexin, α-2-glycoprotein 1, β-2-glycoprotein I, vitamin D-binding protein and so on proved to be new proteins of high abundance and had obvious masking effect on the low-abundance ones. Furthermore, low-abundance proteins were difficult to identify due to the low extraction efficiency of peptides from gels. In the shotgun strategies, the disadvantages of the offline SCX approach were the more time-consuming sample handling and the lack of automation. In addition to the above-mentioned points, all three strategies had similar distribution characteristics of mass range, pI and hydrophobicity of the identified proteins (see Fig. 14), indicating that serum proteins in general had properties appropriate for analysis by all three strategies.

It does appear that 2-D HPLC of intact proteins works best for proteins representing extremes in these properties. 2-DE strategy is difficult in profiling the proteins with low molecular weight, or proteins with extremely acidic, basic or high

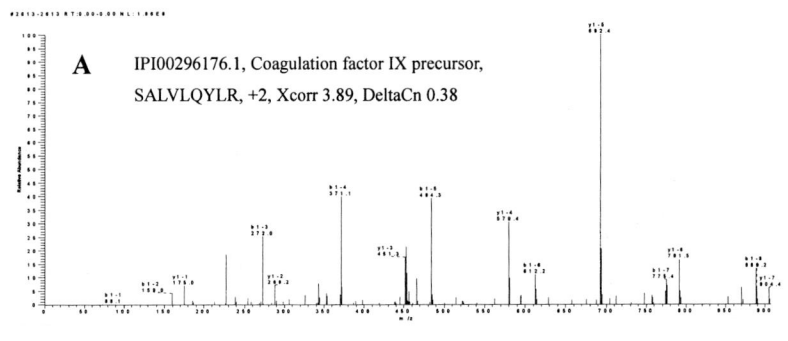

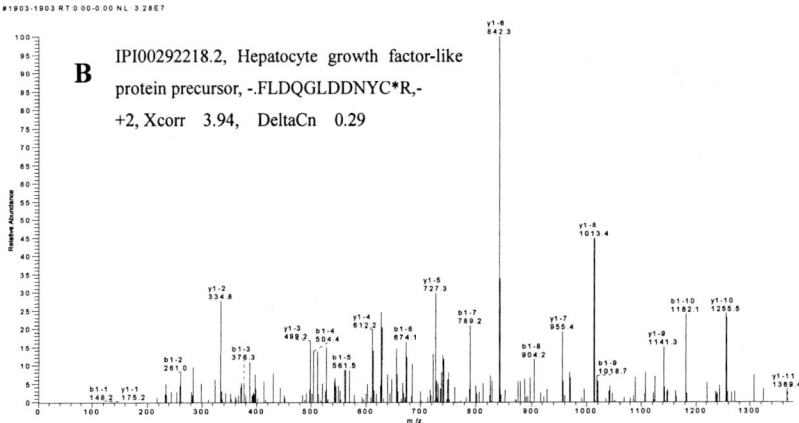

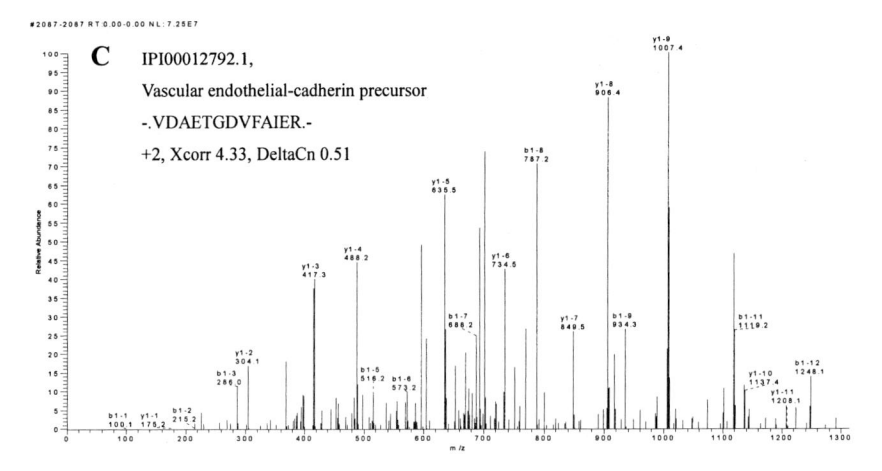

Fig. 13. Tandem spectra of peptides from low-abundance proteins (A) IPI00296176.1, coagulation factor IX precursor, SALVLQYLR, +2, X_{corr} 3.89, ΔC_n 0.38; (B) IPI00292218.2, hepatocyte growth factor-like protein precursor; –.FLDQGLDDNYC*R.-, +2, X_{corr} 3.94, ΔC_n 0.29; (C) IPI00012792.1, vascular endothelial-cadherin precursor, –.VDAETGDVFAIER.-, +2, X_{corr} 4.33, ΔC_n 0.51.

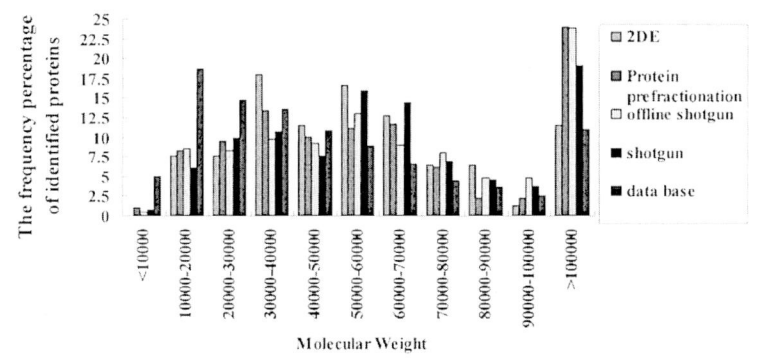

The M_r distribution of identified proteins in four approaches

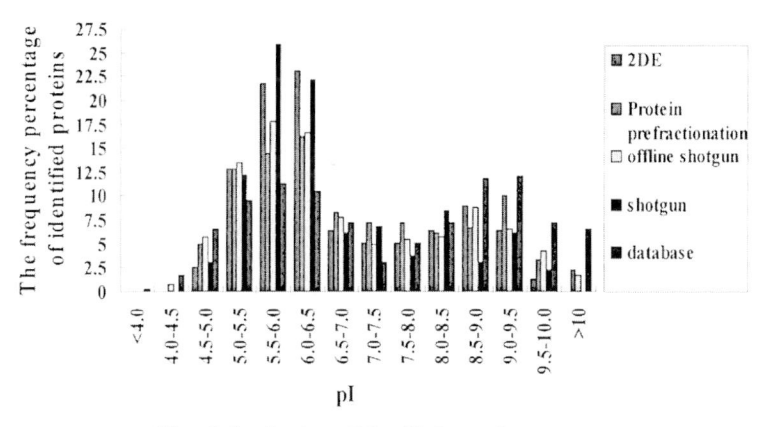

The pI distribution of identified proteins in four approaches

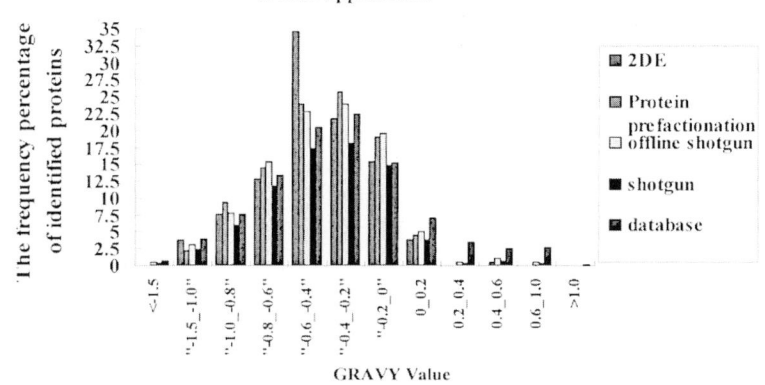

The hydrophobicity distribution of identified proteins in four approaches

Fig. 14. Comparison of the p*I*, M_r and GRAVY value distribution of the proteins identified by 2-D HPLC protein fractionation, shotgun (offline + online) and 2-DE, with the whole data from the database as control.

hydrophobic characters. At last, when the whole data set was compared with the high-confident proteins list of HUPO PPP ($N = 3020$), about 257 unique proteins were overlapped, 139 proteins of which were identified with two or more peptides; While compared with the reported nonredundant list [4], in total 169 unique proteins were overlapped, 117 of which were identified with two or more peptides.

10.5
Concluding remarks

Comprehensive characterisation of serum proteins is a very challenging task for any available technological strategy. All five strategies applied in this study yielded complementary protein identifications. The degree of overlap in identifications among the five strategies was higher than that of the four data sets previously reported [4]. HPLC protein fractionation was useful not only for the gel-based approach but also for the 2-D chromatography-based strategy for characterising the serum proteome. Altogether, the combined five approaches resulted in a reasonably comprehensive serum proteome with a total of 560 proteins identified. HPLC fractionation of serum proteins or peptides was beneficial for the enrichment of the relatively lower-abundance proteins. In addition, careful preparation of the samples, optimising the sensitivity of the MS system by the nanospray setup and improving the separation efficiency of peptides in the offline 2-D LC mode were the key steps in characterising the human serum proteome.

We especially acknowledge the research fund PPP-SP/04–12, the small grants fund of HUPO PPP. This work was supported in part by grants from the National High Technologies R&D Program of China (2001CB510201, 2002BA711A11), the National Natural Science Foundation of China (30370713, 30171049, 30321003, 20275044 and 20275046) and the Science and Technology Foundation of Beijing Municipality (H030230280190). We thank Agilent Technologies for providing the Multiple Affinity Removal System® column for the sample preparation. We are also indebted to Dr. Andrew Alpert (PolyLC) for his generous gift of the PolyWAX LP (weak anion-exchange) column used for the intact serum proteins fractionation. We gratefully acknowledge Dr. Andrew Alpert and Professor Gilbert Omenn for critical reading of the manuscript. Multiple Affinity Removal System® is a trademark of Agilent Technologies. BioBasic® is a trademark of Thermo-Keystone.

10.6
References

[1] Hanash, S., Celis, J. E., *Mol. Cell. Proteomics* 2002, *1*, 413–414.

[2] Omenn, G. S., *Proteomics* 2004, *4*, 1235–1240.

[3] Anderson, N. L., Anderson, N. G., *Mol. Cell. Proteomics* 2002, *1*, 845–867.

[4] Anderson, N. L., Polanski, M., Pieper, R., GatlinT. *et al.*, *Mol. Cell. Proteomics* 2004, *3*, 311–326.

[5] Steel, L. F., Trotter, M. G., Nakajima, P. B., Mattu, T. S. *et al.*, *Mol. Cell. Proteomics* 2003, *2*, 262–270.

[6] Rothemund, D. L., Locke, V. L., Liew, A., Thomas, T. M. et al., Proteomics 2003, 3, 279–287.

[7] Wang, Y. Y., Cheng, P., Chan, D. W., Proteomics 2003, 3, 243–248.

[8] Chertov, O., Biragyn, A., Kwak, L. W., Simpson, J. T. et al., Proteomics 2004, 4, 1195–1203.

[9] Tirumalai, R. S., Chan, K. C., Prieto, D. A., Issaq, H. J. et al., Mol. Cell. Proteomics 2003, 2, 1096–1103.

[10] Adkins, J. N., Varnum, S. M., Auberry, K. J., Moore, R. J. et al., Mol. Cell. Proteomics 2002, 1, 947–955.

[11] Pieper, R., Gatlin, C. L., Makusky, A. J., Russo, P. S. et al., Proteomics 2003, 3, 1345–1364.

[12] Aebersold, R., Mann, M., Nature 2003, 422, 198–207.

[13] Vollmer, M., Horth, P., Nagele, E., Anal. Chem. 2004, 76, 5180–5185.

[14] Marshall, J., Jankowski, A., Furesz, S., Kireeva, I. et al., J. Proteome Res. 2004, 3, 364–382.

[15] Shen, Y., Jacobs, J. M., Camp, D. G., Fang, R. et al., Anal. Chem. 2004, 76, 1134–1144.

[16] Rabilloud, T., Proteomics 2002, 2, 3–10.

[17] Lescuyer, P., Hochstrasser, D. F., Sanchez, J. C., Electrophoresis 2004, 25, 1125–1135.

[18] Heukeshoven, J., Dernick, R., Electrophoresis 1988, 9, 28–32.

[19] Wu, S. L., Amato, H., Biringer, R., Choudhary, G. et al., J. Proteome Res. 2002, 1, 459–465.

[20] Eng, J. K., McCormack, A. L.,Yates, J. R., III, J. Am. Soc. Mass Spectrom. 1994, 5, 976–989.

[21] Washburn, M. P., Wolters, D., Yates, J. R., III, Nat. Biotechnol. 2001, 19, 242–247.

[22] Butler, M., Quelhas, D., Critchley, A. J., Carchon, H. et al., Glycobiology 2003, 13, 601–622.

[23] Zhang, Z., Smith, D. L., Smith, J. B., Proteomics 2001, 1, 1001–1009.

[24] McDonald, W. H., Yates, J. R., III, Dis. Markers 2002, 18, 99–105.

[25] Wolters, D. A., Washburn, M. P., Yates, J. R., III, Anal. Chem. 2001, 73, 5683–5690.

[26] Liu, H., Sadygov, R. G., Yates, J. R., III, Anal. Chem. 2004, 76, 4193–4201.

[27] Smith, R. D., Shen, Y., Tang, K., Acc. Chem. Res. 2004, 37, 269–278.

[28] Oosterkamp, A. J., Gelpi, E., Abian, J., J. Mass Spectrom. 1998, 33, 976–983.

[29] Stapels, M. D., Barofsky, D. F., Anal. Chem. 2004, 76, 5423–5430.

11
A proteomic study of the HUPO Plasma Proteome Project's pilot samples using an accurate mass and time tag strategy*

Joshua N. Adkins, Matthew E. Monroe, Kenneth J. Auberry, Yufeng Shen, Jon M. Jacobs, David G. Camp II, Frank Vitzthum, Karin D. Rodland, Richard, C. Zangar, Richard D. Smith and Joel G. Pounds

Characterization of the human blood plasma proteome is critical to the discovery of routinely useful clinical biomarkers. We used an accurate mass and time (AMT) tag strategy with high-resolution mass accuracy cLC-FT-ICR MS to perform a global proteomic analysis of pilot study samples as part of the HUPO Plasma Proteome Project. HUPO reference serum and citrated plasma samples from African Americans, Asian Americans, and Caucasian Americans were analyzed, in addition to a Pacific Northwest National Laboratory reference serum and plasma. The AMT tag strategy allowed us to leverage two previously published "shotgun" proteomics experiments to perform global analyses on these samples in triplicate in less than 4 days total analysis time. A total of 722 (22% with multiple peptide identifications) International Protein Index redundant proteins, or 377 protein families by ProteinProphet, were identified over the six individual HUPO serum and plasma samples. The samples yielded a similar number of identified redundant proteins in the plasma samples (average 446 ± 23) as found in the serum samples (average 440 ± 20). These proteins were identified by an average of 956 ± 35 unique peptides in plasma and 930 ± 11 unique peptides in serum. In addition to this high-throughput analysis, the AMT tag approach was used with a Z-score normalization to compare relative protein abundances. This analysis highlighted both known differences in serum and citrated plasma such as fibrinogens, and reproducible differences in peptide abundances from proteins such as soluble activin receptor-like kinase 7b and glycoprotein m6b. The AMT tag strategy not only improved our sample throughput but also provided a basis for estimated quantitation.

* Originally published in Proteomics 2005, 13, 3454–3466

Exploring the Human Plasma Proteome. Edited by Gilbert S. Omenn
Copyright © 2006 WILEY-VCH Verlag GmbH & Co. KGaA, Weinheim
ISBN: 3-527-31757-0

11.1
Introduction

The rapid expansion of proteomic analyses in recent years due to the development of powerful enabling technologies parallels the ongoing genomics expansion. Based on the experience gained from the large-scale genomics projects, there is widespread recognition that the proteomics field needs to invest early in significant multilaboratory efforts devoted to improving data quality [1], to making cross-laboratory and cross-platform data comparisons [2], and to developing data standards [3]. One such effort is the Plasma Proteomics Project (PPP), one of the first endeavors of the HUPO [4], which also includes liver and brain initiatives [5].

The Plasma Proteomics Project is timely, as blood plasma and serum are widely recognized as body fluids of great promise for human health diagnostics, *e.g.*, disease prognostics and clinical monitoring [2, 4, 6–18]. Two of the most compelling reasons for studying human plasma are (1) the universal availability of sufficient blood plasma and serum for method development and validation and (2) the long-standing use of plasma and serum as a source of clinically relevant information [6, 19]. The union between the venerable applications of clinical chemistry and powerful new technologies in the form of proteomics is driving a renaissance in human blood plasma and serum analysis.

To gain maximum utility and understanding from this renaissance, many issues need to be addressed. One of these issues is to encourage open and direct comparisons of methods and technologies, using sample(s) made commonly available to many laboratories [3–5] and facilitating the development of a central repository for results with unified data standards. With such a centralized and unified data system in place, a much greater impact of proteomic efforts will be realized, such as more effective mining of results, development of better data analysis tools, more confident protein identifications, and a deeper understanding of the relative strengths and weaknesses of various technologies. The development of a single comprehensive protein database for peptide/protein identifications will lead to better coverage and allow differences among plasma analyses to be dealt with directly [2]. Another advantage is that the results from these combined efforts will also lead to overall improvements in the confidence of identifications that are obtained by different techniques.

A recent approach for obtaining comprehensive high-throughput proteomics is the accurate mass and time (AMT) tag technology pioneered at Pacific Northwest National Laboratory (PNNL), which is based on high-resolution LC separations and high-mass accuracy measurement and is extensively described elsewhere [20–22]. With this approach, normalized LC elution profiles are determined for peptides identified by extensive traditional ion trap (IT) shotgun proteomics experiment(s). Note, the present work employed peptide identifications from two previous studies [15, 17]. The peptide identifications (and thus their known exact masses) and complementary RP capillary LC-elution times are stored in a database to serve as peptide markers (*i.e.*, mass and time tags) for future studies. These tags are then used with the relatively high-throughput, high-mass accuracy, and high resolution of cLC-FT-ICR MS to identify peptides based on matching elution and mass values [20–22], *i.e.*, establishing AMT tags. Ideally, once peptides have been identified with an IT-MS, routine MS/MS measurements are replaced with cLC-FT-ICR MS measurements. This approach provides improved quantitation as well as much higher throughput. The major limitation is that peptides must have been previously identified in an MS/MS spectrum [20, 21] or, alternatively, be subjected to additional experiments for identification [23].

Here, we report on the results obtained by using the AMT tag technology to analyze six Becton-Dickinson Diagnostics-prepared HUPO PPP sera and citrated plasmas [4] along with an in-house reference serum [15] and citrated plasma. Our approach involved using the high-throughput technology to assess the reproducibility and relative confidence of peptide/protein identifications, and make intersample comparisons. The resulting comparisons were clustered to calculate and illustrate correlations among samples and triplicate analyses. The clustered data were also used to demonstrate the possibility of using this technique to detect known differences between serum and citrated plasma and discover new differences. Where possible peptide/protein abundances defined by ion current values were compared to protein concentrations determined by a certified clinical analyzer, the Behring Nephelometer II (BN II), from Dade Behring (Haab *et al.* [54], Omenn *et al.* [55]).

The cLC-FT-ICR MS analysis of eight samples in triplicate was performed in only 4 days total instrument analysis time. These results demonstrate that the AMT tag approach can be used to compare samples at levels comparable to a more typical multidimensional separation using an IT mass spectrometer [55]. Furthermore, these results demonstrate the basis for a more quantitative and higher-throughput global proteomics approach.

11.2
Materials and methods

All reagents were purchased from Sigma, unless noted otherwise. All samples were approved by our IRB and conform to federal and HIPAA regulations.

11.2.1
Human blood serum and plasma

Human sera and plasmas prepared specifically for HUPO PPP by BD Diagnostics were received on dry ice and placed directly at $-80°C$ upon receipt for long-term storage. The PNNL reference serum and citrated plasma consist of different aliquots of the same anonymous female source described previously (Golden West Biologicals, Temecula, CA, USA) [15]. Upon receipt, these samples were aliquoted into 1 mL units and were also stored long-term at $-80°C$ (Tab. 1).

Tab. 1 Sample abbreviation table. Abbreviation, source, and lot number for the plasma and sera analyzed in triplicate by the AMT tag approach for the HUPO PPP

Abbreviation	Source	Lot no.
B3-CIT	HUPO (BD Diagnostics)	BDAA01 CIT-plasma
B3-SERUM	HUPO (BD Diagnostics)	BDAA01 serum
B2-CIT	HUPO (BD Diagnostics)	BDAFA01 CIT-plasma
B2-SERUM	HUPO (BD Diagnostics)	BDAFA01 serum
B1-CIT	HUPO (BD Diagnostics)	BDCA02 CIT-plasma
B1 SERUM	HUPO (BD Diagnostics)	BDCA02 serum
RefP	PNNL ref. (Golden West Biologicals)	M99869 CIT-plasma
RefS	PNNL ref. (Golden West Biologicals)	M99869 serum

11.2.2
Depletion of Igs and trypsin digestion

The Igs were depleted by affinity adsorption chromatography using protein A/G. Two hundred microliters of each sera/plasma was diluted with an equal amount of 20 mM sodium phosphate, pH 7.5, and added to an equal volume of UltraLink Immobilized protein A/G beads (Pierce Chemical, Rockford, IL, USA) that had been preequilibrated with 20 mM sodium phosphate, pH 7.8 as a 50% slurry. This slurry was incubated with gentle rocking for 20 min at 4°C. Ig-depleted serum was separated from the protein A/G beads by centrifugation using an Eppendorf microcentrifuge at 13 000 rpm. The beads were then washed twice with 5 vol of PBS (150 mM NaCl, 10 mM sodium phosphate, pH 7.3) and the washes pooled with the decanted Ig-depleted serum. Bradford protein assays were performed with the equivalent of 1 mg of sample that was denatured with 6 M urea and 1 mM DTT. The denatured protein samples were diluted 1:6 with PBS and then digested with 20 µg modified trypsin (Promega, Madison, WI, USA). Depletion and digestion quality was assessed by SDS-PAGE using a 4–12% gradient NuPage gel (Invitrogen, Carlsbad, CA, USA) and stained with GelCode Blue (Pierce) (Fig. 1).

11.2.3
Peptide cleanup

Eight LC-18 SPE columns (Supelco, Bellefonte, PA, USA) were placed on a Supelco Vacuum manifold (Supelco). The columns were wetted three times with 1 mL vol of methanol and equilibrated three times with 1 mL vol of water. Each trypsin-digested peptide mixture was applied to an individual column and washed three times with 1 mL of water. The peptides were eluted from each column three times with 0.5 mL 80% ACN, 0.1% acetic acid, 0.01% TFA, and water, and collected in a 2 mL Eppendorf tube. The eluted peptides were lyophilized using a SpeedVac and reconstituted to 5 µg/µL with 0.05% TFA and 0.2% acetic acid in water.

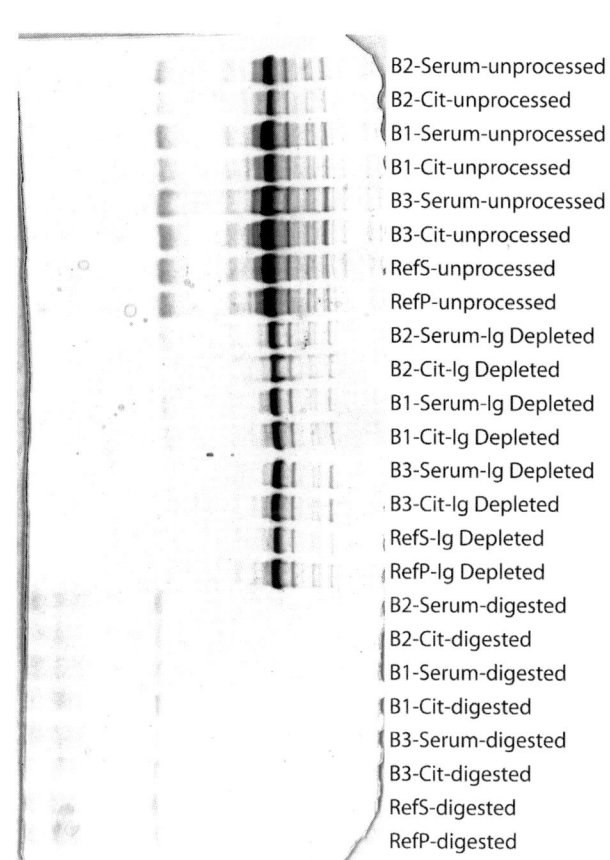

Fig. 1 SDS-PAGE for quality assessment of Ig-depletion and trypsin digestion of plasma samples. A 4–12% NuPage gel stained with GelCode Blue illustrating the sera and plasmas before and after Ig depletion and trypsin digestion. The gel was run from right to left.

Lanes (right to left):
- B2-Serum-unprocessed
- B2-Cit-unprocessed
- B1-Serum-unprocessed
- B1-Cit-unprocessed
- B3-Serum-unprocessed
- B3-Cit-unprocessed
- RefS-unprocessed
- RefP-unprocessed
- B2-Serum-Ig Depleted
- B2-Cit-Ig Depleted
- B1-Serum-Ig Depleted
- B1-Cit-Ig Depleted
- B3-Serum-Ig Depleted
- B3-Cit-Ig Depleted
- RefS-Ig Depleted
- RefP-Ig Depleted
- B2-Serum-digested
- B2-Cit-digested
- B1-Serum-digested
- B1-Cit-digested
- B3-Serum-digested
- B3-Cit-digested
- RefS-digested
- RefP-digested

11.2.4
Capillary RP-LC

RP-LC separations were performed using an in-house capillary LC system with a 60 cm long column (150 µm id × 360 µm od; Polymicro Technologies, Phoenix, AZ, USA) packed with 5 µm Jupiter C_{18} particles (Phenomenex, Torrance, CA, USA). The mobile phases consisted of (A) 0.05% TFA and 0.2% acetic acid in water and (B) 90% ACN and 0.1% TFA in water. Peptides were injected on the column using a 10 µL loop at a flow rate of 1.8 µL/min. The column was reequilibrated with 5% B for 20 min and peptides were eluted with a linear gradient from 5 to 70% B over 80 min. The capillary LC column was interfaced to either an IT mass spectrometer (ThermoFinnigan, San Jose, CA, USA) or a PNNL-modified 9.4 T Bruker-FT-ICR mass spectrometer using ESI [24]. The peptide loading quantity was 10 µg or 5 µg for the IT-MS or FT-ICR MS, respectively.

11.2.5
IT-MS

The ThermoFinnigan IT mass spectrometer (ThermoFinnigan) was configured as published previously [15]. Briefly, the mass spectrometer was configured to cover the m/z range of 400–2000 followed by three MS/MS scans on the three most intense precursor masses from the preceding MS scan. The tandem spectra were generated using a collisional energy of 45%. A dynamic mass exclusion window of 3 min was used.

11.2.6
SEQUEST identification of peptides

Tandem mass spectra were analyzed by SEQUEST (Bioworks 3.0, Thermo-Finnigan) [25–29], which performs its analyses by cross-correlating experimentally acquired mass spectra with theoretical idealized mass spectra generated from a database of protein sequences. These idealized spectra are weighted largely with b and y fragment ions, *i.e.*, fragments resulting from the amide-linkage bond from the *N*- and *C*-termini, respectively. For these analyses, no enzyme rule restrictions were applied to the possible cleavage sites available for peptide generation from the protein database. The peptide mass tolerance was 3.0 and the fragment ion tolerance was 0.0. The resulting identifications were then filtered according to the HUPO specific rules (see Section 2.7).

The FASTA protein database was searched against the July 2003 version of the International Protein Index (IPI; http://www.ebi.ac.uk/IPI/IPIhelp.html), generated by the European Bioinformatics Institute, as selected by the HUPO PPP members, to facilitate interlaboratory comparisons of the proteomic analyses [4].

11.2.7
Putative mass and time tag database from SEQUEST results

The data used to establish the mass and time tag database were obtained from multiple sample sources. The raw LC-IT data from our previous multidimensional analysis [15] were reanalyzed for this work along with individual IT analyses of each HUPO sample described herein (Tab. 1). These combined peptide identifications were used to populate the database that was subsequently used for generating the AMT tag results, which were uploaded to the HUPO database in December 2003. This initial set of results, referred to as the HUPO-Dec_Submission set, were derived using liberal SEQUEST filters, as described elsewhere [22, 30]. Additional peptide identifications were added to the mass and time tag database from a second extensive multidimensional effort using an undepleted plasma sample [17]. These results were not available for upload at the December HUPO PPP submission deadline, but were subsequently uploaded to HUPO in July 2004. These later results are used here for the primary data analysis referred to as the HUPO-Jul_-Submission set, internally designated as MT_Human_X112. The HUPO-Jul_Submission set also used SEQUEST data filters modeled after the HUPO filters. These data filters are: tryptic on at least the *N*- or *C*-terminus and a minimum DelCN value of 0.1 with +1, +2, and +3 charge states, using minimum X_{corr} values of 1.9, 2.2, and 3.75, respectively. The major exception was that at the time of data analysis R_{sp} was not captured in our data and therefore was not filtered with the HUPO expected ≤ 4. Subsequently, we determined that ~20% of all AMT tag peptide identifications and ~12% of those that were identified with high-matched confidence had a value of >4 for R_{sp}. The false-positive identification rate for peptide identifications was 32%; if the filter also contained the requirement for an $R_{sp} \leq 4$, the false-positive rate was 15% using the approach described in Qian et al. [31].

11.2.8
FT-ICR-MS

A Bruker Daltonics 9.4 tesla FT-ICR mass spectrometer was modified and configured for high-throughput proteomics use as described by Belov et al. [24]. Briefly, the FT-ICR mass spectrometer was combined with the capillary LC system (Section 2.4) and modified for concurrent internal mass calibration and auto-sampling. Injected samples contained tryptically digested peptides equivalent to 5 µg protein. These analyses typically result in analyzed peptides with <5 ppm mass measurement accuracy (MMA), depending on the dynamic range of the measurements, see example spectrum in Fig. 2 [24]. While the total analysis time was 4 days of instrument time, the analyses were performed as time became available on the cLC-FT-ICR MS. Complete acquisition was performed over a period of less than a month; B1-Cit and B1-serum samples were analyzed first and the others as analysis time was available.

11.2.9
cLC-FT-ICR MS data analysis

Each sample was analyzed in triplicate by cLC-FT-ICR MS. The resultant data were processed using the PRISM Data Analysis system, a series of software tools developed in-house. The first step involved deisotoping the MS data to give the monoisotopic mass, charge, and intensity of the major peaks in each mass spectrum. Following this step, the data were examined in a 2-D fashion to identify the groups of mass spectral peaks that were observed in sequential spectra. Each group, which was generally ascribed to one detected species and referred to as a unique mass class (UMC), has a median mass, central normalized elution time (NET), and abundance estimate, computed by summing the intensities of the MS peaks that comprise the total UMC.

The peptide identities of UMCs were determined by comparing the measured mass and NET of each UMC to the calculated mass and observed NET of each of the 57 578 peptide identifications in the database. Search tolerances were ±6 ppm for the mass and ±5% of the total normalized run time for the elution time. This peak-matching process gives a list of peptide identifications with scaled ion current values for each sample (Fig. 2 is an example of a matched peak). An abundance of each protein (*i.e.*, ORF) was estimated by averaging the peptide abundance values

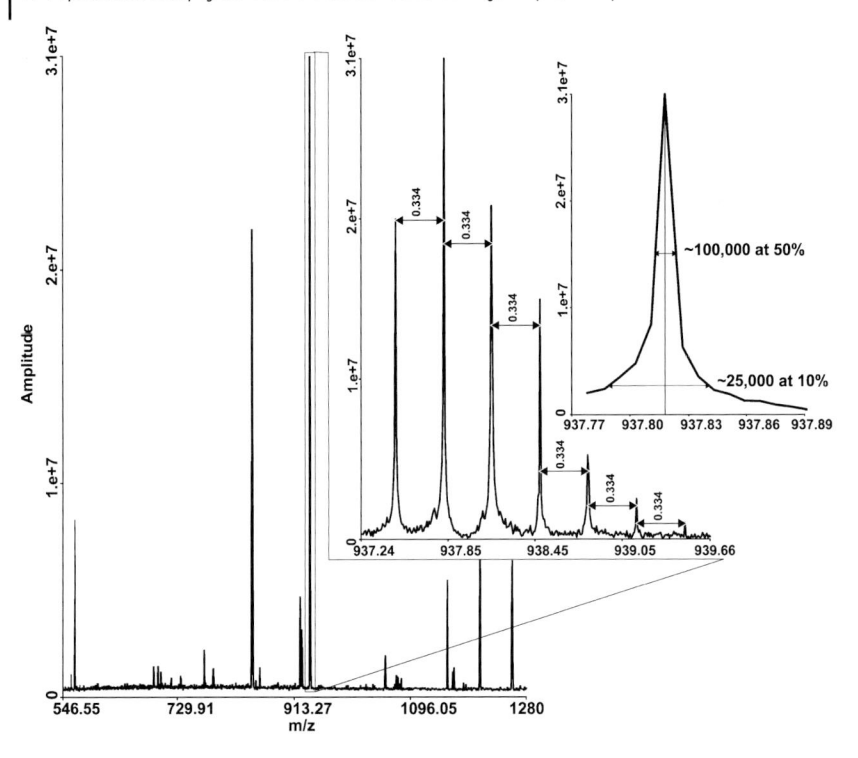

Fig. 2 Example cLC-FT-ICR mass spectrum illustrating resolution and mass accuracy. A peptide identification (VVSMDENFHPLNE-LIPLVYIQ DPK) from alpha-2-macroglobulin with an average mass error of −0.4 ppm compared to the calculated mass. This peptide identification was made in scan 2000 and the NETs for this identification were constrained to scan numbers 1994–2012.

from the ion current. The peak-matching confidence level for these identifications was defined as "high" when an AMT tag peptide was identified in at least two of the three replicates, and referred to throughout the text as high-matched confidence. The identification was defined as "low" confidence if the peak was matched in only one of three replicates. When a peptide was identified in one sample, but not another, e.g., fibrinogen in plasma, but not serum, the protein abundance of the missing protein was set at half the lowest ion current value observed in all 24 cLC-FT-ICR MS runs.

The false-positive rate of the AMT tag process was estimated by peak matching against a combined forward and reverse databases. This approach estimated the overall false-positive rate as 8% for this data set (calculated with combined databases as performed by Gygi et al. [32]), a substantial reduction from the 32% estimated using the HUPO SEQUEST filters without the R_{sp} filter. A second method to evaluate false-positive error of the AMT tag approach was to use the mass error plot to estimate "false-positive background", i.e., those measurements that contribute to a "noise", baseline. The number of identifications that fall below the baseline are false positives, and the "true-positive identifications", are those that contribute to the peak centered at 0 ppm MMA. This value was calculated to be ~10% for this data set (data not shown). To identify the level of degeneracy in the protein identifications ProteinProphet [33] was used as described in Qian et al. [34]. This analysis which provides a compressed number of identifications distinguished 377 protein families.

11.2.10
OmniViz cluster and visual analysis

The AMT tag results were exported into comma-delimited files that contained IPI reference number, protein annotation, peptide (when prepared by peptide), and ion current for peptides or average ion current for proteins. This file was imported into OmniViz 3.6 (OmniViz, Maynard, MA, USA) [35]. OmniViz was then configured to cluster the normalized data values for the triplicate data sets. As stated above, missing data were filled with a value half the lowest detected ion current value for an identified peptide (0.017 for this data set, while the maximum scaled ion current was 282 475). These values were normalized by using the "Standard by Row" normalization in OmniViz to enable comparisons between samples. This normalization is also referred to as the Z-score and is calculated as follows:

$$Z_z = \frac{X - \mu_x}{\sigma_x}$$

where X is the individual value, μ_x is the mean of the values from that row, and σ_x is the SD of the values from that row. Generally, Z-score differences between samples of at least a value of 2 or greater (*i.e.*, more than two SDs different) was considered significant. Z-score is a commonly used normalization method for microarrays [36], in addition to being used for broader applications. The resulting data were then clustered using hierarchical clustering with 24 clusters by magnitude and shape. Additionally, OmniViz was used to visualize and cluster the high matched confidence peptide identifications (rolled up to proteins) and to generate Spearman's pair-wise correlations for peptide abundance by sample.

11.3
Results

11.3.1
PuMT tag database

SEQUEST analysis of μLC-MS/MS data generated previously [15, 17] against the July 2003 IPI database was used to generate the PuMT tag databases used for this work. The numbers of spectra analyzed by SEQUEST are summarized for both the HUPO-Dec_Submission set and the HUPO-Jul_Submission set (269 416 and 653 340, respectively) in Tab. 2. This table also includes the numbers of peptides that passed our standard SEQUEST filters for both data sets [27, 29] and the numbers of PuMT tags with an $X_{corr} > 2.0$. The multidimensional analysis of the PNNL reference serum [15] as used for the HUPO-Dec_Submission set was com-

Tab. 2 PuMT database summary. PuMT tag database statistics for both the 2003 HUPO-Dec_Submission set and the 2004 HUPO-Jul_Submission set (peptides identified two or more times are in parenthesis)

LC-MS/MS data results type	HUPO-Dec_ Submission	HUPO-Jul_ Submission
Spectra analyzed by SEQUEST	269 416	653 340
Filtered peptide identifications[a]	17 018	51 591
Unique peptides identified[a]	2257	4579
Unique peptides identified 2 or more times[a]	1087	2393
Peptide identifications (Washburn et al.)[a]	17 018	51 591
Peptide identifications ($X_{corr} > 2.0$)[b]	34 319	77 284
Peptide identifications (HUPO-like)[c]	19 527	57 578

a) Peptide identifications filtered by routine criteria designed by the developers of SEQUEST [27, 29] with the exception of manual analysis.
b) SEQUEST peptide identification criteria used previously with the AMT approach for microbial proteomics [22, 30], internal designation MT_Human_P79.
c) SEQUEST filters with more stringent cutoffs, at least partially tryptic and a minimum DelCN of 0.1 with +1, +2, and +3 charge states using minimum X_{corr} values of 1.9, 2.2, 3.75, respectively with the exception of $R_{sp} \leq 4$, internal designation MT_Human_X112.

plemented by merging a second similar extensive multidimensional analysis [17] in the HUPO-Jul_Submission set. By merging the two data sets along with the use of stricter SEQUEST filters (see Section 2) for the PuMT tags, the overall confidence of the resulting peptide/protein identifications (*i.e.*, AMT tags) was improved. Combining multidimensional analyses has been shown elsewhere to increase the completeness of a proteomic analysis [32, 37, 38].

Recent analyses of human plasma and other samples provided the basis for estimating the false-positive rates for SEQUEST results, although the filter rules were similar, but not identical to those used here for populating the PuMT tag database [31, 40]. The false-positive values for PuMT identifications were 16–32%, depending on the calculation [31, 40]. In this analysis, using the same approach as previously published [31] we calculated a false-positive rate of 32% for the filters used here for the mass and time tag database.

11.3.2
Summary of peptide/protein identifications by AMT tags

Any peptide identified using the AMT tag approach was mapped to all proteins that contained the specific peptide to allow HUPO PPP the greatest latitude in making decisions with regard to reducing redundancy. For example, by eliminating redundancy, the total number of proteins identified were reduced by about half using ProteinProphet [33] (722 redundant proteins to 377 nonredundant proteins).

Tab. 3 AMT tag data summary. AMT tag database statistics for both the 2003 HUPO-Dec_Submission set and the 2004 HUPO-Jul_Submission set. High-matched confidence was defined here as an AMT tag measured in at least two of three triplicates. Unique peptide-protein pairs are those peptides that map to a single protein

AMT tag date results type	HUPO-Dec_ Submission[a]	HUPO-Jul_ Submission[b]
All possible peptide-protein pairs	4473	2806
High-matched confidence peptide–protein pairs	2811	2298
Unique peptide-protein pairs	2802	1493
High-matched confidence unique peptides	1704	1225
All putative proteins	2625	842
High-matched confidence proteins	1366	585

a) SEQUEST peptide identification criteria used previously with the AMT tag approach for microbial systems [22, 30] with peptide identifications from [15].
b) SEQUEST filters with more stringent cutoffs, at least partially tryptic and a minimum DelCN of 0.1 with +1, +2, and +3 charge states using minimum X_{corr} values of 1.9, 2.2, 3.75, respectively with peptide identifications from [15] and [17].

All peptide-protein pairs (*i.e.*, every peptide possible source from the IPI database was included in the redundant protein numbers) were counted and the 2806 identifications from 1493 unique peptides represented the largest possible number of peptide-protein identifications for the 2004 HUPO-Jul_Submission set (Tab. 3). The high-matched confidence, operationally defined as those found in at least two of three triplicates, peptide-protein pairs reduced to 2298 peptide-protein pairs from 1225 high-confidence peptide identifications for the 2004 HUPO-Jul_Submission set (Tab. 3). Each AMT tag peptide mapped to an average of two proteins due to the degeneracy of the human protein FASTA file and the nature of human proteins. Those peptide identifications that mapped to multiple proteins tended to be of higher confidence (by reproducibility) than those mapped to a single protein entry (data not shown).

The false-positive peptide identifications were high (32%) for the PuMT mass and time tag database based on SEQUEST results. However, the use of high MMA and NETs afforded by the AMT tag approach significantly improved the confidence in peptide identifications over SEQUEST analyses alone [21, 22]. The estimated false-positive errors for these AMT tag identifications are estimated to be between 8 and 10% depending on the method used.

The total instrument time required for cLC-FT-ICR MS analysis of eight samples in triplicate (*i.e.*, 24 cLC-FT-ICR runs) was 4 days. The results from our high-throughput analyses are comparable to protein identifications obtained at most of the other laboratories participating in the HUPO plasma pilot project [55]. Because our results were analyzed in triplicate, we were able to increase the general confidence in an AMT tag peptide identification. Using the HUPO-Jul_Submission set PuMT tag database, 585 proteins were identified with high confidence by reproducibility (Tab. 3).

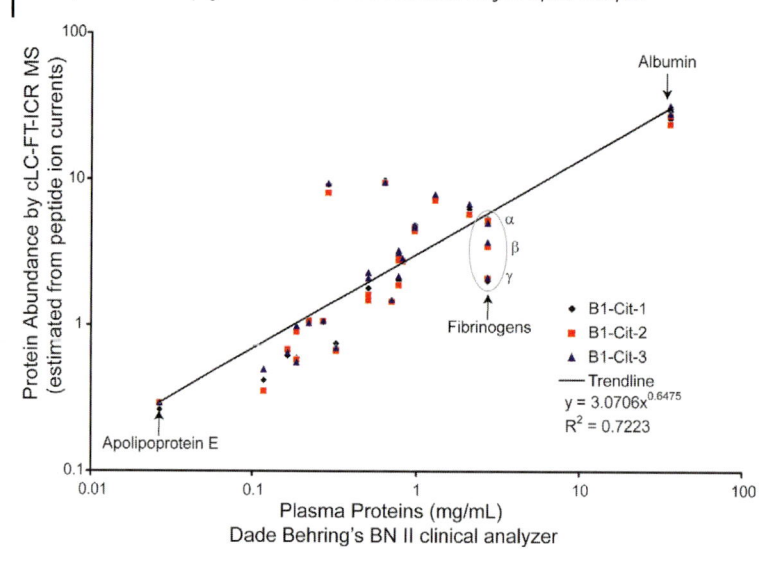

Fig. 3 Correlation between protein abundance estimate from cLC-FTI-CR-MS and protein concentrations measured by standard clinical chemistry methods. Average peptide ion currents from cLC-FT-ICR MS and measurements performed on BN II clinical analyzer (Dade Behring) were compared for 18 proteins in the B1-CIT sample. Triplicate ion current values for each analysis were plotted with multiple protein entries where appropriate (e.g., fibrinogen was mapped to multiple IPI numbers representing fibrinogen α, β, and γ). Selected proteins are also shown on the graph for perspective. Proteins on the graph in order of increasing abundance are: apolipoprotein E, plasminogen, complement C4, ceruloplasmin, transthyretin, antithrombin-III, apolipoprotein A-II, fibronectin, alpha-1-acid glycoprotein 1, hemopexin, complement component 3, haptoglobin, alpha-2-macroglobulin, alpha-1-antitrypsin, apolipoprotein A-I, transferrin, fibrinogen, albumin.

11.3.3
Protein concentration estimates from ion current

The average values of the ion current for all the peptides identified for a particular protein were compared with the concentrations determined by certified assays performed on the BN II at Dade Behring ([54]; www.hupo.org) (Fig. 3). A linear correlation on a log-log plot was used to describe the relationship between MS "abundance" defined by ion current and the concentration measured by BN II immunoassay technology. Note that the correlation between proteins measured by MS *versus* immunoassay techniques will be imperfect, due in part to factors such as the ambiguity in the IPI protein entries actually measured by the BN II, mapping of multiple peptides to multiple proteins, variations in ionization efficiencies, epitope specificity of immunoassays, and multiple subunits and isoforms, *e.g.*, fibrinogens, *etc.* (Fig. 3).

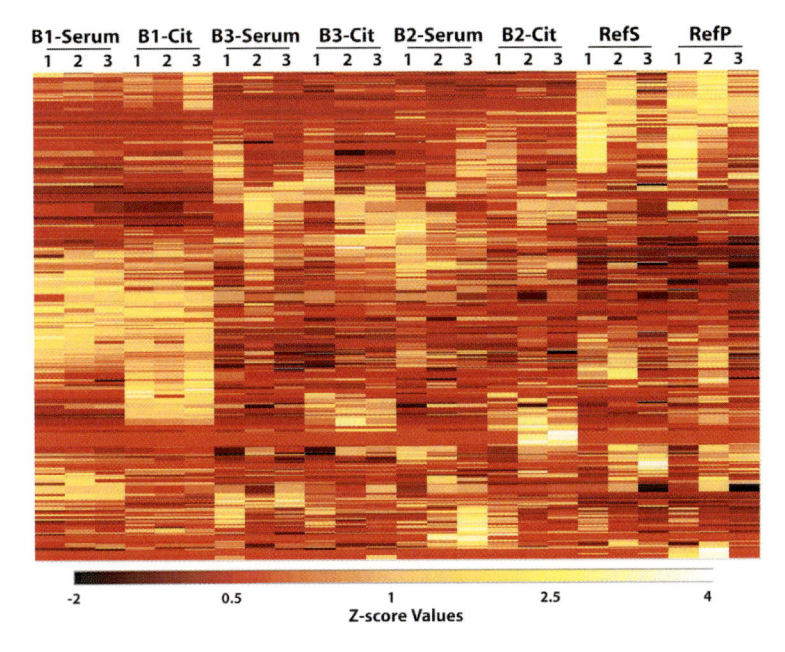

Fig. 4 Corescape view of the 842 highmatched confidence proteins normalized by Z-score. This view was generated from a cluster analysis of Z-score normalized protein abundances using triplicate data for each sample. A number of proteins that appear to be present in relatively higher amounts can be observed in the B1-SERUM and B1-CIT samples. This view shows a high similarity with the results from the B3-SERUM, B3-CIT, B2-SERUM, and B2-CIT samples. RefS and RefP are more similar to each other than to the HUPO samples.

11.3.4
Global protein analysis

The high-confidence protein identifications from the HUPO-Jul_Submission set were analyzed using hierarchical cluster analysis with the OmniViz™ program. Normalization by Z-score on the calculated protein abundance values was used to facilitate cross-comparisons between samples based on the ion current measure of protein abundance used in Fig. 3. Although differences in ion currents were observed for different samples, in general the samples are very similar and the Z-score values reflect this general similarity (Fig. 4). A cluster containing reproducible differences between serum and plasma samples is shown in Fig. 5. This cluster partly illustrates the obvious difference in abundance of fibrinogens in plasma and serum; the clotting process removes fibrinogen from serum. A number of proteins, including some hypothetical proteins, appeared to be present at relatively higher concentrations in plasma than in serum, *e.g.*, zonadhesin. This protein has some known functions related to sperm and would appear unlikely to be present in plasma, but interestingly, this protein contains five von Willebrand D domains, which are common in blood proteins and are involved in clotting [41]. Zonadhesin

Tab. 4 Summary of peptide/protein identifications using the AMT tag approach. A summary of peptide and protein identification statistics by sample

	High-matched confidence Unique		All unique		CV[a]		Peptide dynamic range[b]
	Peptide	Protein	Peptide	Protein	Mean	Median	
B3-CIT	750	435	939	634	33.9	29.3	4585
B3-SERUM	682	379	926	633	39.7	36.4	3524
B2-CIT	792	455	1008	664	37.9	34.8	4361
B2-SERUM	713	428	915	634	38.7	35.2	3717
B1-CIT	864	441	975	572	19.0	13.5	3260
B1-SERUM	847	425	937	542	19.9	15.4	4234
RefP	860	447	1104	706	41.9	37.6	5028
RefS	787	433	1016	701	41.7	37.6	5168

a) Coefficient of variation was calculated from average of the ion current values of all high confident protein identifications.
b) Dynamic range was based on the highest and lowest observed ion current values of an identified peptide.

was also reported by Hefta *et al.* (laboratory 12) [55]. Activin-like receptors appeared to be found preferentially in plasma and were reported by us, as well as by Hefta *et al.* (laboratory 12), and Wang *et al.* (laboratory 12). For each sample, 787 ± 68 high-confidence peptides were identified and an average of 430 ± 23 redundant high-confidence proteins putatively identified (Tab. 4).

The protein composition of plasma *versus* serum revealed both expected and unexpected results. Fig. 5 shows a group of proteins differentially abundant in plasma and serum identified by nonsupervised clustering. A more detailed view of individual plasma/serum pairs is constructed by calculating the abundance ratio (ion current) for all "high-matched confidence" protein identifications common to the sample pair. For example, the African American plasma-serum (B2-CIT and B2-serum) contained 365 high-confidence proteins common to both samples (Fig. 6). Of these, 59 (16%) were two-fold, or more abundant in plasma than in serum, representing proteins that are retained in the clot. In addition to the anticipated fibrinogens and other clotting proteins, 18/59 were consistently depleted in all four plasma/serum sample pairs analyzed. This consistency suggests that the depletion of these proteins (*e.g.*, melanoma inhibitory activity protein 2; sodium/calcium exchanger 2 precursor, titin, zonadhesin precursor, *etc.*) in serum is a real phenomenon.

More unexpectedly, approximately 30 proteins (8%) were two-fold more abundant in serum than in plasma. Nine proteins were consistently more abundant in serum in all four plasma/serum sample pairs analyzed. These serum-enriched proteins include cell division cycle protein 91-like 1, phosphorylase kinase (alpha 1), splice form 2 of P46020 phosphorylase B kinase alpha regulatory chain, and a hypothetical protein. This phenomenon was validated independently by

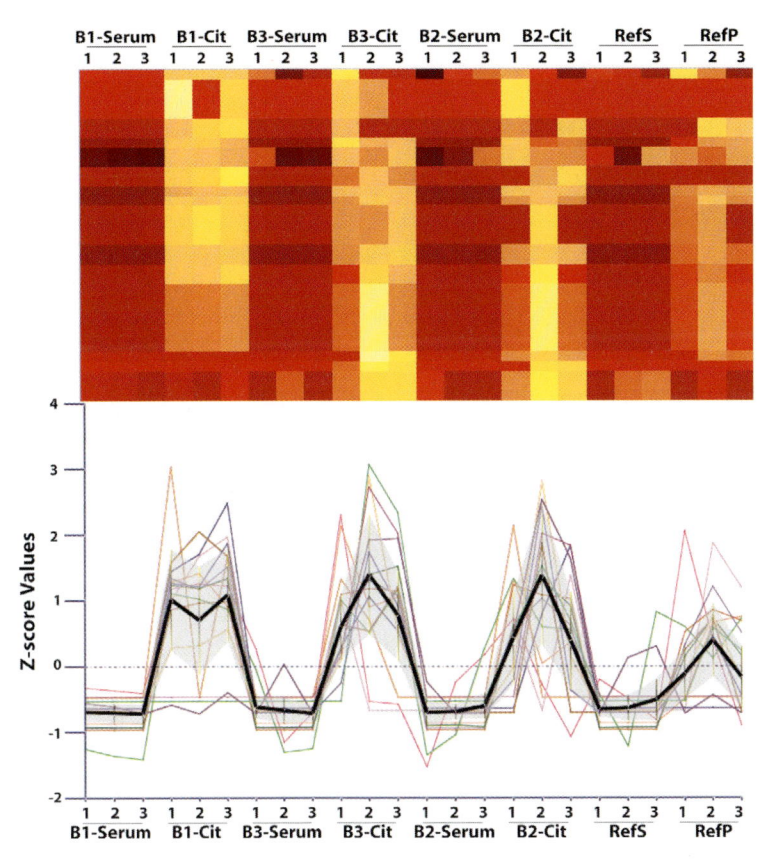

Fig. 5 Identification of proteins differentially abundant in citrated-plasma and serum. This illustration shows a heat map of a serum/plasma specific cluster from results in Fig. 4 and a profile plot. Black profile shows the average value of this cluster and the gray region above and below that profile represents one SD from the measurements.

micro-ELISA analysis of several cytokines in these same HUPO samples, which revealed that RANTES and PDGF-AA were consistently more abundant in serum in all four plasma/serum sample pairs (Richard Zangar, personal communication). This apparent enrichment in serum is more difficult to explain than depletion in serum, although several processes could be proposed. First, removal of clot-associated proteins simplifies the composition of the serum compared to plasma. The resultant simplification in the peptide mixture could facilitate more effective measurement of certain peptides in serum than in plasma. Second, it is conceivable that the serum-enriched proteins are derived from platelet activation. We found 13 proteins previously associated with platelets [42, 43]. These 13 proteins include cytoskeleton, and protein processing, and other proteins not specifically expressed in platelets. Unfortunately, none of these 13 proteins was enriched in serum and thus the potential role of platelets to the phenomena of serum enrich-

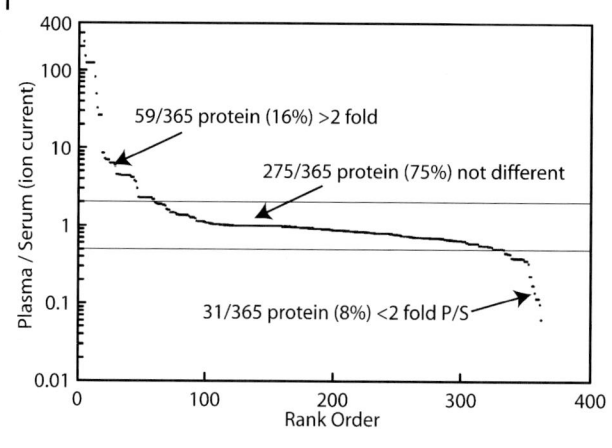

Fig. 6 Abundance ratio of proteins in B2-CIT to B2-serum. Three hundred sixty-five proteins were identified with high-matched confidence and ranked by the plasma/serum abundance (ion current) ratio. Ratios greater than two-fold identify plasma proteins that were depleted in serum.

ment is not clear. Third, the apparent serum enrichment could be an artifact of differential PTMs in plasma *versus* serum. The AMT tag approach identifies peptides, including modified peptides, which were previously identified. For example, if serum contained higher phosphorylase activity than plasma, the unphosphorylated peptide would be higher in serum than in plasma resulting in the observation that the protein was enriched in serum.

A pairwise Spearman's correlation of peptide abundance was performed and interesting visible features resulted (Fig. 7). The columns were allowed to associate with the most closely correlated counterpart, *i.e.*, the analyses that are most similar are next to each other in Fig. 7. Interestingly, the HUPO sera grouped together, which shows higher correlation, as did the HUPO plasma analyses. The PNNL reference samples correlated into a separate group, appropriately revealing that serum and plasma were from a different female-only commercial source (Fig. 7).

11.4
Discussion

11.4.1
Application of FT-ICR MS as a proteomic technology bridge

Over the past decade, proteomics has largely focused on technology development. Many of the major proteomic technologies have specific niches, *e.g.*, MudPIT or similar shotgun proteomics for discovering new proteins in samples with some quantitation [27], SELDI-MS for searching new drug targets and disease markers along with MS spectra algorithms for potentially identifying individuals with disease or a propensity for disease [44], and protein microarrays for measuring the concentrations of known proteins for research applications and potentially disease diagnosis (see another HUPO PPP-related paper by Haab *et al.*, [54]). For our study, we used the AMT tag technology that bridges the gaps among some of these other major technologies.

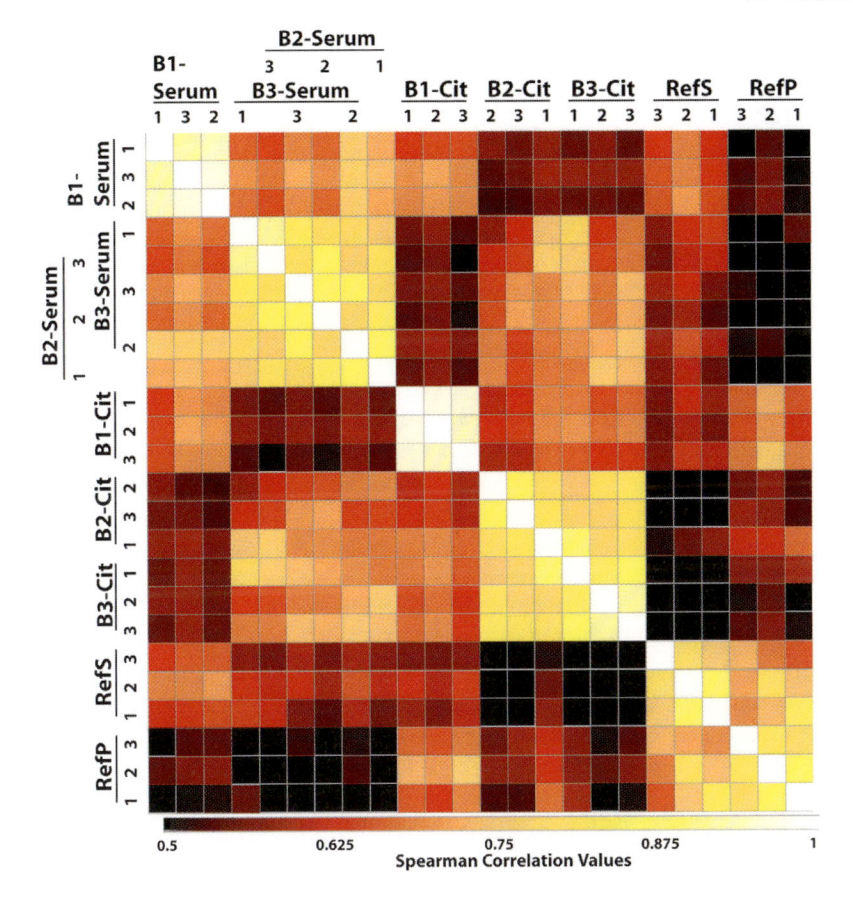

Fig. 7 Pairwise Spearman's correlations of peptide abundance by analysis. Pairwise correlations [53] of the peptide abundances with the columns and rows are set to pair up by the highest correlation. Darker squares represent the least correlated and the lightest the most highly correlated (scale is set to low value of 0.5 to give a greater visual range to the various #values). Highest correlations are within individual samples except in the case of Asian-American and African-American serum samples that exhibited intermixed similarity. In general, the sera and the plasmas were more closely correlated with the exception of the PNNL Ref samples that were from a different source and contained only female serum/plasma.

Advantages of this approach include an estimate of concentration and putative identification of proteins with patterns of interest. This method and similar approaches are likely to be of greater use in the coming years with the recent additional commercialization and competition of FT-ICR instrumentation. Some instruments, such as those with linear ITs with FT-ICR mass analyzers, are particularly exciting in that the AMT tag approach could be undertaken with simultaneous enrichment of a PuMT tag database. MALDI approaches could also be used to attain similar types of information determined by the AMT tag approach.

The recent introduction of linear ITs combined with FT-ICR mass spectrometers provides new opportunities to take advantage of the high-mass accuracy and faster scan times of these instruments. Especially exciting are approaches that may potentially revive many top-down proteomics approaches, including the work of Speicher *et al.* and Karger *et al.* in this special issue [45]. A second approach complementary to the AMT tag approach taken here is to extend the confidence of peptide identifications by MS/MS/MS or MS^3 [46]. These authors illustrated that MS^3 combined with the high MMA improved peptide identifications by MASCOT.

11.4.2
Confidence in any MS-based proteomic approach

Considerable research effort has been directed at determining appropriate data filters, protease cleavage states, MMA, and the number of peptide identifications required for a confident protein identification [15, 17, 27, 32, 46–50]. Blood serum and plasma are increasingly proving to be more difficult to fully characterize with traditional proteomic technologies, as shown by both HUPO and other efforts. Critical assumptions involving data filters and peptide identifications that have been used effectively in other proteomic efforts likely need to be modified for plasma and other body fluids and tissues [31, 51].

Importantly, the identification confidence for peptides does not directly correlate to the identification confidence for proteins identified from these peptides. This distinction has not been sufficiently emphasized in the past. Most methodologies have attempted to limit false-positive peptide identifications to typically <1% of the total number of peptide identifications; however, the false-positive estimates for human proteomics are still regarded as higher. The false-positive incidences for proteins are often considered to be of similar magnitude as the peptides. However, protein misidentifications are actually greater because even correct peptide identifications can result in proteins with multiple identifications. Thus, the false-positive incidence for proteins identified by single peptide identifications should be viewed as having a higher false-positive rate than that for peptide identifications.

A second issue in establishing the confidence of protein identifications concerns the use of a protease cleavage state in peptide identifications. One school of thought has been that nearly all peptides result from highly specific digestion by the exogenous protease, typically trypsin, and thus all confident peptide identifications should conform to fully tryptic digestion patterns (*e.g.*, [10, 48]). Alternatively, complex protein mixtures processed by endogenous proteases may contain unexpected amino- and carboxy-termini, resulting in nontryptic cleavage states (*e.g.*, [15, 18, 27]). Regardless, there is general agreement that identifications based on partially digested, missed digestion sites, and digestion inconsistent with the exogenous protease of choice are much less confident than peptide identifications that completely conform to the expected specificity of the exogenous protease used for a proteomic analysis. At the same time there are clearly many examples of peptides found in plasma, which are not derived by the most common exogenous protease and would result in, at best, "partially" tryptic peptides; examples with trypsin include angiotensinogen I, angiotensinogen II, and vasopressin (G.DRVYIHPFHL.V, G.DRVYIHPF.H, and A.CYFQNCPRG.G, respectively). These partially-tryptic peptides arise from the normal biological processing of inactive precursor molecules. Thus, peptides that do not conform to the protease used should be carefully considered in the context of the underlying biology and the goals of a proteomic experiment.

In order to maintain consistency with the other laboratories analyzing HUPO samples, our present peptide identifications require at least a partially-tryptic state. It should be noted that HUPO criteria also had an $R_{sp} \leq 4$ filter, which tends to offset the partially-tryptic misidentifications. At the time of this analysis our peak-matching process was not setup to use R_{sp} criteria.

A third issue related to protein identification confidence is the application of various analysis tools and data filters to MS/MS data for peptide identification. For example, numerous filtering methods for SEQUEST data are available [15, 27, 32, 47, 49, 50]; all try to set the best balance for sensitivity and specificity. In the development of our AMT tags, we conformed to the data analysis approaches used by other laboratories involved with HUPO plasma samples with the exception of using $R_{sp} \leq 4$ (see Section 2).

11.4.3
Peptide/protein redundancy

Due in part to the large number of redundant entries in the July 2003 IPI protein database, as well as the presence of conserved peptide and protein sequences in the human genome, each peptide was mapped, on average, to two different protein entries. In many cases it was impossible or undesirable to "identify" a specific protein from the peptides observed. For example, haptoglobin alleles 1 and 2, and haptoglobin-related protein all contain a great deal of sequence identity. The three proteins haptoglobin-1, haptoglobin-2, and haptoglobin-related protein are used as examples of the information that would be lost if protein identifications required peptides unique to a single protein (Tab. 5). Haptoglobin-1 or haptoglobin-2 would be considered nondetected over the entire analysis because no specific unique peptides are present (Tab. 5). Furthermore, only the B1-CIT sample contained a peptide that was unique to only haptoglobin-2 (only in the HUPO-Dec_Submission set). This observation corresponded to other measurements that showed the B1-CIT/B1-SERUM sample had the highest haptoglobin-2 concentration relative to the B3-CIT/B3-SERUM samples and that the B2-CIT/B2-SERUM samples contained no haptoglobin-2 (Alex Rai, personal communication).

11.4.4
Identification sensitivity *versus* specificity

A constant concern with all MS-based proteomics approaches involves false-positive and -negative identifications. Another aspect of the same problem stems from the need to balance sensitivity and specificity in various proteomic analyses. Following direction from the HUPO PPP group, both high- and low-confidence (based

Tab. 5 Example of peptide overlap between human proteins, using haptoglobin. Haptoglobin-1, haptoglobin-2, and haptoglobin-related protein derived peptide identifications. High-matched confidence peptide identifications were those that were found to be identified by the AMT tag approach in at least one specific sample in two or three triplicates. NP under confidence shows peptides not present in the 2004 HUPO-Jul_Submission set, but were found with the previous 2003 HUPO-Dec_Submission set.

Confidence	Peptide	Haptoglobin-1	Haptoglobin-2	Haptoglobin-related protein
High	SPVGVQPILNEHTF	√	√	√
High	AVGDKLPECEAVCGKPK	√	√	√
High	DIAPTLTLYVGK	√	√	√
High	DIAPTLTLYVGKK	√	√	√
High	ILGGHLDAK	√	√	√
High	KQLVEIEK	√	√	√
High	NPANPVQR	√	√	√
High	QLVEIEK	√	√	√
High	SCAVAEYGVYVK	√	√	√
High	VTSIQDWVQK	√	√	√
High	VMPICLPSK	√	√	√
High	TSIQDWVQK	√	√	√
High	GSFPWQAK	√	√	√
High	LRTEGDGVYTLNNEK	√	√	
High	YQEDTCYGDAGSAFAVHDLEEDTWYATGILSFDK	√	√	
High	VVLHPNYSQVDIGLIK	√	√	
High	HYEGSTVPEK	√	√	
High	HYEGSTVPEKK	√	√	
High	VMPICLPSKDYAEVGR	√	√	
High	SPVGVQPILNEHTFCAGMSK	√	√	
High	TEGDGVYTLNNEK	√	√	
High	VGYVSGWGR	√	√	
High	YVMLPVADQDQCIR	√	√	
High	DYAEVGR	√	√	
High	GYVSGWGR	√	√	
High	PPEIAHGYVEHSVR	√	√	
High	TEGDGVYTLNDK		√	√
High	TEGDGVYTLNDKK		√	√
High	LRTEGDGVYTLNDK		√	√
High	LRTEGDGVYTLNDKK		√	√
High	VGYVSGWGQSDNFK			√
High	VVLHPNYHQVDIGLIK			√
Low	AVGDKLPECEAVCGK	√	√	√
Low	NYAEVGR			√
NP	AVGDKLPECEADDGCPKPPEIAHGYVEHSVR		√	
NP	LPECEADDGCPKPPEIAHGYVEHSVR		√	
NP	SPVGVQPILNEHTFCVGMSK			√

on reproducibility) results were included in the data analysis of the HUPO-Jul_Submission set. "High-matched confidence" identifications were based upon detection *via* our peak-matching software of at least two of three triplicates. As a result of revised preferences of the HUPO PPP group, we used stringent SEQUEST rules for peptide identifications in our PuMT tag database. Because the AMT tag analysis approach is constrained by the MS/MS data in the PuMT tag database, our reported identifications represent only a subset of the SEQUEST identifications from the IPI July 2003 protein database. In a preliminary analysis, there were approximately 1000 high abundance and highly reproducible UMCs (mass and time features) in the B1-CIT sample that did not map to PuMT tags (data not shown).

11.4.5
Throughput and differential analysis

A significant advantage of the AMT tag strategy is increased throughput compared with other shotgun proteomics approaches. Additionally, this high throughput strategy provides a basis for comparative quantitative analysis of the results. Depending on the fractionation approach and MS conditions used, a multidimensional analysis (*i.e.*, strong cation exchange followed by LC-MS/MS) typically requires the minimum of a day to many weeks. However, by employing the AMT tag approach, we were able to leverage peptide identifications from two previous multiweek multidimensional analyses [15, 17] to serve as the reference mass and time tag database to identify peptides in triplicate analyses from eight samples in only 4 days total instrument time The throughput of this analysis is significantly higher than traditional shotgun proteomics. Another advantage of the rapid turnover of the chromatography and instrumentation is that replicate sample results are more comparable, albeit there were some differences in unnormalized peptide abundances when significant time lapsed during analysis (Fig. 4).

A major disadvantage of the AMT tag approach used here is that peptides must be observed before they show up as in the subsequent analysis. Data-directed methodologies would also need to be applied to adjust to new relevant features found in clinical samples. Also, analysis of the plasma and serum samples in this study makes it clear that even the AMT tag method will not be completely successful without protein depletion or enrichment procedures. A recent enrichment example using the AMT tag approach in another system [52] shows promise when used with depletion strategies.

The authors would like to thank Richard Zangar, Mary Lipton, Gordon Anderson, Eric Strittmatter, BD Diagnostic and their blood donors, Gil Omenn, and Sam Hanash who made specific contributions that helped lead to this publication. This work was performed in the Environmental Molecular Sciences Laboratory, a U.S. Department of Energy (DOE) national scientific user facility at PNNL, with special thanks to Harold Udseth and Ronald Moore. We would also like to acknowledge Mark Gritz and David Koppenaal for their support and encouragement. Portions of this work were supported by Battelle Columbus Operations (IR&D), the National Institutes of Health (NCI CA78722) and the Prote-

omics National Center for Research Resources (RR18522). PNNL is operated by Battelle for the DOE under contract no. DE-AC06-76RLO 1830.** Supplemental results are provided on the public HUPO PPP website (http://www.bioinformatics.med.umich.edu/hupo.org), Results for Laboratory Number 28

11.5
References

[1] Baldwin, M. A., *Mol. Cell. Proteomics* 2003, *3*, 1–9.

[2] Anderson, N. L., Polanski, M., Pieper, R., Gatlin, T., Tirumalai, R. S., Conrads, T. P., Veenstra, T. D. et al., *Mol. Cell. Proteomics* 2004, *3*, 311–326.

[3] Taylor, C. F., Paton, N. W., Garwood, K. L., Kirby, P. D., Stead, D. A., Yin, Z., Deutsch, E. W. et al., *Nat. Biotechnol.* 2003, *21*, 247–254.

[4] Omenn, G. S., *Proteomics* 2004, *4*, 1235–1240.

[5] Hanash, S., *Mol. Cell. Proteomics* 2004, *3*, 298–301.

[6] Anderson, N. L., Anderson, N. G., *Mol. Cell. Proteomics* 2002, *1*, 845–867.

[7] Rose, K., Bougueleret, L., Baussant, T., Bohm, G., Botti, P., Colinge, J., Cusin, I. et al., *Proteomics* 2004, *4*, 2125–2150.

[8] Wu, S. L., Choudhary, G., Ramstrom, M., Bergquist, J., Hancock, W. S., *J. Proteome Res.* 2003, *2*, 383–393.

[9] Wu, S.-L., Amato, H., Biringer, R., Choudhary, G., Shieh, P., Hancock, W. S., *J. Proteome Res.* 2002, *1*, 459–465.

[10] Marshall, J., Jankowski, A., Furesz, S., Kireeva, I., Barker, L., Dombrovsky, M., Zhu, W. et al., *J. Proteome Res.* 2004, *3*, 364–382.

[11] Anderson, L., Anderson, N. G., *Proc. Natl. Acad. Sci. USA* 1977, *74*, 5421–5425.

[12] Liotta, L. A., Ferrari, M., Petricoin, E., *Nature*, 2003, *425*, 905.

[13] Merrick, B. A., *EHP Toxicogenomics* 2003, *111*, 1–5.

[14] Graham, R., *J. Proteome Res.* 2004, *3*, 163.

[15] Adkins, J. N., Varnum, S. M., Auberry, K. J., Moore, R. J., Angell, N. H., Smith, R. D., Springer, D. L., Pounds, J. G., *Mol. Cell. Proteomics* 2002, *1*, 947–955.

[16] Pieper, R., Gatlin, C. L., Makusky, A. J., Russo, P. S., Schatz, C. R., Miller, S. S., Su, Q. et al., *Proteomics* 2003, *3*, 1345–1364.

[17] Shen, Y., Jacobs, J. M., Camp, D. G, 2nd., Fang, R., Moore, R. J., Smith, R. D., Xiao, W. et al., *Anal. Chem.* 2004, *76*, 1134–1144.

[18] Tirumalai, R. S., Chan, K. C., Prieto, D. A., Issaq, H. J., Conrads, T. P., Veenstra, T. D., *Mol. Cell. Proteomics* 2003, *2*, 1096–1103.

[19] Burtis, C. A., Ashwood, E. R., *Tietz Fundamentals of Clinical Chemistry*, fifth edn., W. B. Saunders Company, Philadelphia 2001.

[20] Pasa-Tolic, L., Masselon, C., Barry, R. C., Shen, Y., Smith, R. D., *Biotechniques* 2004, *37*, 626–633.

[21] Anderson, K. K., Monroe, M. E., Daly, D. S., *Proc. Intern. Conf. METMBS*, 2004, 151–156.

[22] Smith, R. D., Anderson, G. A., Lipton, M. S., Pasa-Tolic, L., Shen, Y., Conrads, T. P., Veenstra, T. D., Udseth, H. R., *Proteomics* 2002, *2*, 513–523.

[23] Masselon, C., Pasa-Tolic, L., Tolic, N., Anderson, D. J., Bogdanov, B., Vilkov, A. N., Shen, Y. et al., *Anal. Chem.* 2005, *76*, 400–406.

[24] Belov, M. E., Anderson, G. A., Wingerd, M. A., Udseth, H. R., Tang, K., Prior, D. C., Swanson, K. R. et al., *J. Am. Soc. Mass Spectrom.* 2004, *15*, 212–232.

[25] Yates, J. R, 3rd, Carmack, E., Hays, L., Link, A. J., Eng, J. K., *Methods Mol. Biol.* 1999, *112*, 553–569.

[26] Yates, J. R, 3rd, McCormack, A. L., Eng, J. K., in: *Analytical and Chemistry News & Features*, Anal. Chem.1996, 534–540.

[27] Washburn, M. P., Wolters, D., Yates, J. R, 3rd, *Nat. Biotechnol.* 2001, *19*, 242–247.

[28] Link, A. J., Eng, J., Schieltz, D. M., Carmack, E., Mize, G. J., Morris, D. R., Garvik, B. M., Yates, J. R, 3rd, *Nat. Biotechnol.* 1999, *17*, 676–682.

11.5 References

[29] Eng, J. K., McCormack, A. L., Yates, J. R., *J. Am. Soc. Mass Spectrom.* 1994, *5*, 976–989.

[30] Lipton, M. S., Pasa-Tolic, L., Anderson, G. A., Anderson, D. J., Auberry, D. L., Battista, J. R., Daly, M. J. et al., *Proc. Natl. Acad. Sci. USA* 2002, *99*, 11049–11054.

[31] Qian, W.-J., Liu, T., Monroe, M. E., Strittmatter, E. F., Jacobs, J. M., Kangas, L. J., Petritis, K. et al., *J. Proteome Res.* 2005, *4*, 53–62.

[32] Peng, J., Elias, J. E., Thoreen, C. C., Licklider, L. J., Gygi, S. P., *J. Proteome Res.* 2003, *2*, 43–50.

[33] Nesvizhskii, A. I., Keller, A., Kolker, E., Aebersold, R., *Anal. Chem.* 2003, *75*, 4646–4658.

[34] Qian, W.-J., Monroe, M. E., Liu, T., Jacobs, J. M., Anderson, G. A., Shen, Y., Moore, R. J. et al., *Mol. Cell. Proteomics* 2005, *5*, 1263–1273.

[35] Saffer, J. D., Burnett, V. L., Chen, G., van der Spek, P., *IEEE Comput. Graph. Appl.* 2004, *24*, 10–15.

[36] Cheadle, C., Vawter, M. P., Freed, W. J., Becker, K. G., *J. Mol. Diagn.* 2003, *5*, 73–81.

[37] Durr, E., Yu, J., Krasinska, K. M., Carver, L. A., Yates, J. R., Testa, J. E., Oh, P., Schnitzer, J. E., *Nat. Biotechnol.* 2004, *22*, 985–992.

[38] Jacobs, J. M., Mottaz, H. M., Yu, L. R., Anderson, D. J., Moore, R. J., Chen, W. N., Auberry, K. J. et al., *J. Proteome Res.* 2004, *3*, 68–75.

[39] Qian, W.-J., Liu, T., Monroe, M. E., Strittmatter, E. F., Jacobs, J. M., Kangas, L. J., Petritis, K. et al., *J. Proteome Res.* 2005, *4*, 53–61.

[40] Strittmatter, E. F., Kangas, L. J., Petritis, K., Mottaz, H. M., Anderson, G. A., Shen, Y., Jacobs, J. M. et al., *J. Proteome Res.* 2004, *3*, 760–769.

[41] Voorberg, J., Fontijn, R., van Mourik, J. A., Pannekoek, H., *EMBO J.*, 1990, *9*, 797–803.

[42] Garcia, A., Prabhakar, S., Brock, C. J., Pearce, A. C., Dwek, R. A., Watson, S. P., Hebestreit, H. F., Zitzmann, N., *Proteomics* 2004, *4*, 656–668.

[43] O'Neill, E. E., Brock, C. J., von Kriegsheim, A. F., Pearce, A. C., Dwek, R. A., Watson, S. P., Hebestreit, H. F., *Proteomics* 2002, *2*, 288–305.

[44] Petricoin, E. F., Ardenkani, A. A., Hitt, B. A., Levine, P. J., Fusaro, V. A., Steinberg, S. M., Mills, G. B. et al., *Lancet* 2002, *359*, 572–577.

[45] Wu, S. L., Jardine, I., Hancock, W. S., Karger, B. L., *Rapid Commun. Mass Spectrom.* 2004, *18*, 2201–2207.

[46] Olsen, J. V., Mann, M., *Proc. Natl. Acad. Sci. USA* 2004, *101*, 13417–13422.

[47] Wolters, D. A., Washburn, M. P., Yates, J. R, 3rd, *Anal. Chem.* 2001, *73*, 5683–5690.

[48] Olsen, J. V., Ong, S.-E., Mann, M., *Mol. Cell. Proteomics* 2004, *3*, 608–614.

[49] MacCoss, M. J., Wu, C. C., Yates, J. R., *Anal. Chem.* 2002, *74*, 5593–5599.

[50] Keller, A., Nesvizhskii, A. I., Kolker, E., Aebersold, R., *Anal. Chem.*, 2002, *74*, 5383–5392.

[51] Lopez-Ferrer, D., Martinez-Bartolome, S., Villar, M., Campillos, M., Martin-Maroto, F., Vazquez, J., *Anal. Chem.* 2004, ASAP 11-07-04, *76*, 6853–6860.

[52] Liu, T., Qian, W. J., Strittmatter, E. F., Camp, D. G, 2nd., Anderson, G. A., Thrall, B. D., Smith, R. D., *Anal. Chem.* 2004, *76*, 5345–5353.

[53] Hays, W. L., *Statistics*, fifth edn., Harcourt Brace College Publishers, Austin 1994.

[54] Haab, B. B., Geierstanger, B. H., Michailidis, G., Vitzthum, F. et al., *Proteomics* 2005, *5*, DOI: 10.1002/pmic.200400470.

[55] Omenn, G. S., State, D. S., Adaminski, M., Blackwell, T. W. et al., *Proteomics* 2005, *5*, this issue.

12
Analysis of Human Proteome Organization Plasma Proteome Project (HUPO PPP) reference specimens using surface enhanced laser desorption/ionization-time of flight (SELDI-TOF) mass spectrometry: Multi-institution correlation of spectra and identification of biomarkers*

Alex J. Rai, Paul M. Stemmer, Zhen Zhang, Bao-ling Adam, William T. Morgan, Rebecca E. Caffrey, Vladimir N. Podust, Manisha Patel, Lih-yin Lim, Natalia V. Shipulina, Daniel W. Chan, O. John Semmes, and Hon-chiu Eastwood Leung

We report on a multicenter analysis of HUPO reference specimens using SELDI-TOF MS. Eight sites submitted data obtained from serum and plasma reference specimen analysis. Spectra from five sites passed preliminary quality assurance tests and were subjected to further analysis. Intralaboratory CVs varied from 15 to 43%. A correlation coefficient matrix generated using data from these five sites demonstrated high level of correlation, with values >0.7 on 37 of 42 spectra. More than 50 peaks were differentially present among the various sample types, as observed on three chip surfaces. Additionally, peaks at ~9200 and ~15 950 m/z were present only in select reference specimens. Chromatographic fractionation using anion-exchange, membrane cutoff, and reverse phase chromatography, was employed for protein purification of the ~9200 m/z peak. It was identified as the haptoglobin alpha subunit after peptide mass fingerprinting and high-resolution MS/MS analysis. The differential expression of this protein was confirmed by Western blot analysis. These pilot studies demonstrate the potential of the SELDI platform for reproducible and consistent analysis of serum/plasma across multiple sites and also for targeted biomarker discovery and protein identification. This approach could be exploited for population-based studies in all phases of the HUPO PPP.

12.1
Introduction

Plasma has the unique property of providing a window to events throughout the body. The general circulation contains proteins from every tissue that are released through regulated secretion or *via* cell death and lysis [1, 2]. Reliable and reproducible

* Originally published in Proteomics 2005, 13, 3467–3474.

Exploring the Human Plasma Proteome. Edited by Gilbert S. Omenn
Copyright © 2006 WILEY-VCH Verlag GmbH & Co. KGaA, Weinheim
ISBN: 3-527-31757-0

detection and quantitation of proteins and protein fragments in the circulation will provide insight into normal and pathological processes throughout the body [3–5].

A serious challenge to harvesting the information in the plasma proteome is the complexity of the material [6]. In addition to the thousands of soluble proteins, lipids, and metabolic intermediates, plasma contains intact cells, platelets, and microparticles that are normal components of blood [7]. The thousands of different proteins in plasma are present at concentrations that vary over 10 log units [6, 8, 9]. Given the abundance of proteins and their widely different concentrations, it is reasonable to ask what fraction of the total we can detect and quantify with a particular methodology. Other important questions are the ability to differentiate between different samples (*e.g.*, patient and normal), the reproducibility of the findings in different laboratories, and the sample throughput that can be achieved.

Technical limits in MS make it essential that complex samples be fractionated prior to MS-based protein analysis [1, 10]. One of these limits is that protein and nonprotein components in samples suppress the signals from peptides and proteins that could be of interest. When the plasma is used as a conduit to examine tissues and processes that occur in less accessible parts of the body, the proteins from those sites are likely to be in low abundance [6]. Identifying and quantifying low-abundance proteins in plasma continues to be challenging. In some instances, a metabolic process or cellular "spillage" may act on abundant proteins to produce unique products, *e.g.*, proteolytic fragments, which are easier to quantify [11]. In either case the plasma must be fractionated before mass spectrometric analysis.

The SELDI approach to fractionation is to selectively bind a subset of proteins and peptides directly on the MALDI target [12]. Target surfaces are based on standard chromatographic matrices such as cation- and anion-exchange, metal affinity, and hydrophobic interaction, among others. When multiple surfaces are employed with step-gradient style washing procedures, a single sample can readily be dispersed over many target spots. This limited fractionation allows the detection of additional proteins in plasma, increasing the number of peaks to the hundreds range [12]. In the current study, we have used a single stringency wash for each surface and no prefractionation of samples. This is a strategy that is intended to reduce variability between participating sites but will also limit the total number of peaks detected so that low-abundance proteins are less likely to be represented.

SELDI analysis provides information on a subset of proteins and peptides in the plasma, but pertinent questions remain, including: are the proteins detected by SELDI an important subset of the plasma proteome, *i.e.*, is this an information-rich fraction, and, can the SELDI analysis be done reproducibly for dozens or hundreds of samples in a single laboratory or across several laboratories so that large population-based studies can be reliably performed? Here, we present the results of a pilot study addressing these questions. We also used SELDI to identify differences among samples in the HUPO reference specimens, and demonstrated that at least one differentially expressed protein can be purified and identified using additional proteomic methodologies.

12.2
Materials and methods

12.2.1
Sample preparation

All samples (four aliquots of 250 µL) were shipped directly to participating laboratories by BD Biosciences or the Chinese Academy of Medicine (CAM), according to the HUPO PPP protocol. Samples from BD Biosciences included a lyophilized plasma specimen from the UK National Institute of Biologics Standards and Control (NIBSC), and serum, potassium-EDTA plasma, citrated plasma, and heparinized plasma from donors of three different ethnicities: BD-B1, Caucasian-American, BD-B2, African-American, BD-B3, Asian-American. The fourth specimen set, CAMS, was that of Chinese volunteers from the Peoples Republic of China. All samples were collected using the same protocol and each pool is the combined serum or plasma from two volunteers.

12.2.2
Sample preprocessing

The number of samples analyzed by individual laboratories varied from 4 to 17 and was dependent on multiple factors including the availability of the BD-B1 specimens for late joining laboratories. The analysis was focused on spectra obtained from the CM10 (weak cation exchanger) biochips as this surface was common to the analysis by the various laboratories. Serum or plasma samples, 20 µL, were denatured by the addition of 30 µL of U9 buffer (9 M urea + 2% CHAPS, pH 9), and were incubated at 4°C for 30 min on a shaker. Five microliters of this mixture was then diluted 1:40 in binding buffer (100 mM sodium acetate, pH 4) to 200 µL, which was used for application to the bioprocessor.

12.2.3
Target (CM10) chip preparation and sample incubation

CM10 chips were assembled in a bioprocessor, followed by the addition of 200 µL binding buffer (100 mM sodium acetate, pH 4) to each well. Samples were placed on a shaker for 5 min at 100 rpm on an orbital shaker and the buffer was decanted. The process was repeated once before sample addition. Diluted sample, 200 µL, was applied to CM10 surface directly. The bioprocessor was placed in an orbital shaker and shaken at 100 rpm for 30 min at room temperature. Solutions in wells were discarded and 150 µL of buffer was added to each well as a washing step. The bioprocessor was allowed to shake for 5 min in orbital shaker before the buffer was decanted. This wash process was repeated and the bioprocessor was disassembled. The biochips were briefly rinsed with distilled water and then air-dried. Saturated SPA (0.5 µL in 50% ACN and 0.1% TFA) was added twice onto each spot and air-dried before the chips were read using the Ciphergen PBSII or PBSIIc ProteinChip Reader.

12.2.4
Scanning protocol

A mass range of up to 20 000 m/z was scanned for low-molecular-weight proteins. The optimized range was set between 1000 and 16 000 m/z. The laser intensity was optimized for each laboratory with the range being from 170 to 195. Detector sensitivity was set to 10, with shots being focused by optimization at the center. SELDI quantitation was chosen as the data acquisition method. Four warm-up shots were fired at a laser intensity of 10 greater than acquisition and the data from these warming shots were not included. A total of 40 shots were acquired in each spot before proceeding to the next spot.

12.2.5
Data processing

The region below 1000 m/z of each spectrum was eliminated because of the high noise level. Automatic baseline correction, *i.e.*, background subtraction, was used and data filtering was set as default at 0.2 times expected peak width; the signal enhance feature was not applied. Data were normalized to total ion current (TIC) from 1000 to 20 000 m/z.

12.2.6
Bioinformatics analysis of data and correlation coefficient matrix

SELDI allows for the selective analysis of a subset of the sample proteome. Differences in analytical protocols can result in very different mass spectra. In this study, only spectra spotted using the carboxymethyl chip surface were used for analysis. Spectra from different sites that, due to protocol differences, failed to share a majority of common peaks (in m/z locations, but not in peak intensities) with other sites were excluded from the final analysis. Protocol differences that led to data rejection included running chips on instruments that did not meet resolution specification or deviations from wash and/or binding stringency. To assess the concordance of SELDI-based proteomic profiling data among multiple sites, qualified peaks were identified manually in the m/z range of 1500–20 000. The selection is based on the overall quality of the peaks and their presence in more than a single site. The peak intensity data were then used to estimate a correlation coefficient matrix. For easy visualization, the correlation coefficient data were plotted as a pseudo-color image.

12.2.7
Protein purification, SDS-PAGE analysis, and extraction of proteins

After chromatographic fractionation of 45 µL of plasma, fractions were screened for peaks of interest using the same CM10 protocol. Selected fractions were dried in a SpeedVac, redissolved in 20 µL of Laemmli sample buffer, and loaded on SDS-PAGE gel. The gel was stained using the Colloidal Blue Staining Kit (Invitrogen). Selected protein bands were excised for further processing and identification. Gel pieces containing no protein were processed alongside the protein bands as a negative control. Selected protein bands were excised from the polyacrylamide gel with Pasteur pipettes (six to seven gel plugs *per* band). The gel pieces were washed with 200 µL of 50% methanol/10% acetic acid for 30 min, dehydrated with 100 µL of ACN for 15 min, and extracted with 70 µL of 50% formic acid, 25% ACN, 15% isopropanol for 2 h at room temperature with vigorous shaking. Two microliters of the extracts were then applied to NP-20 ProteinChip Arrays and allowed to dry. SPA (1 µL of a 50% solution) was applied twice to the spots, and the arrays were analyzed on a PBSII or PBSIIc ProteinChip Reader.

12.2.8
Peptide mass fingerprinting (PMF)

A 1 µL aliquot of each protease digest was spotted on an NP-20 ProteinChip Array and air-dried. Subsequently, a 1 µL aliquot of 20% saturated CHCA in 50% ACN, 0.5% TFA was applied twice to each spot. Data were collected in the peptide mass range ($<10\,000$ m/z) on a PBSII ProteinChip Reader. Spectra were externally calibrated with mass calibrants (All-in-1 Peptide standard, Ciphergen). m/z values of peptides that were unique to each protein of interest were submitted for a search of the NCBI and/or Swiss-Prot protein databases, using the ProFound search algorithm (http://129.85.19.192/prowl-cgi/ProFound.exe) as the database-mining tool.

12.2.9
MS/MS analysis

Single MS and MS/MS spectra were acquired on a Qq-TOF mass spectrometer equipped with a Ciphergen PCI-1000 ProteinChip Interface. A 1 µL aliquot of each protease digest was spotted on an NP-20 ProteinChip Array. Saturated CHCA (1 µL of 50% ACN/0.5% TFA) was immediately applied to the spot and the two solutions were mixed by pipetting. Spectra were collected from 900 to 3500 m/z in single MS mode. After reviewing the spectra, specific ions were subjected to MS/MS analysis. The MS/MS spectra were submitted to the database-mining tool MASCOT (Matrix Sciences) for identification.

12.2.10
Western blot analysis

Serum/plasma samples were resuspended in Laemmli sample buffer and separated by 4–20% SDS-PAGE. The gel was transferred onto an NC membrane, blocked with 5% nonfat dry milk in PBS-T (*i.e.*, PBS + 0.5% Tween-20), probed with anti-haptoglobin alpha subunit primary antibody (Rai *et al.*, unpublished observations at 1:10 000 dilution in PBS-T buffer, washed three times in PBS-T, probed with horseradish peroxidase-donkey antirabbit secondary antibody (GE Healthcare, Piscataway, NJ, USA), washed three times with PBS-T, then developed on film after incubation with ECL reagent (Pierce Biotechnology, Rockford, IL, USA).

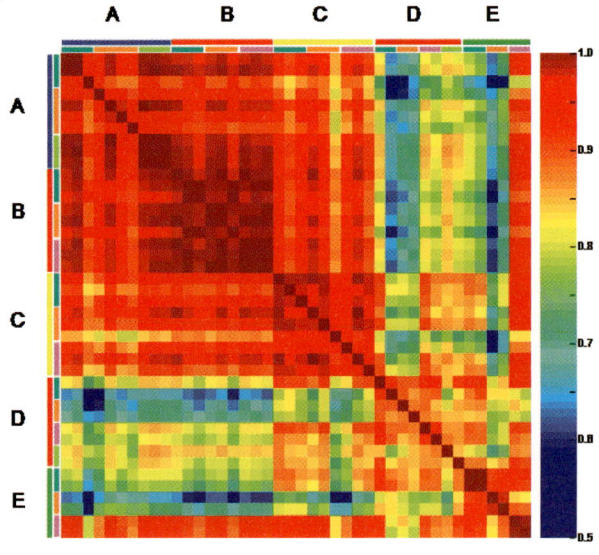

Fig. 1 Correlation coefficient matrix-analysis of qualified SELDI spectra from five sites. Data were used to estimate a correlation coefficient matrix. Letters correspond to anonymized sites and numbers correspond to correlation percentage in pseudo-color view. For additional details see Section 2.

12.3 Results

We solicited data from all sites participating in the HUPO PPP using SELDI analysis, and obtained eight data submissions. Of the eight, only five met criteria for inclusion in group analysis, as detailed in Section 2. The laboratories that produced data meeting criteria for acceptance routinely run SELDI analysis and have internal laboratory criteria to qualify both their instruments and target preparation procedures before analysis of test samples. The data sets that did not meet criteria were all from new or infrequent users of SELDI that did not have procedures in place to qualify their instruments and sample preparation.

A total of 42 spectra generated at five sites using the carboxymethyl surface were submitted and all were included in the final correlation analysis. Spectra from each site were generated using reference plasma samples from three or four HUPO sample sets with a minimum of two replicates. A total of 60 peaks were identified by automatic peak detection of all peaks in the spectra with $S/N > 5$ (second pass = 2) and used for the estimation of correlation coefficient matrix, which is displayed in Fig. 1. It shows that despite the different machine conditions and slight variations in protocols under which the SELDI experiments were conducted, the peak intensity data of the selected peaks with the m/z range of 1500–20 000 demonstrate a remarkably high level of concordance with all but five spectra having minimum correlation coefficients >0.7. In addition, to determine the precision of the data sets, we calculated CVs for all peaks with $S/N > 5$ (Tab. 1). Intralaboratory CV ranged from 15 to 43%, and was similar in the processing of serum or plasma. This discrepancy in CVs among the four sites is likely due to the use of robotic instrumentation for sample processing, which was used by two of the four sites.

Tab. 1 Intralaboratory precision comparison-SELDI analysis of HUPO serum and EDTA plasma reference specimens

Specimen type	Donor source (laboratory no.)	CV of replicates, %	Average CV (laboratory no.), %
Serum	BD-B2 (1)	43.1	43.1 (1)
	BD-B3 (1)	44.1	
	CAM (1)	41.6	
	BD-B2 (2)	26.2	19.4 (2)
	BD-B3 (2)	17.1	
	BD-B1 (2)	14.8	
	BD-B2 (3)	24.6	37.6 (3)
	BD-B3 (3)	39.1	
	BD-B1 (3)	49.2	
	BD-B2 (4)	16.4	19.9 (4)
	BD-B3 (4)	24.5	
	BD-B1 (4)	20.6	
	CAM (4)	17.9	
EDTA plasma	BD-B2 (1)	40.1	38.1 (1)
	BD-B3 (1)	37.2	
	CAM (1)	36.9	
	BD-B2 (2)	17.1	15.3 (2)
	BD-B3 (2)	12.3	
	BD-B1 (2)	16.6	
	BD-B2 (3)	30.6	35.1 (3)
	BD-B3 (3)	40.7	
	BD-B1 (3)	33.8	
	BD-B2 (4)	23.4	26.2 (4)
	BD-B3 (4)	33.5	
	BD-B1 (4)	24.1	
	CAM (4)	24.1	
	BD-B2 (5)	13	18.8 (5)
	BD-B3 (5)	19.2	
	BD-B1 (5)	24.2	

Note: comparison of serum samples is based on analysis at four laboratories, whereas comparison of plasma is based on analysis at five laboratories.

We selected one of these data sets for further in-depth analysis. A careful inspection of the spectra identified multiple peak differences between the different sample types within a donor set (Tab. 2). These were noted on several different chip surfaces, including cation- and anion-exchange, and metal affinity capture. Most of the differences were noted to be peaks that were differentially present between sample types, *i.e.*, plasma *versus* serum. In addition, when samples were processed under reduced conditions, a particular set of peak differences at ~9200 and ~15 950 m/z were noted, distinguishing the different donor source serum samples (Fig. 2).

Tab. 2 Representative peak differences among plasma and serum samples on IMAC, weak cation exchange (WCX), and strong anion-exchange (SAX) surfaces. Absence of a peak is denoted by an empty box, whereas the presence is denoted with a checkmark; number of checkmarks indicates relative peak intensity. An m/z values are approximate

Peak m/z	Citrate	Heparin	K-EDTA	Serum
IMAC				
3100	✓	☐	☐	☐
3400	✓	☐	☐	☐
3900	✓	✓	✓✓	✓✓
4550	✓	✓	☐	☐
4600	☐	☐	☐	✓✓
5100	✓	☐	☐	☐
5200	☐	☐	☐	✓
5300	☐	☐	☐	✓✓
5850	✓	✓	☐	☐
5900	☐	✓	✓	✓✓✓
6100	☐	☐	☐	✓
6200	☐	☐	✓	☐
6500	✓	✓	☐	☐
7750	✓	☐	✓	✓✓
8000	☐	☐	☐	✓
8100	☐	☐	☐	✓
9200	☐	☐	☐	✓✓
9300	☐	✓	✓	✓✓
9500	☐	☐	☐	✓
10 000	✓	✓	✓	☐
10 250	☐	☐	☐	✓
11 700	✓	✓	☐	☐
13 400	✓	✓	☐	✓
13 900	✓	✓	☐	✓
14 600	✓	✓	✓	☐
19 900	✓✓	✓✓	✓✓	☐
28 100	✓✓	✓	☐	✓✓
WCX				
3200	☐	☐	☐	✓
3250	☐	☐	☐	✓
4100	☐	☐	☐	✓
4200	☐	☐	☐	✓✓
5300	☐	☐	☐	✓
5900	☐	☐	☐	✓✓
6100	☐	☐	☐	✓
6150	☐	☐	☐	✓
7550	☐	☐	☐	✓
7700	☐	☐	☐	✓✓
7900	☐	☐	☐	✓
8150	✓	✓	✓	✓✓✓

Tab. 2 Continued

Peak m/z	Citrate	Heparin	K-EDTA	Serum
8900	☐	☐	☐	√√
9200	☐	☐	☐	√√
9300	√	√	√	√√
10 250	☐	☐	☐	√
15 100	☐	☐	√	☐
15 900	☐	☐	√	☐
SAX				
Triplet at 4200	☐	☐	☐	√√
5000	☐	☐	☐	√
Doublet at 5700	☐	☐	☐	√
Doublet at 5800	☐	☐	☐	√√
12 500	☐	☐	☐	√
12 600	☐	☐	☐	√

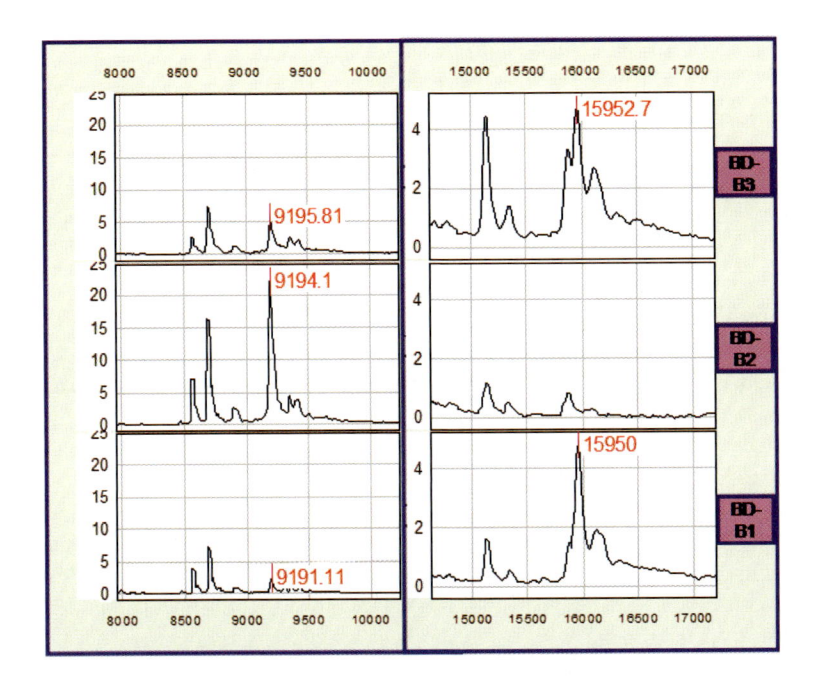

Fig. 2 Detection of differentially expressed peaks in HUPO reference samples using SELDI-TOF MS. Serum samples were reduced with DTT and denatured proteins were solubilized with urea and CHAPS (see Section 2). Two microliters of resulting preparations were profiled using the CM10 (carboxymethyl) ProteinChip Array on (1) B3, (2) B2, and (3) B1 HUPO sample sets. Note that samples were run in triplicate, but representative results from one replicate are shown.

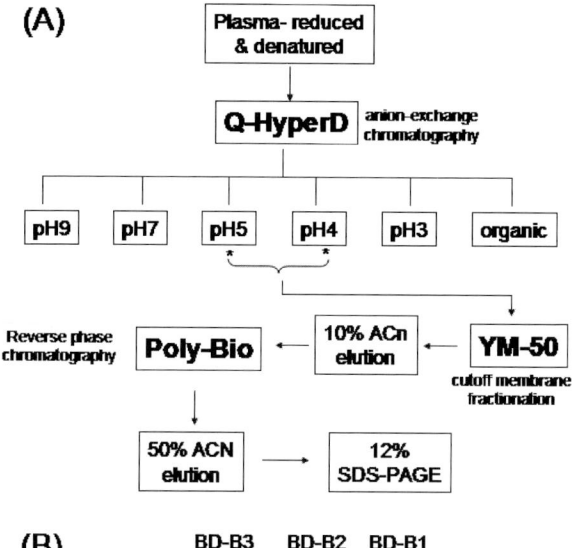

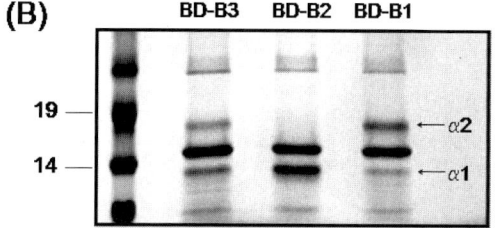

Fig. 3 Purification scheme for ~9200 m/z peak. (A) A scheme for sequential chromatographic separation of serum samples for enrichment of this protein peak. (B) 12% SDS-PAGE analysis of the reference samples after fractionation.

We sought to determine the identity of the protein corresponding to the ~9200 m/z peak. As detailed in Fig. 3, anion-exchange fractionation was used as a first step towards protein purification, and the samples were profiled using carboxymethyl arrays. Additional steps included cut-off membrane fractionation and RP chromatography. We looked for differences in the samples at the appropriate molecular mass range that correlated well between the peaks, as shown by SELDI, with that of the protein bands, as visualized by SDS-PAGE. The final fraction was separated by SDS-PAGE, the band was excised from the gel, trypsin digested, and PMF was performed (Fig. 4). After interrogating the protein databases with seven tryptic peptides, the protein band was identified as the haptoglobin alpha subunit. This identity was confirmed using MS/MS analysis (data not shown).

To verify that the haptoglobin alpha subunit was indeed expressed at different levels in the samples, we employed an antibody specific to the alpha subunit for use in Western blot analysis. This antibody was raised against a peptide that is specific to the alpha subunits of haptoglobin, affinity purified using a peptide column, and its specificity was confirmed by 1- and 2-DE (Rai et al., unpublished observations). The result, as shown in Fig. 5, demonstrates that these various HUPO serum specimen samples exhibit differences in protein levels of both the α1 and α2 subunits. In addition, a second antibody to the haptoglobin protein, used for Western blot analysis, has confirmed the differences that are represented (data not shown).

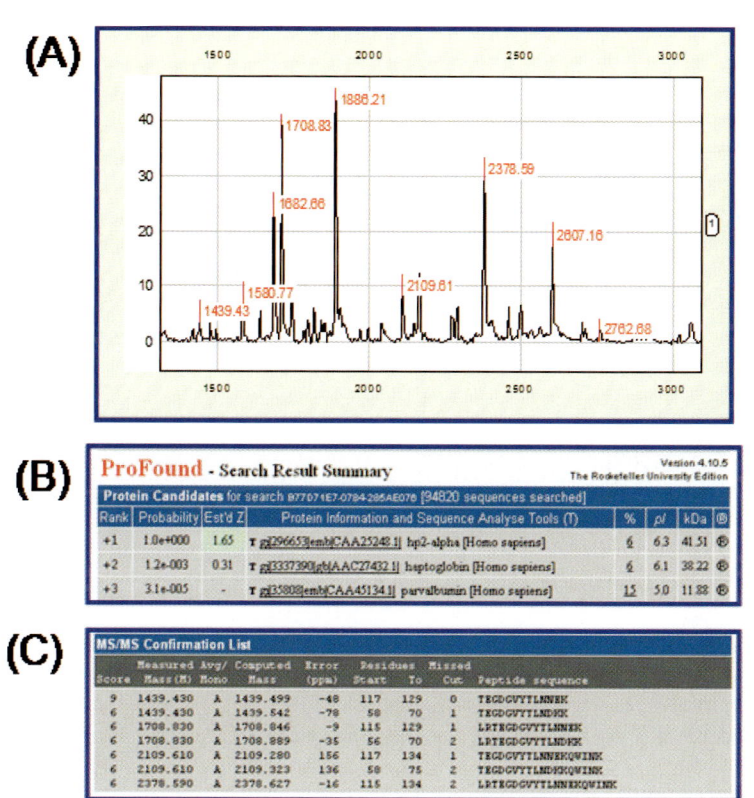

Fig. 4 PMF analysis of the 9.2 kDa protein band. (A) SELDI-TOF analysis after tryptic digest. (B) ProFound result summary after database search analysis. (C) List of peptides matching to the haptoglobin alpha subunit.

12.4 Discussion

SELDI peak intensity data are a measure of the relative abundance of protein expression levels. The absolute value of an individual peak is also affected by sample preparations, instrument settings, and data normalization procedures. For a particular peak, the intensity values in spectra from different sites could be very different; however, the results from this study show that the relative intensities of multiple peaks along the m/z range are very consistent among the spectra from different sites. This indicates the possibility of comparing SELDI results across multiple sites.

During our analysis of the SELDI data submitted by eight participating laboratories, it became clear that stringent steps must be taken to enable meaningful comparisons of interlaboratory data. This is in contrast to the relative ease and accuracy of differential profiling carried out on one instrument in one laboratory.

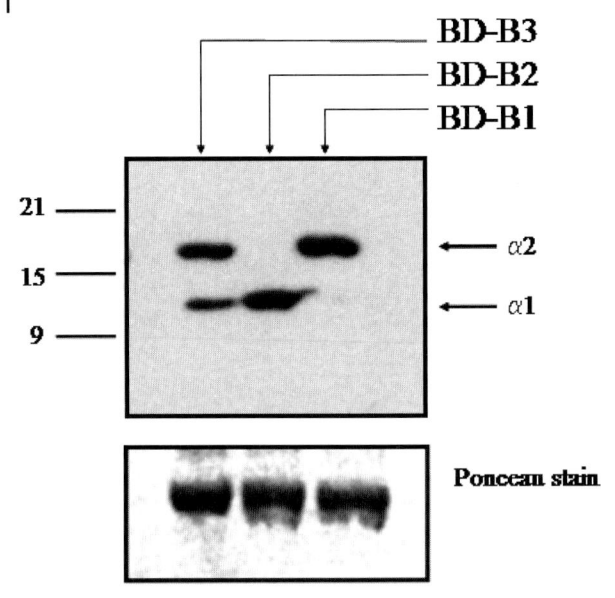

Fig. 5 Confirmation of differences in haptoglobin alpha subunit expression as demonstrated by Western blot analysis of HUPO samples. Serum samples (1 µL each) were electrophoresed on 4–20% SDS-PAGE, blotted to NC, and probed with antihaptoglobin alpha subunit antibody; Ponceau staining of the membrane was used to demonstrate equal protein loading.

Consequently, when interlaboratory results are to be compared, care must be taken to standardize the parameters of analysis. These include identical sample preparation and treatment protocols, identical tuning of the analytical instruments with the same set of external standards for mass accuracy, and checks of sensitivity and reproducibility between runs. This is facilitated by the use of standard quality control chips with IgG and insulin, and procedures for standardization (all are available from Ciphergen Biosystems). We also noted that results obtained by manual sample application to the chips differed significantly from that of sample application using robotic instrumentation, with regard to precision. This demonstrates that sample application protocols as well as instrument parameters must be standardized for reliable interlaboratory comparisons of protein profiles obtained using SELDI. Only in this manner can consistent results be obtained, similar to that demonstrated in our correlation coefficient matrix (Fig. 1).

The samples analyzed by SELDI were done without prefractionation and with a focus on the low-mass components. The portion of the mass spectra <20 000 m/z is considered the low-mass range for this study. Spectra obtained on the PBSII and PBSIIc instruments used in this work typically have a greater number of peaks with higher average intensities in the low-mass range and, therefore, our analysis was focused on that part of the spectra. Other approaches that focus on the low-mass portion of the proteome have been termed "peptidomics" and are presented elsewhere in this issue [13]. These approaches enlist prefractionation of samples to remove the vast majority of protein components and increase the signal intensity of the remaining low-mass species. The SELDI approach can incorporate any method of prefractionation, including size-based separation. By running the samples in this project without prefractionation, we are able to use the same target for both the high- and low-mass components and to compare results of the instrumentation between laboratories without concern for interlaboratory differences introduced during prefractionation.

Focusing on the low-mass range, we compared the spectra resulting from the different HUPO PPP reference specimen sets. Many differences, both between sample types and donor sets, were noted among these spectra, and are detailed in Tab. 2. It is clear from the results that the HUPO PPP reference serum contains a greater number of peaks in the low-molecular-mass range than any of the reference plasma sample types. Analysis of the protein profiles revealed that the serum sample is clearly the most target-rich sample. Each of the plasma samples: citrate, heparin, and EDTA, exhibit relatively fewer protein peaks on the three surfaces tested, as compared to serum. This may be because all three clotting additives have the ability to compete or interfere with the binding of protein species on these surfaces. Although serum contains a plethora of peaks, the identity of the corresponding protein species remains to be identified. It is important to remember that there is only one representative of each sample type and specimen set, with all laboratories receiving that sample. Therefore, these results cannot be considered to be representative of the groups they are associated with either for sample type (heparinized plasma *vs.* EDTA treated plasma) or specimen set (BD-B1 *vs.* BD-B2 *vs.* BD-B3), and additional samples must be run to validate the apparent differences that are observed. In the samples analyzed, we selected one peak which varied between groups to illustrate the potential of the SELDI platform to support the capability of targeted protein identification. This was done using standard chromatographic steps, SDS-PAGE, trypsin digestion of selected protein bands followed by PMF and MS/MS analysis.

The haptoglobin $\alpha1$ and $\alpha2$ subunits were identified as proteins that are differentially present in the various HUPO donor sample sets. Because each sample set was collected from only two individuals and pooled to create a single sample of each type, the differences found in this analysis cannot be generalized to larger groups, *i.e.*, ethnic-specific differences. It is clear, however, that the haptoglobin $\alpha1$ and $\alpha2$ subunits are differentially expressed in these donor samples. It is interesting to note that differences among these samples were also observed using immunoassay technology within the HUPO PPP [14]. However, because the antibodies used could not distinguish $\alpha1$ and $\alpha2$ subunits, only the amount of haptoglobin protein was shown to vary between the donor sets. Such differences, both in haptoglobin concentration and differences in protein composition are logically plausible. The haptoglobin gene is known to exhibit polymorphism in the human population [15]. The encoded protein exists as three major phenotypes (1-1, 2-1, and 2-2) due to differences in the expression of the $\alpha1$ and $\alpha2$ subunit composition [16, 17]. Further, there are minor sequence variants that are characteristic of different ethnic groups. Our identification of haptoglobin as a differentially expressed protein among the HUPO donor sets illustrates a viable approach for the targeted identification of biomarkers. Such a methodology can be generalized and applied for the identification of biomarkers to distinguish between any two different sample groups. However, it is noteworthy that even though the concentration of Hp is in the µg/mL range, making it one of the higher abundance proteins in plasma, the approach that we use for biomarker discovery can be generalized to identify markers of lower abundance. For such markers, profiling will likely require front-end prefractionation to reduce sample complexity. This will allow the detection of lower abundance components. Consequently, the requirement for starting materials for protein purification will be greater, and as long as sufficient quantities of materials are available, identification of lower abundance protein components is feasible.

The SELDI platform, consisting of the target chips and reader, is designed to find differences among the proteomes of sample groups [18]. The data from the HUPO PPP demonstrate that unfractionated samples of serum and plasma yield information on approximately 60 "peaks". This "first pass" analysis presumably reflects the most highly abundant proteins in these samples. Additional sample fractionation and analysis has been shown to expand SELDI proteome coverage by several fold [12]. The obvious question is whether or not the fraction of the proteome observable by SELDI contains information that can establish differences between groups. The answer is unequivocally yes. Even with the limited sample size in the current study, clear differences between serum and plasma and between plasma types were found using the SELDI platform (Tab. 2). These differences are consistent and are reproducibly found by all laboratories submitting qualified data.

SELDI analysis provides information on a subset of proteins and peptides in serum and plasma that can establish differences between sample groups. This analysis can be done reproducibly in multiple laboratories, and the analysis is amenable to simultaneous analysis of dozens or hundreds of samples. In addition to the current work detailed here, similar results have been demonstrated in another recent publication [19], and techniques to further improve data quality for robust peak identification have also been described [20]. These features establish SELDI analysis as a powerful approach to proteomic analysis in population-based studies, and hence the utility of this technology can be exploited in all phases of the HUPO studies.

This work was supported in part by a grant from HUPO to A.J.R. and was assisted by the services of the Protein Interactions and Proteomics Core at Wayne State University, which is supported by NIEHS grant P30 ES06639.

12.5
References

[1] Pieper, R., Su, Q., Gatlin, C. L., Huang, S. T., Anderson, N. L., Steiner, S., *Proteomics* 2003, *3*, 422–432.

[2] Anderson, N. L., Polanski, M., Pieper, R., Gatlin, T. *et al.*, *Mol. Cell. Proteomics* 2004, *3*, 311–326.

[3] Verma, M., Wright, G. L., Jr., Hanash, S. M., Gopal-Srivastava, R. *et al.*, *Ann. NY Acad. Sci.* 2001, *945*, 103–115.

[4] Zhang, R., Barker, L., Pinchev, D., Marshall, J. *et al.*, *Proteomics* 2004, *4*,: 244–256.

[5] Bischoff, R., Luider, T. M., *J. Chromatogr. B.* 2004, *803*, 27–40.

[6] Anderson, N. L., Anderson, N. G., *Mol. Cell. Proteomics* 2002, *1*, 845–867.

[7] VanWijk, M. J., VanBavel, E., Sturk, A., Nieuwland, R., *Cardiovasc. Res.* 2003, *59*, 277–287.

[8] Rabilloud, T., *Proteomics* 2002, *2*, 3–10.

[9] Corthals, G. L., Wasinger, V. C., Hochstrasser, D. F., Sanchez, J. C., *Electrophoresis* 2000, *21*, 1104–1115.

[10] Steel, L. F., Trotter, M. G., Nakajima, P. B., Mattu, T. S. *et al.*, *Mol. Cell. Proteomics* 2003, *2*, 262–270.

[11] Tirumalai, R. S., Chan, K. C., Prieto, D. A., Issaq, H. J. et al., *Mol. Cell. Proteomics* 2003, *2*, 1096–1103.

[12] Merchant, M., Weinberger, S. R., *Electrophoresis* 2000, *21*, 1164–1177.

[13] Tammen, H., Schulte, I., Hess, R., Menzel, C. et al., *Proteomics* 2005, *5*, DOI: 10.1002/pmic.200400419.

[14] Haab, B. B., Geierstanger, B. H., Michailidis, G., Vitzthum, F. et al., *Proteomics* 2005, *5*, DOI: 10.1002/pmic.200400470.

[15] Langlois, M. R., Delanghe, J. R., *Clin. Chem.* 1996, *42*, 1589–1600.

[16] Van Vlierberghe, H., Langlois, M., Delanghe, J., *Clin. Chim. Acta* 2004, *345*, 35–42.

[17] Wassell, J., *Clin. Lab.* 2000, *46*, 547–552.

[18] Tang, N., Tornatore, P., Weinberger, S. R., *Mass Spectrom. Rev.* 2004, *23*, 34–44.

[19] Semmes, O. J., Feng, Z., Adam, B. L., Banez, L. L. et al., *Clin. Chem.* 2005, *51*, 102–112.

[20] Malyarenko, D. I., Cooke, W. E., Adam, B. L., Malik, G. et al., *Clin. Chem.* 2005, *51*, 65–74.

13
An evaluation, comparison, and accurate benchmarking of several publicly available MS/MS search algorithms: Sensitivity and specificity analysis*

Eugene A. Kapp, Frédéric Schütz, Lisa M. Connolly, John A. Chakel, Jose E. Meza, Christine A. Miller, David Fenyo, Jimmy K. Eng, Joshua N. Adkins, Gilbert S. Omenn and Richard J. Simpson

MS/MS and associated database search algorithms are essential proteomic tools for identifying peptides. Due to their widespread use, it is now time to perform a systematic analysis of the various algorithms currently in use. Using blood specimens used in the HUPO Plasma Proteome Project, we have evaluated five search algorithms with respect to their sensitivity and specificity, and have also accurately benchmarked them based on specified false-positive (FP) rates. Spectrum Mill and SEQUEST performed well in terms of sensitivity, but were inferior to MASCOT, X!Tandem, and Sonar in terms of specificity. Overall, MASCOT, a probabilistic search algorithm, correctly identified most peptides based on a specified FP rate. The rescoring algorithm, PeptideProphet, enhanced the overall performance of the SEQUEST algorithm, as well as provided predictable FP error rates. Ideally, score thresholds should be calculated for each peptide spectrum or minimally, derived from a reversed-sequence search as demonstrated in this study based on a validated data set. The availability of open-source search algorithms, such as X!Tandem, makes it feasible to further improve the validation process (manual or automatic) on the basis of "consensus scoring", *i.e.*, the use of multiple (at least two) search algorithms to reduce the number of FPs.

13.1
Introduction

A major goal of the HUPO Plasma Proteome Project (PPP) is a comprehensive analysis of the protein constituents of human plasma and serum [1]. The pilot phase of this project brought together submissions from 47 different laboratories, of which 18 laboratories submitted peptide and protein identification

---

* Originally published in Proteomics 2005, 13, 3475–3490

Exploring the Human Plasma Proteome. Edited by Gilbert S. Omenn
Copyright © 2006 WILEY-VCH Verlag GmbH & Co. KGaA, Weinheim
ISBN: 3-527-31757-0

tables based on MS/MS acquired in either an ion trap (IT) or Q-TOF-like mass spectrometers coupled to multidimensional LC. In order to maximize the discovery of low abundance and potentially interesting peptides and/or proteins, the HUPO-PPP committee emphasized the need for laboratories to submit peptide and protein identification tables along with corresponding protocols, and assigned identifications as either "high-" or "low-confidence". Although this approach is potentially flawed, since the error rate and number of false-positive (FP) protein identification submissions are unknown, it does at least allow the capture of all information; it is then up to informaticians to expertly curate the information such that a reliable analysis of the protein constituents of human plasma and serum can be reported.

It is well recognized that more extensive analysis of LC-MS/MS data is required if data from different experiments, instruments, and laboratories are to be compared [2, 3]. Recent guidelines [4] and issues [5] for the dissemination and publication of large proteomic data sets indicate a growing awareness that a significant number of published protein identifications are indeed incorrect. Hence, an appraisal of MS software and a more informed understanding of the scoring schemes employed by current industry standard MS/MS database search algorithms are warranted [6]. Future literature mining (*e.g.*, Anderson *et al.* [7]) and bioinformatic prediction tools rely heavily on expertly curated data sets so it is imperative that the level of reported FPs remains low, preferably below 1% level.

In MS/MS, gas-phase peptide ions (precursor ions) undergo CID with molecules of an inert gas, such as helium or argon. Under low-energy CID (<100 eV) conditions, typical for ITs, the precursor ion fragments along the peptide backbone bonds give rise to mainly *y*-, *b*-ions, and their neutral losses. Importantly, most of the intensity of the precursor ion is distributed amongst its product ions and depending on the peptides' composition and charge state, might also give rise to selective cleavages, such as enhanced cleavage *N*-terminal to a proline amino acid residue and/or oxidized methionine residues [8, 9], which might hinder its structural elucidation by both *de novo* sequencing and/or database search methods. If an MS/MS spectrum is acquired for a peptide, then its amino acid sequence can be determined by matching the MS/MS spectrum to a known *in silico* generated database of peptide spectra using search algorithms such as SEQUEST [10] and MASCOT [11] in an uninterpreted manner. The rate-limiting step in defining a proteome by these methods is not the capacity to correlate tryptic peptides in this manner, but rather the capacity to accurately interpret such data [12]. Ultimately, investigators aim to determine the protein or gene from which a peptide is derived. This problem is complicated by the fact that a peptide sequence usually does not uniquely define a protein [13]. To this end statistical approaches and models, which attempt to make tandem MS data analysis a consistent and transparent process across research groups, mass spectrometers, and even different MS/MS database search tools, have been developed [14, 15]. These models would undoubtedly benefit from a more informed understanding of the strengths, weaknesses, and limitations of current search algorithms.

Current MS/MS search algorithms scoring functions can essentially be classified into two categories. One category of search algorithms, referred to as heuristic algorithms, correlate the acquired experimental MS/MS spectrum with a theoretical spectrum and calculate a score based on the similarity between the two. These search algorithms are often based on the notion of "shared peak count" (SPC), which simply counts the number of peaks common to the two spectra. Examples of heuristic algorithms include SEQUEST, Spectrum Mill, X!Tandem, and Sonar. Probabilistic algorithms (*e.g.*, MASCOT), on the other hand, model to some extent the peptide fragmentation process (*e.g.*, ladders of sequence ions) and calculate the probability that a particular peptide sequence produced the observed spectrum by chance. A recent review by Sadygov *et al.* [16] provides a useful update and supplement regarding the different models of MS/MS database search algorithms.

13.1.1
Heuristic algorithms

SEQUEST [10] uses a preliminary scoring (Sp) algorithm, based on a variation of the SPC, to select the 500 best candidate peptide sequences for direct cross-correlation. To speed up computations, fast FTs are used to compute the cross-correlation (X_{corr}), but this does not have any influence on the score itself. For each candidate peptide sequence several scores and rankings are determined.

Spectrum Mill allows MS/MS spectra to be filtered prior to searching, which significantly reduces the number of spectra that need to be analyzed. Its scoring concept is similar to that of the SPC in that 25 of the most abundant fragment ions (above noise level) are matched. Bonus points are awarded depending on the ion type (b or y) as well as penalty points for unmatched peaks, which is inversely proportional to the relative peak intensity of the unmatched fragment ion. A "scored peak intensity" (SPI) is also calculated, which is the proportion of the TIC that has been assigned (values less than 70% represent a poor interpretation). Again, empirically determined thresholds are used to indicate the correctness of a match, which are applied in an automated fashion.

Sonar [17] (http://bioinformatics.genomicsolutions.com/service/prowl/sonar.html) ranks the proteolytic peptides from proteins in a sequence collection by calculating a score based on the dot product between the theoretical and experimental tandem mass spectra (similar to clustering approaches [18, 19]). The score is subsequently converted into an expectation value [2]. The expectation value is obtained by collecting statistics during the search to estimate the distribution of scores for random and false identifications. This distribution is hypergeometric, and the expectation value of high scoring peptides can therefore be obtained by extrapolation. The expectation value represents the number of peptides that are expected to get a certain score by random matching.

X!Tandem [20, 21] (http://www.proteome.ca/x-bang/tandem/tandem.html) is an open-source search engine that has been optimized for speed. It generates theoretical spectra for the peptide sequences using knowledge of the intensity patterns associated with particular amino acid residues. These spectra are then correlated with the experimental data using a dot product (similar to Sonar). Subsequently, an expectation value is calculated.

13.1.2
Probabilistic algorithms

Details of the probabilistic MASCOT scoring algorithm have not been published. However, the Matrix Science website (http://www.matrixscience.com) indicates that the MASCOT algorithm incorporates a probability-based implementation of the MOWSE scoring algorithm used for PMF [22] as well as, amongst other things, fragment ion series, mass accuracy, and peptide length. For each peptide, MASCOT reports a probability-based "Ions Score", which is defined as $-10^{*}\log_{10}(p)$, where p is the probability that the observed match between experimental data and the database sequence is a random event. Knowing the size of the sequence database being searched, it becomes possible to provide an objective measure of the significance of a result. MASCOT V2.0 also reports an expectation value, which is similar to those reported by both Sonar and X!Tandem.

Since the majority of search algorithms will always return a score even if the peptide represented by the product ion spectrum is not in the database, it is useful to have an idea of the distribution of the scores for correct or incorrect hits to be able to assess the significance of a particular result. Empirically determined thresholds (filters) have been used [23–25] to indicate the correctness of a match. More recently, the PeptideProphet [14] rescoring algorithm uses Fisher's Linear Discriminant Analysis (LDA) to combine the different SEQUEST scores with other information (*e.g.*, mass difference). The Expectation-Maximization (EM) algorithm as well as Bayes theorem is then used to derive a probability that the peptide hit is correct.

In this paper, we explore the performance of the different MS/MS search algorithms, which were used by the participating HUPO-PPP laboratories, on IT data specially prepared by one of these laboratories. Overall, the main aim of the work is to accurately compare and benchmark the different MS/MS search algorithms based on a validated data set. The more detailed aims of the search algorithm analysis are: (i) to create an expertly curated reference data set that could be used for testing improved MS/MS scoring functions; (ii) to assess the strengths and weaknesses in terms of sensitivity and specificity of the different algorithms; (iii) to accurately benchmark the different algorithms at a specified FP rate; (iv) to assess the effect of database size and different search strategies (tryptic *vs.* nontryptic); (v) to determine the utility of reversed sequence database searches; and (vi) to assess the idea of consensus scoring by combining the results of multiple search algorithms.

13.2
Materials and methods

13.2.1
HUPO-PPP reference specimens

Two reference specimens from BD Diagnostics (citrated plasma (Cit-plasma) and serum) for each of three ethnic groups (B1-Caucasian-American, B2-African-American, and B3-Asian-American) were used in these studies [1]. The B1-serum and B1-Cit-plasma (Caucasian-American ethnic group) as well as B3-serum and B3-Cit-plasma (Asian-American ethnic group) were extensively analyzed including manual MS/MS spectrum validation.

13.2.2
Sample preparation and MS analysis

The HUPO-PPP samples were prepared for MS and run by the PNNL (Adkins and Pounds, see Acknowledgements) as described by Adkins *et al.* [26]. Briefly, serum and plasma were immunoglobulin (Ig) depleted, digested using modified trypsin (Promega), and conditioned by C18 SPE column (Supelco) clean-up. RP separation was performed with an Agilent 1100 capillary column (90 min gradient) interfaced to an LCQ Deca XP IT mass spectrometer (ThermoFinnigan, San Jose, CA, USA) using ESI. The mass spectrometer was operated in the data-dependent mode to automatically switch between MS and MS/MS acquisition, selecting the three most intense precursor ions for fragmentation using CID.

13.2.3
Protein sequence databases

All tandem mass spectra were searched against two protein sequence databases and randomized versions of these databases (forward and reverse): a Ludwig Institute nonredundant database (NR, August 2003, ~1.5 million entries) [27] and the Human International Protein Index database (IPI, version 2.21 July 2003, ~56 000 entries, European Bioinformatics Institute, www.ebi.ac/uk/IPI/) [28]. The randomized versions of these databases were created by taking all protein sequence entries and reversing them, such that the original sequence length and composition were preserved.

13.2.4
MS/MS database search strategy

Since the majority of submissions by HUPO-PPP participating laboratories were based on IT-MS/MS data, it was deemed appropriate to restrict our analysis to search algorithms used by these individual laboratories. Peptide and protein identification lists submitted by the individual participating laboratories to the University of Michigan (central repository) were based on search results from MASCOT, SEQUEST, Sonar, X!Tandem, and Spectrum Mill. Four independent research groups with considerable experience in using one or more of these programs volunteered to analyze the MS data prepared by the PNNL. The JPSL group (Melbourne, Australia) used SEQUEST and MASCOT. Independently, the Agilent team (Jose Meza, Christine Miller, and John Chakel) used Spectrum Mill, David Fenyo at GE Healthcare (formerly Amersham Biosciences) used Sonar and X!Tandem, and Jimmy Eng (ISB, Seattle) used SEQUEST and PeptideProphet to analyze the data. Each group independently decided on their choice of parameters (*i.e.*, data extraction and search parameters) as well as search strategy in order to maximize and optimize at the search algorithm level. Comparisons between search algorithms were carried out using only the subset of spectra common to all the searches (3952 MS/MS spectra, see Section 2.4.1). The MS data as well as protein sequence databases used (where appropriate) were identical for all groups.

13.2.4.1 SEQUEST and MASCOT workflow performed by the JPSL research group

The LCQ_DTA utility, obtained from ThermoFinnigan as part of the SEQUEST package of programs, was used to extract the MS/MS spectra from the raw instrument data files into individual spectra files (.dta file extension). Parameters used to extract MS/MS spectra were: 700–5000 (min–max mass); minimum of 35 peaks and minimum TIC of 1×10^5 counts. Spectra were not merged, and since doubly- and triply-charged precursor ions cannot accurately be distinguished using low-resolution ESI-IT MS, all spectra not calculated as singly charged were extracted as both doubly- and triply-charged spectra. This resulted in the analysis of 3952 MS/MS spectra for the B3-Cit-plasma (Asian-American) sample. Searches were carried out using both algorithms against the IPI and NR database in both forward and reverse directions using the following search parameters: trypsin-constrained (full with two missed cleavages) as well as no-enzyme (unconstrained) searches; no static or differential modifications; 3 Da precursor ion tolerance and 0.5 Da fragment ion tolerance using monoisotopic masses, and ESI-IT selected as instrument setting.

13.2.4.2 SEQUEST and PeptideProphet workflow performed by the ISB research group

ThermoFinnigan LCQ raw instrument data files were first converted to the mzXML file format using the ReAdW program [29]. The mzXML2Other program was used to extract individual spectra from the mzXML files into MS/MS files of the .dta format. For the reasons stated previously, all spectra not extracted as singly charged were extracted as both doubly- and triply-charged and no individual spectra were merged. After extraction, a filtering program, named dtafilter (http://sourceforge.net/projects/sashimi), was used to reduce the data set based on the following parameters: 600–4200 Da peptide mass range and a minimum of six peaks with a minimum intensity of 2. This resulted in the analysis of 5579 MS/MS spectra for the B3-Cit-plasma (Asian-American) sample.

SEQUEST database searches were performed on these spectra against the Human IPI protein sequence database (version 2.21). The search parameters were as follows: average masses used for both the peptide mass and fragment ion calculations, peptide mass tolerance set to 3.0, fragment ion tolerance set to 0.0, variable modification of +16.0 to methionine residues, and a sequence constraint of at least one tryptic cleavage site. All search results were passed to the PeptideProphet algorithm using default parameters for IT data. Based on multiple factors of the search results, including individual database search scores and distribution of peptides exhibiting expected cleavage rules, the PeptideProphet algorithm assigned a probability of being a correct identification to each search result.

13.2.4.3 Spectrum Mill workflow performed by the Agilent group

LCQ instrument data files (*.raw) were extracted with the Spectrum Mill Data Extractor using the following parameters: 600–5000 (min–max mass); sequence tag length on (>1) and off with no spectral merging for two separate sets of search results. Where spectral charge state cannot be determined, no charge state is assigned during extraction and both +2 and +3 charge states are considered during searches. Searches were carried out against the IPI database in both forward and modified reverse directions using the following search parameters: initial search in "multihomology" mode in which combinations of carbamylated lysine, oxidized methionine, and Pyro–Glu modifications were applied; trypsin specific with two missed cleavage; 2.5 Da precursor ion tolerance; 0.7 Da fragment ion tolerance; and ESI-IT as instrument. The initial results were also autovalidated using the following parameters for the "protein details" mode: SPI >70% for matches with score >8 for +1, >7 for +2, and >9 for +3; SPI >90% for score >6 on +1. A second autovalidation step was done in "peptide" mode using criteria of a score >13 and SPI >70%. In addition, both autovalidation steps required a forward–reverse score >1 for +1 and +2 and >2 for +3 peptides. The validated peptides were used to identify a set of proteins from which a result file was created. A second round of searches with unvalidated peptide spectra was performed against the set of proteins in this result file using a no-enzyme (unconstrained) search to identify possible nonspecific or semitryptic peptide fragments. All database matches above the threshold score of 3 were summarized and reported.

13.2.4.4 Sonar and X!Tandem workflow performed by David Fenyo

X!Tandem and Sonar searches were performed by grouping the MS/MS spectra (files with .dta extension) generated by the LCQ_DTA utility (ThermoFinnigan) into single files (with .pkl extension) to speed up the searches. The parameters used in the extraction of the MS/MS spectra were the same settings as for the SEQUEST and MASCOT searches (see Section 2.4.1). The search parameters used were tryptic digestion with a maximum of two missed cleavage sites, parent ion tolerance of 3 Da, fragment ion tolerance of 0.5 Da, no complete or partial modifications, and the ESI-IT settings. The searches with both Sonar and X!Tandem were performed against IPI database in both forward and reverse directions. Perl scripts were written to automate searches and to parse the output of X!Tandem and Sonar. An expectation value cut-off of 1 was used to filter the results.

13.2.5
Web interface for data validation, integration, and cross annotation

Scripts written in Perl (version 5.8.4, http://www.perl.com) were used to manage the different data sets and the results of associated database searches obtained from the four independent groups. To assist with the process of manual validation, the

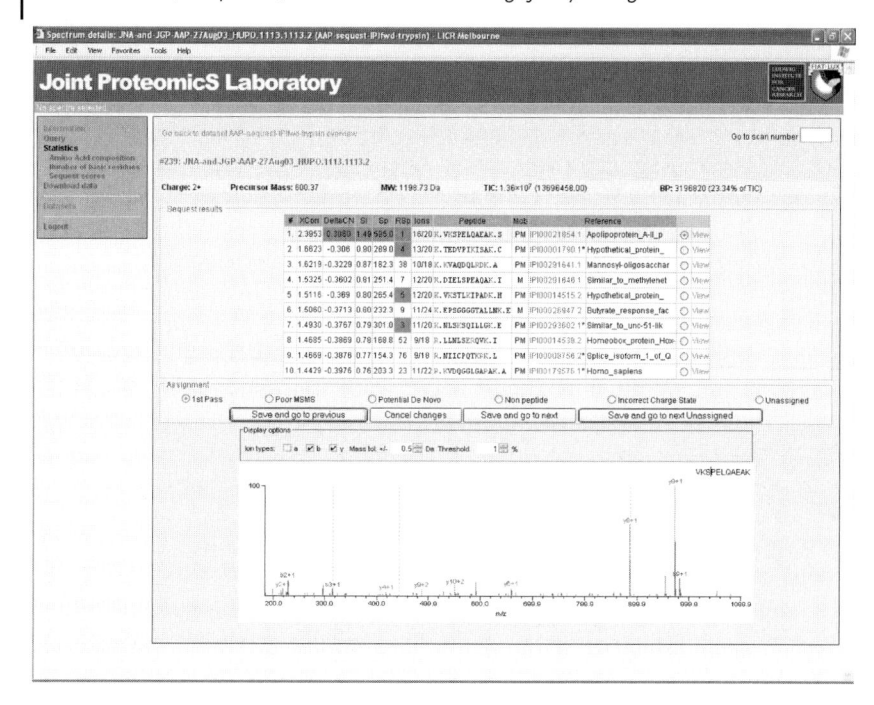

Fig. 1 Web-interface for viewing and manually assigning tandem MS peptide identification results. The top ten SEQUEST search results (scores and ancillary information) for a particular spectrum are shown. The selected top hit is used to annotate the spectrum (java applet) showing matching b and y ions within a user defined threshold and tolerance. Clicking the *View* radio control selects the chosen peptide hit, which is saved in a temporary file if one of the *Save* buttons is selected. Traversing large lists of spectra is made simpler with the "Go to scan number" function at the top of the web page.

Perl scripts also provided the following functionalities: (1) peptide hits with scores above user specified thresholds (cut-points) and/or accepted published cut-offs are highlighted as a visual aid to indicate that a hit is probably correct; (2) a protein summary view (list of inferred proteins) based on correctly identified peptides are sorted by number of matching peptide hits showing all assigned and unassigned peptide spectra matching a particular protein record; (3) options to autovalidate search results based on an already manually validated data set; and (4) highlight and detect inconsistencies between different search algorithms and/or results for the same data set (*i.e.*, same spectrum assigned with two different peptide sequences). Using the Apache web server, a web interface was assembled to allow easy access and manual validation of the data. The annotated spectra were displayed using a Java applet (see Fig. 1). The web interface also allows the user the ability to perform some simple statistics on the data sets, such as comparing numbers of peptide hits which are ranked first or in the top ten for different algorithms. These statistics can be viewed in the form of Venn diagrams and/or concordance plots.

The FP and true-positive (TP) rates can also be calculated based on specified rules. For more sophisticated analysis, the validated data set (list of identifications with their scores) can be exported in tab-delimited format for import into spreadsheet packages (such as Excel) and the R statistical package [30] (http://www.r-project.org). Public access to the web interface, database, and associated search results as well as peaklists (.dta files) and supplementary material can be found at http://www.ludwig.edu.au/archive/.

The informatics strategy employed to achieve accurately validated as well as unbiased data sets consisted of several phases. First, the four groups optionally validated all MS/MS spectra using their chosen search engine and/or analysis tools by a combination of automated as well as manual assignment. All the data sets in the form of spreadsheets were then collated and made accessible *via* the web interface. The SEQUEST and MASCOT search results (JPSL group) for the B3-Cit-plasma (Asian-American) sample consisting of 3952 MS/MS spectra were separately validated by two independent experts according to established protocols [8, 31]. For SEQUEST, cut-off filters, or thresholds developed by Yates *et al.* [23] and others [24, 25] were used as a guide to highlight probable correct identifications. For MASCOT, the peptides Ions Score, ranking, E-value as well as associated protein record were used as a guide to highlight probable correct identifications. All SEQUEST and MASCOT peptide identification search results were therefore independently classified and assigned, using the web interface, as either "1st Pass" (correct), "Poor" (spectra with few ions and/or poor S/N), or "Potential *de novo*" (good quality with many peaks above the noise level). The Perl scripts were then run to first detect inconsistencies (*i.e.*, same spectrum assigned with two different peptide sequences) between the SEQUEST and MASCOT search results. Second, to autovalidate the search results from the other groups based on the already validated SEQUEST and MASCOT assignments (*i.e.*, where peptide sequences were the same for a particular spectrum they were classified as 1st Pass (correct)). Third, peptide identification lists including scores and assignments for all the search algorithms were examined (sorted by descending score or probability) for unassigned spectra as well as conflicts between all search algorithms. Finally, all unassigned as well as conflicts (inconsistencies) were resolved by means of manual inspection by two independent MS experts (a detailed listing of all assignments and peptide sequences returned by the different algorithms for different search strategies can be obtained from http://www.ludwig.edu.au/archive/). An example listing (subset) with explanation is provided in Tab. 1S (supplementary material).

13.2.6
ROC curve generation

Receiver operating characteristic (ROC) plots were generated using the statistical package R (version 2.0.1) and used to measure the sensitivity (*i.e.*, the ability to make a correct identification irrespective of the quality of the data, see Eq. 1) and specificity (*i.e.*, the ability to calculate low-ranking scores for random (incorrect) matches, see Eq. 2) of all the MS/MS search algorithms used in this study. An ROC is a graphical plot of the TP rate *versus* the FP rate for a binary classifier system as its discrimination threshold is varied. For each search carried out using the forward protein sequence database, peptide hits were classified based on their score and whether they were correct or incorrect. If the score for a peptide hit was above the threshold, the hit was assigned as positive, and below the threshold they were assigned as negative. If a specific threshold value was selected, it was therefore possible to assign all peptides as either TP, true negative, FP- or false-negative hits.

$$\text{Sensitivity} = TP/(TP + FN) \tag{1}$$

where TP is the number of "true positives" (correct hits with scores above threshold) and FN is the number of "false negatives" (correct hits with scores below threshold).

$$\text{Specificity} = TN/(TN + FP) \tag{2}$$

where TN is the number of "true negatives" (incorrect hits with scores below threshold) and FP is the number of "false positives" (incorrect hits with scores above threshold).

13.3
Results and discussion

The Cit-plasma and serum samples analyzed as part of this study serve as excellent reference data sets because the acquired MS/MS peptide spectra originate from tryptic digests of plasma and/or serum proteins. Human plasma has a disproportional dynamic range of protein concentrations in that only 22 abundant proteins contribute ~99% of the total protein mass, while an unknown number of relatively low-abundance proteins make up to <1% of the total protein mass [32]. Currently available reference data sets are often mixtures of standard proteins of less dynamic range than that found in human specimens [33]. A particular challenge for MS/MS search algorithms and/or the validation process (automated and/or manual) is whether low-abundance peptides, which presumably originate from low-abundance proteins, are identifiable. Since the currently analyzed samples were not albumin depleted (the most abundant protein (40 mg/mL) in plasma) it is expected that the majority of peptides identified will belong to this protein. The capture and inclusion of lower scoring peptide hits (gray area between correct and incorrect hits), belonging to albumin, should enhance the quality of the reference data set. So even though each peptide hit is validated independently based on its score and annotated spectrum (whether automatically or manually), the inferred protein identity contributes to the overall subjective decision-making process. The inclusion of lower scoring peptide hits that match high-abundance proteins is therefore fundamental in determining the lower detection limits of current MS/MS search algorithms. More often than not, many low-abundance proteins are only identified by a single peptide (the so-called "one-hit wonders" [34]). Irrespective of how these peptides should be dealt with in terms of protein identification, it is important that their spectra are captured so as to facilitate future algorithmic improvements. Finally, all peptide hits were not only validated by a combination of automated as well as manual inspection, but were also cross-validated based on the results of the other search algorithms.

13.3.1
Comparison of MS/MS search algorithms

In order to compare the MS/MS search algorithms effectively one needs to calculate their coverage or sensitivity (*i.e.*, how many correct hits can be found irrespective of score) and their specificity (*i.e.*, whether the correct hit is significant relative to the other hits). Our reference data set enables accurate calculation of both of these metrics since all the hits returned by the various algorithms have been compared against each other as well as being validated by independent investigators.

13.3.1.1 Sensitivity and concordance between MS/MS search algorithms

The sensitivity of a search algorithm demonstrates its ability to make a correct identification using any data, irrespective of the quality of the data. Based on the B3-Cit-plasma reference data set and extensive validation and cross-checking/annotation between search algorithms, the overall number of correct peptide hits that were ranked first (irrespective of score) were tabulated in the form of concordance tables for trypsin-constrained (Tab. 1A) or no-enzyme (unconstrained) (Tab. 1B) searches of the IPI protein sequence database. The total number of correct hits for each search algorithm is indicated in bold text (diagonal line) (Tab. 1A and B) and ordered such that the search algorithm with the most hits appear first. For trypsin-constrained searches (Tab. 1A) it can be seen that SEQUEST identified 526 peptide hits, whilst Spectrum Mill (with tag >1 enabled) identified 397 peptide hits. Based on this observation it is clear that a large number of peptide spectra exhibit incomplete fragmentation patterns (*i.e.*, a less than ideal ladder of sequence ions due to fragmentation kinetics, *etc.*). Nevertheless, over 400 correct peptide hits are identified by at least four different search algorithms indicating reasonable concordance between the different algorithms. The fact that the SEQUEST/PeptideProphet combination (ISB group) identified slightly less hits than that of SEQUEST alone (JPSL group) can probably be attributed to differences in search parameters (*e.g.*, average *vs.* monoisotopic) and/or software versions. For no-enzyme (unconstrained) searches (Tab. 1B) it can be seen that SEQUEST and Spectrum Mill (when used in a less restrictive mode (*i.e.*, "no tag")) are better able to correctly identify peptides from poorer quality spectra (*i.e.*, higher sensitivity) and also identify a higher number of peptides compared with a trypsin-constrained search. All of the additional peptides identified in the no-enzyme mode were confirmed as belonging to already identified protein records (*e.g.*, albumin).

Tab. 1A. Number of correctly identified peptide spectra that are ranked first based on trypsin-constrained searches against the Human IPI v2.21 protein sequence database

	SEQUEST	Peptide-Prophet	MASCOT	Spectrum Mill	Sonar	X!Tandem	Spectrum Mill(tag)
SEQUEST	526	463	463	402	443	424	338
PeptideProphet	463	499	453	390	435	416	327
MASCOT	463	453	492	389	443	431	324
Spectrum Mill	402	390	389	476	389	374	395
Sonar	443	435	443	389	475	422	324
X!Tandem	424	416	431	374	422	457	314
Spectrum Mill(tag)	338	327	324	395	324	314	397

Tab. 1B. Number of correctly identified peptide spectra that are ranked first based on no-enzyme (unconstrained) searches against the Human IPI v2.21 protein sequence database

	SEQUEST	Spectrum Mill	MASCOT	Peptide-Prophet	Spectrum Mill(tag)
SEQUEST	531	422	438	436	352
Spectrum Mill	422	528	388	375	436
MASCOT	438	388	457	388	327
PeptideProphet	436	375	388	455	321
Spectrum Mill(tag)	352	436	327	321	438

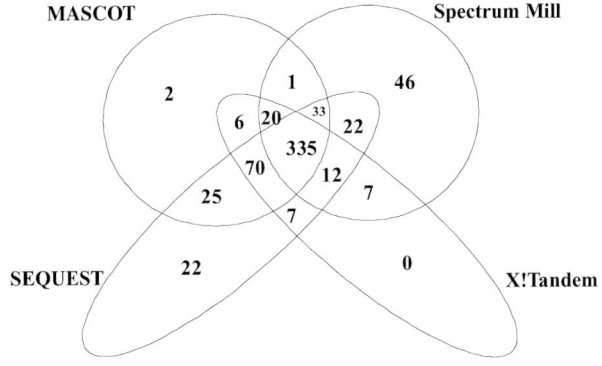

Fig. 2 Four-way Venn diagram showing the overlap between four of the MS/MS search algorithms. The number of correctly identified peptides by one or more algorithms is indicated, e.g., 335 peptide hits are correctly identified based on a consensus of all four algorithms (intersection), whilst 608 peptide hits are correctly identified by one or more algorithms (union).

The overlap, between four of the MS/MS search algorithms, in terms of the number of correctly identified peptide hits that are ranked first is shown in the form of a Venn diagram (Fig. 2) for trypsin-constrained searches of the IPI protein sequence database. Out of a possible 608 hits from the four algorithms (union), 335 peptides are identified by all four algorithms (intersection), whilst 70 peptides are identified by a single algorithm. Almost 75% of these peptide hits are singly charged spectra and 46 of these were independently identified by Spectrum Mill. Upon further inspection, the majority of the 46 hits are either small peptides between 600 and 700 Da in mass or constitute modified peptides (two with methionine oxidations, one with a pyroglutamic residue and three with internal carbamylated lysine residues). The majority of these matches were found to be highly credible upon closer inspection by two independent experts.

13.3.1.2 Specificity and discriminatory power of the primary score statistic for the different MS/MS search algorithms: Distribution of scores and ROC plots

Based on the B3-Cit-plasma reference data set, the distribution of the scores for top-ranking hits obtained from each of the MS/MS search algorithms was plotted for trypsin-constrained as well as no-enzyme (unconstrained) searches of the IPI and/or NR database in both forward and reverse directions (whichever was available). These plots (Fig. 3A–E) illustrate the distribution of scores (highest and lowest) as well as the potential overlap between scores of correct and incorrect peptide hits. For MASCOT searches (Fig. 3A) there is a clear distinction between correct (green) and incorrect (red) peptide hits, especially for trypsin-constrained searches. A search of the reverse databases gives 0 and 6 correct hits for the IPI and NR databases, respectively. The six peptide sequences identified from the reverse NR database are equivalent to real peptides that were also identified in the normal, "forward" search, and all were less than ten residues in length. For SEQUEST (Fig. 3B), it can be seen that there is more overlap between correct and incorrect peptide hits based on the X_{corr} score, especially for no-enzyme (unconstrained) searches (i.e., lower specificity). As for the MASCOT search, a number of correctly identified peptides were obtained when searching the NR database in reverse order. The distribution of scores for Spectrum Mill (based on no-tag search mode) (Fig. 3C) appears to be similar to those of SEQUEST (i.e., slightly more overlap when compared with MASCOT). The distribution of X!Tandem "hyperscores" and Sonar scores for trypsin-constrained searches is displayed on a log-scale (Fig. 3D and E, respectively). A comparison between all of the search algorithms suggests that MASCOT and X!Tandem demonstrate the highest specificity and therefore ability to calculate low-ranking scores for random (incorrect) matches.

ROC plots do not give an indication of the total number of correct hits nor do they illustrate the number of correct hits that might not be ranked first, but they do allow an overall comparison of the sensitivity and specificity of a search algorithm independent of a specific threshold. Based on the B3-Cit-plasma reference data set, ROC plots were generated for the different MS/MS search algorithms for trypsin-constrained (Fig. 4A) and no-enzyme (unconstrained) (Fig. 4B) search results. Since the ROC curve displays the sensitivities and FP rates at all possible cut-off levels, it can be used to assess the performance of the primary score (i.e., X_{corr} for SEQUEST or Hyperscore for X!Tandem), independent of any decision threshold. Therefore, ideal behavior would be a curve that approaches a sensitivity of 1.0 without any FP (i.e., 1-specificity is 0.0). This would indicate that a search algorithm

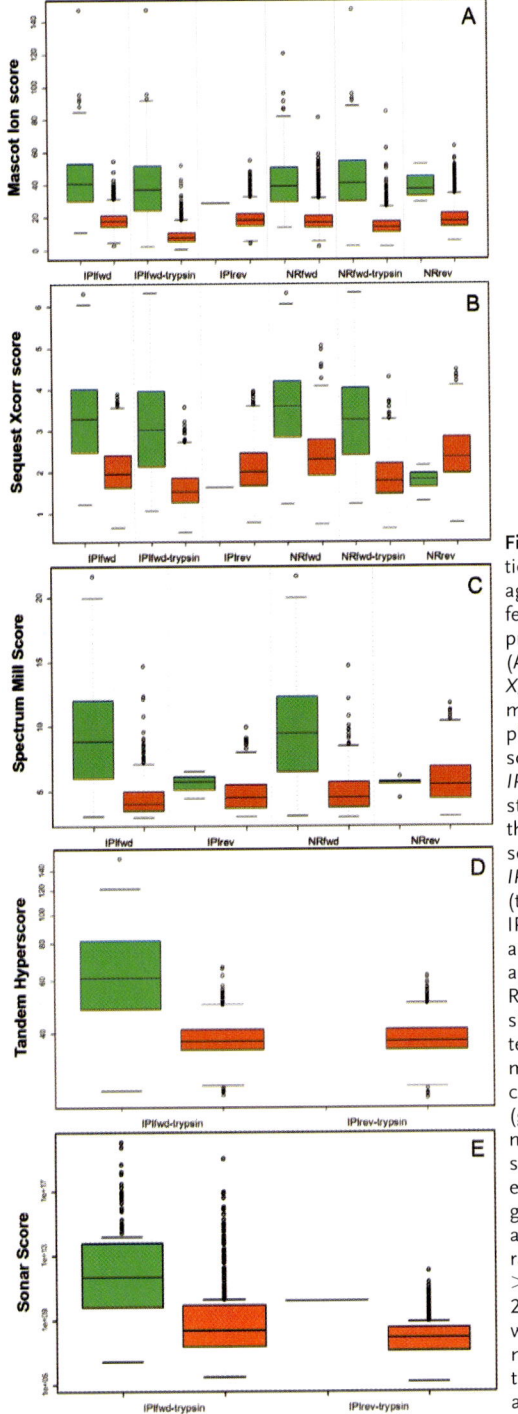

Fig. 3 Box-plots showing the distribution of scores obtained for searches against different databases, using different search parameters and reversed protein sequence databases.
(A) MASCOT Ion Score; (B) SEQUEST X_{corr}; (C) Spectrum Mill score (no tag mode); (D) X!Tandem hyperscore displayed on a log 2 base scale; (E) Sonar score displayed on a log 2 base scale. *IPIfwd* (green) denotes an unconstrained (no enzyme) search against the normal (forward) IPI protein sequence database (~56 000 entries). *IPIfwd-trypsin* (green) denotes a tryptic (two missed cleavages) search of the IPI database and *IPIrev* (red) denotes an unconstrained (no enzyme) search against the reversed IPI database. Reversed databases were created by simply reversing each individual protein sequence entry and as such maintaining the original sequence composition and length. The *NRfwd* (green) denotes a search against the normal (forward) NR protein sequence database (~1.5 million entries). Box-plots were automatically generated using the statistical package R, version 2.0.1 using default parameters (*i.e.*, outliers are scores >1.5X the interquartile range (75–25%), which are indicated by dots (o), whiskers represent the highest score not considered to be an outlier, and the box represents scores between 25 and 75% with median at 50%).

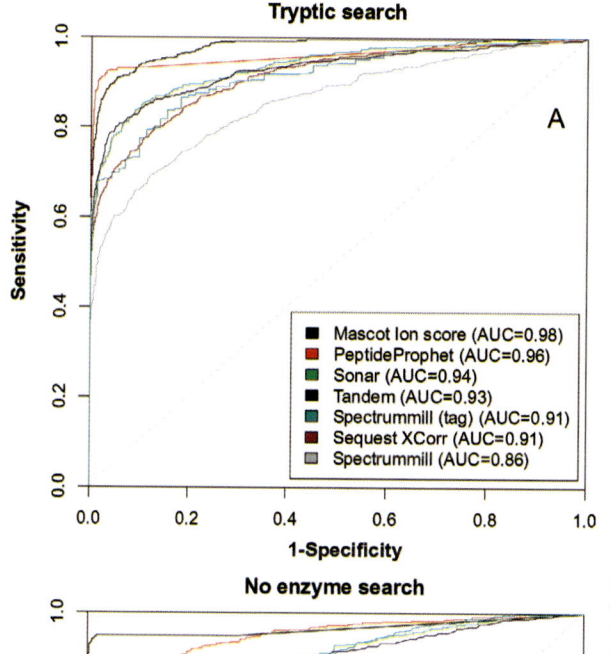

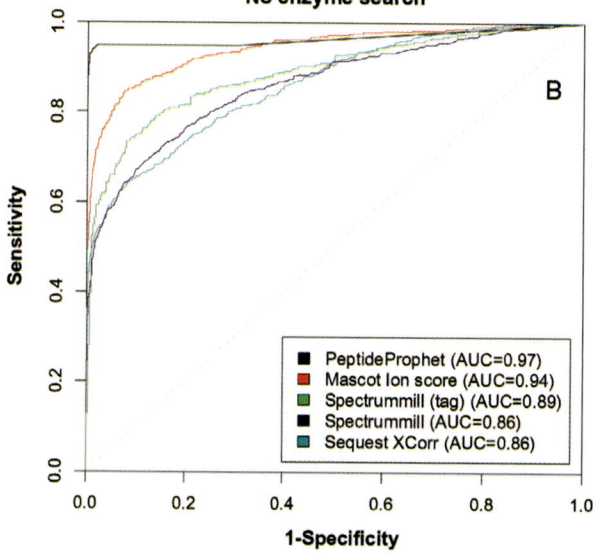

Fig. 4 ROC plot for the different search algorithms based on searches against the IPI protein sequence database: (A) tryptic-constrained (two missed cleavages) and (B) unconstrained (no enzyme). Hundred percent discrimination between correct and incorrect peptide hits would be indicated by a sensitivity of 1.0 and 1-specificity of 0.0 (*i.e.*, a search algorithm is able to identify all TP hits without any FPs). The AUC is indicated for each search algorithm (values of 0.5 would be considered random also called the chance diagonal (dotted line)).

is perfectly able to discriminate between correct and incorrect peptide hits, and the calculated area under the curve (AUC) would be 1.0. The AUC is a measure of the overall performance in terms of separating positives and negatives with values approaching 0.5 indicating random discrimination (*i.e.*, the diagonal line also called the chance diagonal). For trypsin-constrained searches (Fig. 4A) it can seen that the MASCOT Ion Score and SEQUEST/PeptideProphet combination perform better than X!Tandem and Sonar, which again perform better than SEQUEST and Spectrum Mill (tag >1 enabled). From Fig. 4B (no-enzyme searches), it can be seen that all the search algorithms, with the exception of the SEQUEST/PeptideProphet combination, perform worse when compared with the representative trypsin-constrained searches. The fact that the AUC improves slightly for the SEQUEST/PeptideProphet combination for no-enzyme searches (0.97 vs. 0.96) indicates that the number of tryptic termini is an important determinant in deriving the probability for a peptide hit. None of the individual search algorithms take this into account when classifying correct *versus* incorrect peptide hits.

13.3.1.3 Calculation of score thresholds based on specified FP identification error rates

Based on the B3-Cit-plasma reference data set and database search results of known validity, score thresholds (cut-offs) for the different MS/MS search algorithms were calculated at specified FP identification rates (0.1, 1, and 5%). These score thresholds were also calculated with regard to their charge state and various filtering criteria (*e.g.*, $R_{sp} < 5$ and $\Delta C_n >= 0.1$ for SEQUEST) for trypsin-constrained and/or no-enzyme (unconstrained) searches (see Tab. 2A–F). The criteria that give rise to the most TP hits at the 1% FP rate are indicated in bold text in the tables. The calculated score thresholds can first be used to judge the usefulness of a specific criterion (*e.g.*, R_{sp} for SEQUEST) and whether or not its inclusion improves the overall specificity and sensitivity of a particular search algorithm. Second, a sense of what constitutes equivalence between search algorithms can be obtained if one compares score thresholds at a specified FP rate (*i.e.*, what MASCOT score is equivalent to an X!Tandem score, for example). Finally, the score thresholds can be used for autovalidation purposes (*i.e.*, assume all peptide hits to be correct if their scores are above the calculated thresholds) but with the following caveats: that the thresholds be applied to similar data sets (*i.e.*, LCQ-like data) obtained under similar experimental conditions and analyzed using the same search parameters (*i.e.*, searches are performed using 3 Da precursor ion tolerance and against similar sized protein sequence databases).

From Tab. 2A, for trypsin-constrained searches, it can be seen that the $\Delta C_n >= 0.1$ as well as $R_{sp} < 5$ criteria improve the overall specificity of the SEQUEST algorithm. The R_{sp} criterion has largely been ignored in published studies to date but it is clear from Tab. 2A that this filter should be included when analyzing search results from complex protein extracts such as cell lysates and tissues such as blood. For no-enzyme (unconstrained) searches (Tab. 2A), it can be seen that many random (incorrect) matches can be filtered by applying the strict trypsin rule (*i.e.*, peptide must be fully tryptic).

Tab. 2A. SEQUEST X_{corr} thresholds calculated based on different criteria at specified FP error rates for trypsin-constrained and no-enzyme searches against the Human IPI v2.21 protein sequence database

Charge state of precursor ion	Criteria applied to peptide hit	Trypsin-constrained search %FP rate			No-enzyme search %FP rate		
		0.1	1	5	0.1	1	5
All	None	3.56 (37%)	2.64 (59%)	2.28 (71%)	3.89 (30%)	3.3 (49%)	2.95 (60%)
	ΔC_n [a]	3.21 (44%)	2.55 (62%)	2.16 (73%)	3.89 (29%)	3.1 (53%)	2.4 (73%)
	$\Delta C_n + R_{sp}$ [b]	3.21 (44%)	**2.44 (65%)** [d]	1.68 (85%)	3.41 (43%)	2.05 (68%)	0 (70%)
	Tryptic [c]	–	–	–	3.11 (50%)	**2.27 (73%)**	0 (89%)
	$\Delta C_n + R_{sp}$ + tryptic	–	–	–	3.01 (48%)	0[e] (63%)	0 (63%)
	No. of hits [f]	526			531		
Singly-charged peptides (1+)	None	2.41 (18%)	2.05 (41%)	1.73 (75%)	2.82 (9%)	2.41 (28%)	2.11 (48%)
	ΔC_n	2.22 (27%)	1.86 (58%)	1.59 (76%)	2.29 (35%)	**2.01 (49%)**	1.38 (68%)
	$\Delta C_n + R_{sp}$	2.02 (35%)	**1.72 (58%)**	1.35 (64%)	2.03 (26%)	1.37 (31%)	0 (31%)
	Tryptic	–	–	–	2.69 (10%)	2.22 (33%)	1.71 (79%)
	$\Delta C_n + R_{sp}$ + tryptic	–	–	–	0 (24%)	0 (24%)	0 (24%)
	No. of hits	168			127		
Doubly-charged peptides (2+)	None	3.56 (50%)	2.44 (89%)	1.97 (98%)	3.13 (66%)	2.76 (81%)	2.41 (91%)
	ΔC_n	3.01 (70%)	2.35 (90%)	1.81 (97%)	3.08 (66%)	2.52 (84%)	2.07 (92%)
	$\Delta C_n + R_{sp}$	3.01 (70%)	**2.18 (93%)**	1.5 (97%)	3.01 (67%)	2.01 (84%)	0 (85%)
	Tryptic	–	–	–	3.01 (64%)	**2.22 (85%)**	0 (89%)
	$\Delta C_n + R_{sp}$ + tryptic	–	–	–	3.01 (61%)	0 (75%)	0 (75%)
	No. of hits	281			331		
Triply-charged peptides (3+)	None	3.21 (74%)	2.68 (83%)	2.42 (88%)	3.89 (58%)	3.43 (78%)	3.13 (84%)
	ΔC_n	3.21 (73%)	2.68 (83%)	2.34 (90%)	3.89 (56%)	3.3 (78%)	2.75 (86%)
	$\Delta C_n + R_{sp}$	3.21 (73%)	**2.53 (84%)**	2.02 (96%)	3.41 (64%)	2.38 (71%)	0 (71%)

Tab. 2A. Continued

Charge state of precursor ion	Criteria applied to peptide hit	Trypsin-constrained search %FP rate			No-enzyme search %FP rate		
		0.1	1	5	0.1	1	5
	Tryptic	–	–	–	3.11 (82%)	2.41 (95%)	0 (99%)
	$\Delta C_n + R_{sp}$ + tryptic	–	–	–	2.38 (71%)	0 (71%)	0 (71%)
	No. of hits	77			73		

a) ΔC_n criteria $>= 0.1$
b) R_{sp} criteria < 5
c) True (full) tryptic criteria
d) %TP peptide identifications based on the specified criteria and total number of correctly identified peptide hits (see point f below)
e) Negligible score threshold (*i.e.*, almost zero)
f) Total number of correctly identified peptide hits

Tab. 2B. MASCOT Ions Score thresholds calculated based on different criteria at specified FP error rates for trypsin-constrained and no-enzyme searches against the Human IPI v2.21 protein sequence database

Charge state of precursor ion	Criteria applied to peptide hit	Trypsin-constrained search %FP rate			No-enzyme search %FP rate		
		0.1	1	5	0.1	1	5
All	None	51.80 (25%[b])	23.08 (78%)	16.20 (90%)	54.39 (23%)	34.63 (63%)	27.96 (80%)
	Tryptic[a]	–	–	–	47.94 (33%)	22.38 (**79%**)	0[c] (87%)
	No. of hits[d]	493			457		
Singly-charged peptides (1+)	None	51.80 (2%)	27.83 (**35%**)	21.17 (59%)	42.79 (11%)	38.03 (22%)	31.21 (37%)
	Tryptic	–	–	–	42.79 (10%)	32.36 (**29%**)	18.46 (72%)
	No. of hits	125			87		
Doubly-charged peptides (2+)	None	39.02 (62%)	20.00 (**93%**)	13.87 (96%)	54.39 (31%)	33.12 (79%)	24.45 (91%)
	Tryptic	–	–	–	47.94 (43%)	19.06 (**85%**)	0 (88%)
	No. of hits	299			311		
Triply-charged peptides (3+)	None	16.06 (91%)	20.84 (**84%**)	–	40.92 (37%)	33.58 (64%)	28.26 (81%)
	Tryptic	–	–	–	40.77 (37%)	19.39 (**95%**)	0 (97%)
	No. of hits	69			59		

a) True (full) tryptic criteria
b) %TP peptide identifications based on the specified criteria and total number of correctly identified peptide hits (see point d below)
c) Negligible score threshold (*i.e.*, almost zero)
d) Total number of correctly identified peptide hits

Tab. 2C. X!Tandem score thresholds calculated based on specified FP error rates for trypsin-constrained searches against the Human IPI v2.21 protein sequence database

Charge state of precursor ion	Criteria applied to peptide hit	Trypsin-constrained search %FP rate		
		0.1	1	5
All	None	66.4 (42%[a])	50.7 (68%)	45 (81%)
	No. of hits[b]	457		
Singly-charged peptides (1+)	None	54.4 (19%)	53.7 (20%)	44.6 (44%)
	No. of hits	116		
Doubly-charged peptides (2+)	None	66.4 (63%)	53.3 (82%)	46.3 (93%)
	No. of hits	284		
Triply-charged peptides (3+)	None	62.3 (21%)	48.1 (82%)	44.3 (89%)
	No. of hits	57		

a) %TP peptide identifications based on the specified criteria and total number of correctly identified peptide hits (see point b below)
b) Total number of correctly identified peptide hits

Tab. 2D. Sonar score thresholds calculated based on specified FP error rates for trypsin-constrained searches against the Human IPI v2.21 protein sequence database

Charge state of precursor ion	Criteria applied to peptide hit	Trypsin-constrained search %FP rate		
		0.1	1	5
All	None	$5.3e^{13}$ (21%[a])	$3.6e^9$ (63%)	$2.4e^8$ (78%)
	No. of hits[b]	475		
Singly-charged peptides (1+)	None	$3.7e^{12}$ (12%)	$4.5e^{10}$ (33%)	$1.6e^9$ (60%)
	No. of hits	129		
Doubly-charged peptides (2+)	None	$6.7e^{10}$ (54%)	$3.4e^8$ (77%)	$5.3e^7$ (87%)
	No. of hits	281		
Triply-charged peptides (3+)	None	$5.3e^{13}$ (38%)	$4.1e^9$ (77%)	$3.4e^8$ (85%)
	No. of hits	65		

a) %TP peptide identifications based on the specified criteria and total number of correctly identified peptide hits (see point b below)
b) Total number of correctly identified peptide hits

The MASCOT Ions Score thresholds for both trypsin and no-enzyme searches (Tab. 2B) indicate that thresholds are higher for singly-charged peptide ions compared with doubly- and triply-charged peptides. A comparison of the thresholds with the reported MASCOT "identity score ($p < 0.05$)" of 43 for trypsin-constrained

Tab. 2E. Spectrum Mill (tag >1) score thresholds calculated based on different criteria at specified FP error rates for trypsin and hierarchical iterative searches against the Human IPI v2.21 protein sequence database

Charge state of precursor ion	Criteria applied to peptide hit	Trypsin and no-enzyme iterative search %FP rate		
		0.1	1	5
All	None	14.68 (13%[b])	9.42 (53%)	7.96 (66%)
	Tryptic[a]	8.46 (58%)	**7.67 (63%)**	5.39 (79%)
	No. of hits[c]	438		
Singly-charged peptides (1+)	None	8.6 (16%)	8.6 (16%)	5.77 (51%)
	Tryptic	7.65 (21%)	**7.65 (21%)**	5.39 (46%)
	No. of hits	104		
Doubly-charged peptides (2+)	None	10.78 (50%)	10.78 (50%)	8.66 (71%)
	Tryptic	8.46 (69%)	**8.46 (69%)**	7.96 (73%)
	No. of hits	268		
Triply-charged peptides (3+)	None	14.68 (23%)	12.13 (45%)	7.95 (85%)
	Tryptic	7.51 (86%)	**6.38 (91%)**	4.03 (97%)
	No. of hits	66		

a) True (full) tryptic criteria
b) %TP peptide identifications based on the specified criteria and total number of correctly identified peptide hits (see point c below)
c) Total number of correctly identified peptide hits

Tab. 2F. SEQUEST/PeptideProphet thresholds calculated based on specified FP error rates for trypsin-constrained searches against the Human IPI v2.21 protein sequence database

Charge state of precursor ion	Criteria applied to peptide hit	Trypsin-constrained search %FP rate		
		0.1	1	5
All	None	0.96 (56%[a])	0.11 (88%)	0 (93%)
	No. of hits[b]	499		
Singly-charged peptides (1+)	None	0.49 (57%)	0.29 (67%)	0 (76%)
	No. of hits	126		
Doubly-charged peptides (2+)	None	0.96 (76%)	0.17 (94%)	0.01 (99%)
	No. of hits	301		
Triply-charged peptides (3+)	None	0.86 (68%)	0 (93%)	0 (97%)
	No. of hits	72		

a) %TP peptide identifications based on the specified criteria and total number of correctly identified peptide hits (see point b below)
b) Total number of correctly identified peptide hits

and 60 for no-enzyme searches reveals the following: for trypsin-constrained searches a cut-off score of 43 gives an FP rate of 0.03% and TP rate of 38% whilst applying the reported homology score gives an FP rate of 0.23% and a TP rate of 70.38% (data not shown); for the no-enzyme searches a cut-off score of 60 gives an FP rate of 0% and a TP rate of 14.22% whilst applying the reported homology score gives an FP rate of 0.34% and TP rate of 60.39% (data not shown).

A comparison of trypsin-constrained and no-enzyme (unconstrained) searches for SEQUEST and MASCOT searches (Tab. 2A and B, respectively) indicates that score thresholds are considerably higher at all predefined FP rates for no-enzyme searches. Indeed for both SEQUEST and MASCOT, similar score thresholds are obtained for a no-enzyme search against the IPI protein sequence database compared with a trypsin-constrained search of the NR database (which is comprised of ~1.5 million entries) (data not shown). This clearly highlights the "distraction effect" as a result of an effective increase in database size due to the increased number of peptides that must be queried.

The hyperscore thresholds for X!Tandem and Sonar score thresholds, based on trypsin-constrained searches, are shown in Tab. 2C and D, respectively. For X!Tandem (Tab. 2C), the thresholds are constant across all charge states, indicating that singly-charged spectra do not have such a negative effect on the X!Tandem scoring function. Also, based on these calculations, a score of 50 (~1% FP) would be more appropriate than previously suggested (unpublished) score cut-offs of 45 which equates to ~5% FP rate under these conditions. The Spectrum Mill (tag >1 mode) results (see Tab. 2E) are based on a five-phase iterative search strategy. Again (similar to SEQUEST), it can be seen that the scores are dependent on the charge state of the precursor ion. Finally, the probability thresholds at the different FP rates for SEQUEST/PeptideProphet are shown in Tab. 2F.

It is clear from Tab. 2A–F that the number of TP peptide identifications is lowest for singly-charged peptide spectra. This is perhaps not surprising when one considers that singly-charged precursor ions are inherently smaller, fragment in a less predictable manner, and generate less fragment ions. Approximately, 30% of the low-mass ions are not observed on a 3-D IT due to the low-mass cut-off. Doubly- and triply-charged peptide spectra are less affected by the low-mass cut-off, and since the majority of tryptic peptide spectra are doubly-charged under electrospray conditions and have a mobile proton [35], more ideal fragmentation is facilitated and hence identified by current search algorithms.

13.3.1.4 Benchmarking of the different MS/MS search algorithms at 1% FP error rate

Based on the results from Tab. 2A–F, Tab. 3 provides an overall comparison and accurate benchmark of the search algorithms evaluated in this study in terms of the number of correctly identified peptide spectra (TP) at 1% FP rate. Overall, taking into account all charges (first row Tab. 3) it can be seen that PeptideProphet when applied to SEQUEST results identifies 439 peptides whilst Spectrum Mill (used with tag >1) identifies 276 peptides. However, at the individual charge state level, especially singly-charged, there appears to be much variation between the different search algorithms.

Tab. 3 Number of correctly identified peptide spectra (TP rate) based on a 1% FP rate (benchmark) for the different search algorithms for trypsin-constrained searches against the Human IPI v2.21 protein sequence database

Charge state of precursor ion	SEQUEST/ Peptide-Prophet	MASCOT	SEQUEST ($\Delta C_n + R_{sp}$)	X!Tandem	Sonar	Spectrum Mill (tag >1)
All	439	385	342	311	299	276
Singly-charged peptides (1+)	84	44	97	23	43	22
Doubly-charged peptides (2+)	283	278	261	233	216	185
Triply-charged peptides (3+)	67	58	65	47	50	60

13.3.1.5 Effect of database size and search strategy

In order to investigate the effect of database size as well as optimal search strategy, the total number of correct hits (ranked first or in the top ten) reported by both MASCOT and SEQUEST was tabulated based on searches against the IPI and NR databases using trypsin-constrained as well as no-enzyme (unconstrained) searches (see Tab. 4). First, it can be seen that the search algorithms lose sensitivity as the search space is increased (i.e., more peptides have to be queried) and that MASCOT is affected more than that of SEQUEST since the correct peptide hit appears more often in the top ten hits rather than being ranked first. This indicates that the SEQUEST scoring function is slightly more sensitive (i.e., better able to rank poorer quality peptide spectra) compared with that of MASCOT, especially when large protein sequence databases are used and/or unconstrained searches are carried out. Second, of the 581 correctly identified peptides (top ten considered, no-enzyme search) for SEQUEST, 89% are true-tryptic, 11% are semitryptic, and none are nonspecific, whereas of the incorrectly identified peptides, 2% are true-tryptic, 20% are semitryptic, and 78% are nonspecific. These values are in close agreement with those calculated by Keller et al. [33] based on tryptic digestion of an 18 standard protein mixture. Our findings therefore support and confirm the observation that trypsin is a very specific protease [36, 37]. In fact, the majority of semitryptic peptides identified in this analysis were derived from human albumin, the most abundant protein in these samples.

13.3.1.6 Utility of reversed sequence searches

The utility of reversed sequence searches to restrict the number of FP peptide identifications has been explored by various groups [25, 38, 39]. The idea is to analyze a particular data set and identify peptides using both the "normal" forward and "random" reversed protein sequence database searches. The random database could be appended to the normal database or searched separately. Our protocol consisted of the following steps: (1) reversed sequence searches were carried out separately; (2) the search results were then filtered so as to remove correct matches

Tab. 4 Number of correctly identified peptide spectra for SEQUEST and MASCOT based on different search strategies and protein sequence databases

		Top hit[a]		Top ten hits[b]	
		Trypsin[c]	No-enzyme[d]	Trypsin	No-enzyme
SEQUEST	IPI[e]	526	531	535	581
	NR[f]	498	418	552	481
MASCOT	IPI	492	457	539	526
	NR	425	363	508	446

a) Only the correct peptide hits that are ranked first are considered
b) Correct peptide hits ranked amongst the top ten are considered
c) Trypsin-constrained search (full tryptic) with two missed cleavages
d) No-enzyme (unconstrained) search
e) Human IPI v2.21 protein sequence database comprising ~56 000 entries
f) Ludwig Institute NR (nonredundant) protein sequence database comprising ~1.5 million entries

based on the validated normal forward search (see Section 3.1.2); (3) the scores were then sorted in descending order and the threshold determined based on the *n*th ranked score depending on the specified (acceptable) FP rate. For example, if the FP rate is 1% and 1000 peptide spectra are scored, the tenth highest score would be the score threshold. Our findings indicate that similar score thresholds, albeit slightly higher thresholds, were obtained compared with those from the normal forward search (Tab. 2A–F). This appears to be in agreement with others regarding the estimation of FP rates based on the reverse database model [39]. In order for this approach to be effective it would have to be repeated for each experiment. The obvious disadvantage of the reverse database model is the number of false-negative peptide hits (*i.e.*, the correct peptide identifications below the threshold) but it demonstrates an improvement on empirically derived published score cut-offs.

13.3.1.7 Consensus scoring between MS/MS search algorithms

The idea of consensus scoring has previously been raised [40] and briefly explored here. The basic idea is to merge search results from two or more algorithms and combine the scores for peptide spectra where there is consensus between different algorithms. The top ranking peptide hit or top ten peptide hits for each spectrum, from the different algorithms, could be considered. Interestingly, based on the data sets used in this study, when one compares all the top ranked peptide sequences returned by both MASCOT (trypsin search) and SEQUEST (trypsin search), 646 peptide sequences are found to be identical, and of these, 465 have been validated as correct. However, when one compares the top ranked peptide sequences returned by both MASCOT (trypsin search) and SEQUEST (no-enzyme search), 470 peptide sequences are found to be identical, and of these, 450 have been validated as correct (data available from website, see Section 2.5). A closer inspection of these 20 peptide spectra (identical sequences but not 1st Pass) reveals that they are mostly poorer quality (singly-charged) spectra with low scores and exhibiting less than ideal ladders of sequence ions. Further examination and observation regarding consensus amongst at least three algorithms reveal that the MASCOT scoring function generally performs poorer on singly-charged spectra and/or spectra exhibiting few ions or spectra exhibiting many ions but with a few very intense peaks. Indeed when one compares all the top ranked peptide sequences returned by all the search algorithms and filter out nonidentical sequences, we find that the remaining peptides have all been classified by the investigators as correct. This suggests that the consensus approach based on multiple scoring functions definitely has merit and that the scores could be considered as independent and orthogonal. Further work needs to be carried out to determine exactly how many search algorithms (or independent scoring functions) are required so as to allow confident and automated validation of peptide identifications and therefore accurate protein identifications.

13.4
Concluding remarks

Our aims in this paper were to assess the strengths and weaknesses of different MS/MS search algorithms on IT data, and to provide guidelines to help assess the significance of peptide identification results obtained from the individual HUPO-PPP participating laboratories. Important considerations when carrying out MS/MS database searches are the specified search parameters (*i.e.*, mass tolerance which is dependent on the instrument and calibration), search strategy (*i.e.*, semitryptic *vs.* tryptic), chosen protein sequence database to query (*i.e.*, IPI *vs.* NCBI NR which is dependent on the particular experiment), and chosen search engine. The choice of search engine should not only be guided by the range of mass spectrometers available but also whether or not it is restrictive regarding the above choices as well as its overall sensitivity and specificity, which we have addressed in this study.

It is clear from this study that the number of correctly identified peptides that are ranked first by the different algorithms decreases (less sensitive) as the search space is increased (*i.e.*, no-enzyme search and/or large protein sequence database). This is particularly notable for MASCOT compared with SEQUEST, on the basis of the number of correctly identified peptides that are no longer ranked first but appear in the top ten. SEQUEST and Spectrum Mill (using no tag filter) are more sensitive than the other algorithms but MASCOT, Sonar, and X!Tandem are more specific (*i.e.*, better able to discriminate between correct and incorrect peptide hits). Overall, calculating the TP rate at a specified FP rate shows that MASCOT performs better than the other algorithms used in this study. Application of a rescoring algorithm, such as PeptideProphet, improves the specificity of the SEQUEST algorithm and based on these results should also improve the results of the other algorithms.

Score thresholds, if used, can be determined based on reverse sequence searches as demonstrated in this study. For high-confidence peptide identifications these thresholds could be combined with orthogonal scoring information, such as scores from other search algorithms. The availability of open-source algorithms, such as X!Tandem as well as OMSSA [41] make this process feasible. In this respect, an algorithm that demonstrates high sensitivity should be used in conjunction with an algorithm that demonstrates high specificity. Thresholds, if used, should also be calculated on a *per*-experiment basis because the number of spectra generated and the detectable dynamic range of proteins have a major influence on the number of potential FP identifications. For example, at a predefined score threshold, the number of FP identifications will be higher if a large number of spectra are generated that do not correctly match anything in the protein sequence database. This scenario is typical of human specimens, such as plasma, which exhibits a disproportional dynamic range of protein concentrations.

The MS data, generated for this study, were performed in the Environmental Molecular Sciences Laboratory, a US national scientific user facility sponsored by the Department of Energy's Office of Biological and Environmental Research and located at Pacific Northwest National Laboratory. We thank Joel Pounds, Dick Smith, and Ron Moore for access to the MS data; James Eddes for the mass spectrum applet used in the web interface; Robert Moritz for access to the JPSL MASCOT server. Funding was provided, in part, by the HUPO-PPP and by the Australian National Health and Medical Research Council (program grant no. 280912).

13.5
References

[1] Omenn, G. S., *Proteomics* 2004, 4, 1235–1240.

[2] Fenyo, D., Beavis, R. C., *Anal. Chem.* 2003, 75, 768–774.

[3] Nesvizhskii, A. I., Aebersold, R., *Drug Discov. Today* 2004, 9, 173–181.

[4] Clauser, K., Negvizhskii, A., Carr, S., Aebersold, R., Baldwin, M., Burlingame, A. et al., *Mol. Cell. Proteomics* 2004, 3, 531–533.

[5] Baldwin, M. A., *Mol. Cell. Proteomics* 2004, 3, 1–9.

[6] Simpson, R. J., *Eur. J. Pharm. Sci. Rev.* 2004, 9, 25–36.

[7] Anderson, N. L., Polanski, M., Pieper, R., Gatlin, T. et al., *Mol. Cell. Proteomics* 2004, 3, 311–326.

[8] Kapp, E. A., Schutz, F., Reid, G. E., Eddes, J. S. et al., *Anal. Chem.* 2003, 75, 6251–6264.

[9] Reid, G. E., Roberts, K. D., Kapp, E. A., Simpson, R. J., *J. Proteome Res.* 2004, 3, 751–759.

[10] Eng, J. K., McCormack, A. L., Yates III, J. R., *J. Am. Soc. Mass Spectrom.* 1994, 5, 976–989.

[11] Perkins, D. N., Pappin, D. J., Creasy, D. M., Cottrell, J. S., *Electrophoresis* 1999, 20, 3551–3567.

[12] Marshall, J., Jankowski, A., Furesz, S., Kireeva, I. et al., *J. Proteome Res.* 2004, 3, 364–382.

[13] Kearney, P., Thibault, P., *J. Bioinform. Comput. Biol.* 2003, 1, 183–200.

[14] Keller, A., Nesvizhskii, A. I., Kolker, E., Aebersold, R., *Anal. Chem.* 2002, 74, 5383–5392.

[15] Nesvizhskii, A. I., Keller, A., Kolker, E., Aebersold, R., *Anal. Chem.* 2003, 75, 4646–4658.

[16] Sadygov, R. G., Cociorva, D., Yates III, J. R., *Nat. Methods* 2004, 1, 195–202.

[17] Field, H. I., Fenyo, D., Beavis, R. C., *Proteomics* 2002, 2, 36–47.

[18] Tabb, D. L., MacCoss, M. J., Wu, C. C., Anderson, S. D., Yates, J. R., 3rd, *Anal. Chem.* 2003, 75, 2470–2477.

[19] Beer, I., Barnea, E., Ziv, T., Admon, A., *Proteomics* 2004, *4*, 950–960.
[20] Craig, R., Beavis, R. C., *Bioinformatics* 2004, *20*, 1466–1467.
[21] Craig, R., Beavis, R. C., *Rapid Commun. Mass Spectrom.* 2003, *17*, 2310–2316.
[22] Pappin, D. J., Hojrup, P., Bleasby, A. J., *Curr. Biol.* 1993, *3*, 327–332.
[23] Link, A. J., Eng, J., Schieltz, D. M., Carmack, E. et al., *Nat. Biotechnol.* 1999, *17*, 676–682.
[24] Washburn, M. P., Wolters, D., Yates, J. R., 3rd, *Nat. Biotechnol.* 2001, *19*, 242–247.
[25] Peng, J., Elias, J. E., Thoreen, C. C., Licklider, L. J., Gygi, S. P., *J. Proteome Res.* 2003, *2*, 43–50.
[26] Adkins, J. N., Varnum, S. M., Auberry, K. J., Moore, R. J. et al., *Mol. Cell. Proteomics* 2002, *1*, 947–955.
[27] Moritz, R. L., Ji, H., Schutz, F., Connolly, L. M. et al., *Anal. Chem.* 2004, *76*, 4811–4824.
[28] Kersey, P. J., Duarte, J., Williams, A., Karavidopoulou, Y. et al., *Proteomics* 2004, *4*, 1985–1988.
[29] Pedrioli, P. G., Eng, J. K., Hubley, R., Vogelzang, M. et al., *Nat. Biotechnol.* 2004, *22*, 1459–1466.
[30] R Developement Core Team, *A Language and Environment for Statistical Computing*, R Foundation for Statistical Computing, Vienna, Austria 2005, http://www.R-project.org.
[31] Simpson, R. J., Connolly, L. M., Eddes, J. S., Pereira, J. J. et al., *Electrophoresis* 2000, *21*, 1707–1732.
[32] Anderson, N. L., Anderson, N. G., *Mol. Cell. Proteomics* 2002, *1*, 845–867.
[33] Keller, A., Purvine, S., Nesvizhskii, A. I., Stolyar, S. et al., *Omics* 2002, *6*, 207–212.
[34] Veenstra, T. D., Conrads, T. P., Issaq, H. J., *Electrophoresis* 2004, *25*, 1278–1279.
[35] Wysocki, V. H., Tsaprailis, G., Smith, L. L., Breci, L. A., *J. Mass Spectrom.* 2000, *35*, 1399–1406.
[36] Keil, B., *Specificity of Proteolysis*, Springer-Verlag, Berlin 1992.
[37] Olsen, J. V., Ong, S. E., Mann, M., *Mol. Cell. Proteomics* 2004, *3*, 608–614.
[38] Cargile, B. J., Bundy, J. L., Stephenson, J. L., Jr., *J. Proteome Res.* 2004, *3*, 1082–1085.
[39] Qian, W. J., Liu, T., Monroe, M. E., Strittmatter, E. F. et al., *J. Proteome Res.* 2005, *4*, 53–62.
[40] Resing, K. A., Meyer-Arendt, K., Mendoza, A. M., Aveline-Wolf, L. D. et al., *Anal. Chem.* 2004, *76*, 3556–3568.
[41] Geer, L. Y., Markey, S. P., Kowalak, J. A., Wagner, L. et al., *J. Proteome Res.* 2004, *3*, 958–964.

14
Human Plasma PeptideAtlas*

Eric W. Deutsch, Jimmy K. Eng, Hui Zhang, Nichole L. King, Alexey I. Nesvizhskii, Biaoyang Lin, Hookeun Lee, Eugene C. Yi, Reto Ossola and Ruedi Aebersold

Peptide identifications of high probability from 28 LC-MS/MS human serum and plasma experiments from eight different laboratories, carried out in the context of the HUPO Plasma Proteome Project, were combined and mapped to the EnsEMBL human genome. The 6929 distinct observed peptides were mapped to approximately 960 different proteins. The resulting compendium of peptides and their associated samples, proteins, and genes is made publicly available as a reference for future research on human plasma.

The protein content of human plasma is considered important for medical diagnosis and has the potential to provide a complete snapshot of the health of an individual. In addition to proteins that carry out their function within the circulatory system, plasma contains proteins that are secreted or leaked from cells and organs throughout the body. As a diagnostic tool, plasma is even more valuable by virtue of its accessibility, with millions of samples stored in clinical archives and even more obtained every year from patients.

Human plasma is thought to contain a large number of proteins, perhaps nearly all human proteins on account of low-level tissue leakage [1]. Further, human plasma also contains proteins from foreign organisms as well as millions of distinct immunoglobulins. However, a mere 22 proteins make up 99% of the mass of protein in human serum [2], and thus an investigation of the thousands of very low-abundance proteins is difficult.

Several recent studies have sought to provide a preliminary definition of the human plasma proteome [3–6]. Adkins *et al.* [3] performed LC-MS/MS experiments on immunoglobulin-depleted samples and reported 490 distinct proteins. Pieper *et al.* [4] identified 325 distinct proteins from samples with eight high-abundance proteins removed *via* immunoaffinity chromatography. Anderson *et al.* [5] provided

---

* Originally published in Proteomics 2005, 13, 3497–3500

Exploring the Human Plasma Proteome. Edited by Gilbert S. Omenn
Copyright © 2006 WILEY-VCH Verlag GmbH & Co. KGaA, Weinheim
ISBN: 3-527-31757-0

a nonredundant list from four separate sources (previous literature and three other published experiments) of 1175 proteins. Chan et al. [6] published a list of 1444 distinct serum proteins from a large-scale LC-MS/MS experiment. A comparison of the data in these reports has shown limited overlap between studies and raised the question of how data from different plasma proteome studies could be evaluated and represented to facilitate meaningful comparisons.

HUPO has undertaken the Plasma Proteome Project (PPP), which aims to provide a comprehensive analysis of the proteins of human plasma and serum, including the analysis of variation within individuals as well as across individuals [7, 8]. As part of this project, various samples have been sent to over 40 laboratories for local analysis using a variety of protocols and platforms. Further information about this project can be found in other reports in this issue.

We previously developed the PeptideAtlas process [9] to create and make public a genome-mapped atlas of peptides observed in a set of LC-MS/MS proteomics experiments, initially for human and *Drosophila melanogaster*, with processing of data from additional organisms underway. Here, we present a PeptideAtlas build derived solely from human plasma (including serum) sample experiments, mostly generated for the HUPO PPP. Although the experiments were performed in different laboratories with varying protocols and platforms, the raw MS data have all been processed through the pipeline of tools developed at the Institute for Systems Biology with the goal of analyzing peptide MS/MS data consistently and with known error rates. The pipeline includes a step that assigns a probability of correctness for all putative peptide identifications. This uniform statistical validation ensures a consistent and high-quality set of peptide and protein identifications.

We assembled 28 MS/MS experiments, collectively representing 1001 LC-MS/MS runs, as summarized in Tab. 1. Of these experiments, 20 were the analysis of HUPO PPP standard samples, which are described elsewhere in this issue. The other eight are unpublished serum experiments, mostly performed at the Institute for Systems Biology (ISB) as part of other work. Nearly all the ISB data employ the glycopeptide capture technique [10] to mitigate the effects of the extremely abundant proteins.

The mass spectra were searched using SEQUEST [11], and then each possible top identification was assigned a probability of being the correct identification using the PeptideProphet software [12]. The results of this automated searching and validation with an error model were loaded into an instance of the SBEAMS – Proteomics database (http://www.sbeams.org/). All peptides with a PeptideProphet probability of being correct p greater or equal to 0.90 were combined in the database to form a master list of observed peptides across all these experiments. This list of peptides was then mapped to the EnsEMBL human proteome and genome, and the results are loaded into the PeptideAtlas database [9].

Beginning with over 1.9 million spectra in 1001 MS runs, the PeptideProphet analysis yielded 87 209 spectra with a probability of $p \geq 0.90$. This resulted in 6929 distinct peptides with $p \geq 0.90$. By combining the error rates in all the individual experiments, we calculated an overall false positive rate of 14% for the 6929 distinct peptides. Of these, 6342 peptides were successfully mapped to the EnsEMBL

Tab. 1 Summary of the contribution to the Plasma PeptideAtlas from each experiment

Search ID	Sample tag	HUPO laboratories	No. of $p \geq 0.90$ spectra	No. of distinct peptides	No. of new distinct peptides	Is HUPO?
411	b1-CIT_glyco_lcq	2	5832	740	740	Y
412	NIBSC_glyco_lcq	2	10054	1190	726	Y
414	b1-CIT_glyco_qstar	2	1379	306	61	Y
453	HUPO12_run31	12	731	235	187	Y
454	HUPO12_run32	12	1014	191	68	Y
455	HUPO12_run33	12	1037	293	149	Y
456	HUPO12_run34	12	810	169	40	Y
436	HUPO22_M_CA_S	22	9078	1578	1434	Y
399	HUPO28_b1-CIT	28	386	230	76	Y
400	HUPO28_b1-SERUM	28	514	289	64	Y
401	HUPO28_b2-CIT	28	1922	470	98	Y
402	HUPO28_b2-SERUM	28	1604	385	29	Y
403	HUPO28_b3-CIT	28	558	326	24	Y
404	HUPO28_b3-SERUM	28	556	307	15	Y
408	HUPO29_b1-CIT_1	29	417	88	15	Y
409	HUPO29_b1-CIT_win1	29	3008	549	155	Y
410	HUPO29_b1-CIT_win2	29	593	183	34	Y
407	HUPO34_b1-HEP	34	8805	1562	650	Y
413	HUPO37_b1-HEP_2LCQ	37	24	23	6	Y
422	HUPO40	40	5645	697	190	Y
254	Serum_peo_peptides		7154	1026	663	N
275	Breakfast_qtof08		334	117	10	N
278	Caex_qtof08		905	255	38	N
281	cat_ex_qtof		3514	1040	300	N
283	Cation_ex_lcq		15751	2238	963	N
368	PID_serum		4861	557	187	N
405	HUPO28_Ref-CIT		373	257	5	N
406	HUPO28_Ref-SERUM		337	224	2	N

Columns 1 and 2 provide an internal SBEAMS search batch numeric identifier and a short name (tag) for each experiment, respectively. The sample tags include the official HUPO sample names (*e.g.*, b1-SERUM) if known. Column 3 provides the HUPO laboratories from which the data are derived. The last eight experiments are serum experiments not from HUPO-provided samples, although the last two were provided by HUPO laboratories. Columns 4–6 tabulate the number of spectra identified with PeptideProphet $p \geq 0.9$, number of distinct peptides therein, and number of new distinct peptides added to the cumulative total (as plotted in Fig. 1). Clearly, the early experiments (arbitrarily sorted by HUPO laboratories number here) will have the greatest contribution to the cumulative list as nearly every peptide is new. The final column indicates if the full experimental raw data are part of the official HUPO PPP.

Tab. 2 Summary of HUPO-only and all data Plasma PeptideAtlas builds

HUPO only	ALL data	Statistic
20	28	Experiments (samples) included in build
727	1001	MS runs (mass spectrometer output files)
1568528	1943440	MS/MS spectra searched with SEQUEST
53976	87209	MS/MS spectra scored $p \geq 0.9$ by PeptideProphet
4761	6929	Distinct peptide sequences
4416	6342	Distinct peptides that mapped to EnsEMBL 29.35b
1058	1606	Possible proteins implicated in mapping
755	1131	Possible genes implicated in mapping
622	960	Simple reduced proteins (correction for ambiguous mappings)
436	666	Unambiguously mapped proteins (contain nondegenerate peptide)

Columns 1 and 2 list the statistics for the HUPO-only data and all plasma/serum experiment PeptideAtlas builds, respectively. These statistics are discussed further in the text.

29.35b genome build. The remainder of the peptides were identified *via* SEQUEST searching against the IPI v2.21 database [13] with sequences that are not exactly in the EnsEMBL build. This has been observed in other PeptideAtlas builds [9].

This list of 6342 distinct peptides mapped to 1606 different EnsEMBL proteins and 1131 different EnsEMBL genes; however, in many cases a single peptide mapped ambiguously to several proteins. A simple strategy for reducing the multiple mappings [9] suggested that approximately 960 proteins have been identified in these samples. There were 666 distinct proteins to which a peptide was unambiguously mapped. See Tab. 2 for a summary of these statistics for both the HUPO PPP sample only build and the build for all 28 experiments.

The accumulation of new distinct peptides as additional identified MS/MS spectra were added to the process is summarized in Fig. 1. Each point represents the addition of another experiment, arbitrarily sorted as shown in Tab. 1. The initial experiments contributed greatly to the cumulative numbers of distinct peptides, but the trend did become somewhat shallower as expected. The curve will asymptotically approach the total number of detectable peptides (with the used technologies and techniques). However, this level is far from being reached. At this point, approximately 65 new distinct peptides are being added for every 1000 new $p \geq 0.90$ spectra. This is a rate somewhat smaller than that observed in the main PeptideAtlas build [9].

We compared the results of the Plasma PeptideAtlas build with the compendium of plasma proteins of Anderson *et al.* [5] derived from four other sources. We mapped the proteins in that source to EnsEMBL proteins and then determined which of those proteins are in the Plasma Peptide-Atlas. Some proteins from Anderson *et al.* did not map to EnsEMBL readily with the accession numbers given, and were excluded for the purpose of this comparison. Of the proteins

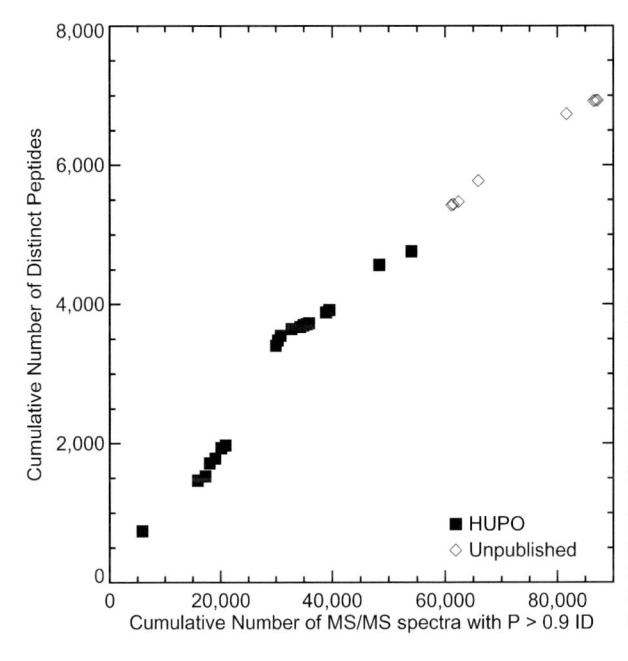

Fig. 1 Cumulative number of distinct peptides as a function of the addition of more MS/MS spectra identified with $p \geq 0.9$. Eventually the pattern is expected to show saturation as most observable peptides are cataloged. However, at present, it still appears that ~65 new peptides are still cataloged *per* 1000 identified spectra added.

found in all the four sources, all are found in the Plasma PeptideAtlas. For the proteins found in at least three, two, and one sources, we find in the Plasma PeptideAtlas 96, 76, and 27%, respectively.

The collaborative analysis of all the HUPO samples obtained from 18 laboratories yielded a total of 3020 proteins for which at least two peptides were reported in two different analyses [14]. We compared this set to the results of our Plasma PeptideAtlas build based purely on the HUPO samples with $p \geq 0.90$ (false positive rate ~14%), and found that our build contains 479 of the 3020 proteins.

We have set up the Plasma PeptideAtlas data as a DAS source that can be browsed using the EnsEMBL genome browser. Instructions on configuring the EnsEMBL browser to view these data can be found on the PeptideAtlas website.

The compendium of peptides, derived from this large set of LC-MS/MS experiments on human plasma and serum samples, is publicly available for future studies. As part of the PeptideAtlas project, we will continue to accept submission of raw MS data derived from human plasma samples and publicly release new builds of the Human Plasma Peptide-Atlas at our website http://www.peptideatlas.org/ with an increasing set of experiments. In addition to the build results, the raw mass spectrometer output for all published or otherwise public datasets are downloadable in mzXML [15] format from our repository.

This work has been funded in part with federal funds from the National Heart, Lung, and Blood Institute, National Institutes of Health, under contract no. N01-HV-28179. We gratefully acknowledge HUPO laboratories 12, 22, 28, 29, 34, 37, and 40 for allowing us to use these data in the Plasma PeptideAtlas.

14.1
References

[1] Anderson, N. L., Anderson, N. G., *Mol. Cell. Proteomics* 2002, *1*, 845–867.

[2] Tirumalai, R. S., Chan, K. C., Prieto, D. A., Issaq, H. J., Conrads, T. P., Veenstra, T. D., *Mol. Cell Proteomics* 2003, *2*, 1096–1103.

[3] Adkins, J. N., Varnum, S. M., Auberry, K. J., Moore, R. J., Angell, N. H., Smith, R. D., Springer, D. L., Pounds, J. G., *Mol. Cell. Proteomics* 2002, *1*, 947–955.

[4] Pieper, R., Gatlin, C. L., Makusky, A. J., Russo, P. S., Schatz, C. R., Miller, S. S, Su, Q. *et al.*, *Proteomics* 2003, *3*, 1345–1364.

[5] Anderson, N. L., Polanski, M., Pieper, R., Gatlin, T., Tirumalai, R. S., Conrads, T. P., Veenstra, T. D. *et al.*, *Mol. Cell. Proteomics* 2004, *3*, 311–326.

[6] Chan, K. C., Lucas, D. A., Hise, D. *et al.*, *Clin. Proteomics* 2004, *1*, 101–225.

[7] Omenn, G. S., *Dis. Markers* 2004, *20*, 131–134.

[8] Omenn, G. S., *Proteomics* 2004, *4*, 1235–1240.

[9] Desiere, F., Deutsch, E. W., Nesvizhskii, A. I., Mallick, P., King, N., Eng, J., Aderem, A. *et al.*, *Genome Biol.* 2004, *6*, R9.

[10] Zhang, H., Yi, E. C., Li, X.-J., Mallick, P., Kelly-Spratt, K. S., Masselon, C. D., Camp, I. D. G. *et al.*, *Mol. Cell. Proteomics* 2005, *4*, 144–155.

[11] Eng, J., McCormack, A. L., Yates, J. R., *J. Am. Soc. Mass Spectrom.* 1994, *5*, 976–989.

[12] Keller, A., Nesvizhskii, A. I., Kolker, E., Aebersold, R., *Anal. Chem.* 2002, *74*, 5383–5392.

[13] Kersey, P. J., Duarte, J., Williams, A., Karavidopoulou, Y., Birney, E., Apweiler, R., *Proteomics* 2004, *4*, 1985–1988.

[14] Omenn, G. S., States, D. J., Adamski, M., Blackwell, T. W., Menon, R., Hermjakob, H., Apweiler, R. *et al.*, *Proteomics* 2005, *5*, DOI: 10.1002/pmic.2005–00358.

[15] Pedrioli, P. G. A., Eng, J. K., Hubley, R., Vogelzang, M., Deutsch, E. W., Pratt, B., Nilsson, E. *et al.*, *Nat. Biotechnol.* 2004 *22*, 1459–1466.

15
Do we want our data raw? Including binary mass spectrometry data in public proteomics data repositories*

Lennart Martens, Alexey I. Nesvizhskii, Henning Hermjakob, Marcin Adamski, Gilbert S. Omenn, Joël Vandekerckhove and Kris Gevaert

With the human Plasma Proteome Project (PPP) pilot phase completed, the largest and most ambitious proteomics experiment to date has reached its first milestone. The correspondingly impressive amount of data that came from this pilot project emphasized the need for a centralized dissemination mechanism and led to the development of a detailed, PPP specific data gathering infrastructure at the University of Michigan, Ann Arbor as well as the protein identifications database project at the European Bioinformatics Institute as a general proteomics data repository. One issue that crept up while discussing which data to store for the PPP concerns whether the raw, binary data coming from the mass spectrometers should be stored, or rather the more compact and already significantly processed peak lists. As this debate is not restricted to the PPP but relates to the proteomics community in general, we will attempt to detail the relative merits and caveats associated with centralized storage and dissemination of raw data and/or peak lists, building on the extensive experience gained during the PPP pilot phase. Finally, some suggestions are made for both immediate and future storage of MS data in public repositories.

The completion of the human genome project, with the corresponding rise of the field of proteomics, led to the creation of the HUPO projects as the next major collaborative scientific enterprise in the life sciences [1]. In order to achieve the high-aiming goals of these projects in a reasonable time frame, collaborations between multiple labs around the world have been set up, with each of these labs analyzing standard samples using distinct protocols and hardware. The Plasma Proteome Project (PPP), as the pioneering project in the larger HUPO consortium, is the first of these to have amassed a large body of proteomics data during its recently completed pilot phase [2]. Centralized data storage and subsequent dissemination of these data to the scientific community has been addressed through the initial data

---

* Originally published in Proteomics 2005, 13, 3501–3505

Exploring the Human Plasma Proteome. Edited by Gilbert S. Omenn
Copyright © 2006 WILEY-VCH Verlag GmbH & Co. KGaA, Weinheim
ISBN: 3-527-31757-0

collection and management work of Marcin Adamski at the University of Michigan, Ann Arbor [3] and the protein identification database (PRIDE) [4] project of the European Bioinformatics Institute (EBI). During the construction of these resources, a lot of discussion was attributed to the storage of the MS data. In particular the storage of the raw, binary data that the machines report has been discussed thoroughly.

As the question of storing raw data has recently been taken up by editors of proteomics journals as well [5], and furthermore affects the proteomics community at large [6], we here present a series of advantages and limitations inherent to the publication of raw data compared to processed peak lists, building on the unique experiences obtained through the PPP.

There seems to be a general consensus in the proteomics community today to request submission of the source data on which reported identifications are based [5]. This will allow other researchers to verify and validate the published conclusions independently. Publishing source data also has the benefit of allowing additional (computational) analyses by other researchers, which could lead to the uncovering of new, biologically relevant information that was missed in the original analysis.

These source data can take a number of forms, but by far the most common representations are either the proprietary, binary "raw" formats that the mass spectrometers churn out during their analyses or the text-based, processed peak lists that are typically submitted to search engines for identification of the peptides that produced those spectra. In the case of fragmentation spectra, the peak lists contain the parent peptide m/z and charge (if the charge is known) and a listing of measured m/z values and their intensities for the fragment peaks. Search engines then attempt to match these fragment peaks to *in silico* generated fragmentation spectra of all peptides in a search database. The peak lists are often called MS/MS spectra and due to the extensive automation of acquisition software, they are often the only format encountered by researchers. These files can take a variety of formats, yet all are essentially text-based, small (a few kilobytes *per* file), readily readable by both humans and software programs and easily compressible (two-fold to three-fold compression ratios are routine using GNU ZIP (GZIP) (GNU – GNU's Not Unix)). Additionally, each of these peak list formats can conveniently be transcribed in any other format. A few common examples are SEQUEST files (dta), Micromass peak lists (pkl), and MASCOT Generic Format files (mgf). There is a slight variability in the amount of information these different formats can accommodate, but in general conversion between formats tends to be conservative. Furthermore, the mzData format, a community standard recently developed by the HUPO Proteomics Standards Initiative (PSI) [7] that elicits broad support among both instrument and software vendors, will ultimately eliminate the need for these format conversions.

As noted above, peak lists present an already processed view on the originally recorded data. Typically proprietary, vendor-supplied software is used to extract these peak lists from the raw data. Frequently applied processing techniques during this extraction phase include noise-filtering, centroiding, deconvolution, and deisotoping of the peaks. As there is no standard protocol for these processing steps, problems often arise because what one scientist regards as standard processing might seem "lossy" conversion to another, leading some to label these peak lists as an unfit distribution medium for MS data.

The raw data formats in contrast are much larger in size (typically well above 10 MB *per* file) and are usually stored in a proprietary, binary format. This makes the files impractical to read for both users and third-party software programs, all the more so because the exact format description is typically not disclosed by the vendors. Since the binary format can already be a compressed representation of the data, standard compression algorithms such as GZIP do not always reduce the size of these files. A simple analysis was performed to illustrate both size differences and the effects of data compression (Fig. 1). The much larger size of the raw data does, however, allow these files to contain much more information than peak lists. Raw files contain all the individual peaks as registered by the instrument detector and, for LC-MS machines, can store elution profiles and times for the LC part. Depending on the vendor and make of the machine, other useful instrument-related information can be stored in these files as well.

Recently, several interesting developments have been described that can put this wealth of additional information in raw files to good use [8, 9]. The key to interpreting these raw data directly has been the development of specific software to parse the binary content of these raw files into intelligible data, a tedious and time-consuming task that typically needs to be redone each time a new machine or a new version of an existing machine or its operating software appears. Furthermore, this reverse-engineering of a proprietary format is typically frowned upon by vendors. Next to the above-mentioned caveats associated with proprietary raw data formats, there is also the very real problem of "aging" that comes with any binary formatted data. As time goes by, support for certain formats tends to evaporate and within the space of several years, readers can no longer be found for the format. A detailed review of the issues concerning proprietary data formats and science can be found in [10].

The mzXML format of the Institute of Systems Biology [11], designed as an intermediate format between raw data and peak lists, could bring some solace if it were supported by vendors, but a more pervasive effort on behalf of the entire community to standardize raw data formats is more likely to succeed in eliciting such global support.

When it comes to storing mass spectrometric data in proteomics data repositories, the discussion tends to focus on an "either-or" decision. Most proponents for the storage of raw data currently have (limited) facilities to parse this kind of data, and are therefore able to exploit the richer information therein. The other camp, which advocates the storage of the processed peak lists, tends to lack this software, making the raw data essentially inaccessible to them (unless they happen to possess the particular, proprietary instrument software that allows the transformation to peak lists). It is our opinion that the choice should not be an exclusive one. In fact, we are convinced that both formats have a distinct and additive value at this time and as such fulfill complementary roles.

When a reevaluation of the peak lists using a different search algorithm or using a newer sequence database as search base is the scope of the research done with the original data, peak lists typically are the most readily accessible and efficient sources of MS data. For more advanced purposes however, such as obtaining large training sets for machine learning approaches for the prediction of peptide elution times [12] or, in the case of quantitative proteomics experiments based on stable isotope labeling [8], the raw formats present the only data source rich enough for these analyses.

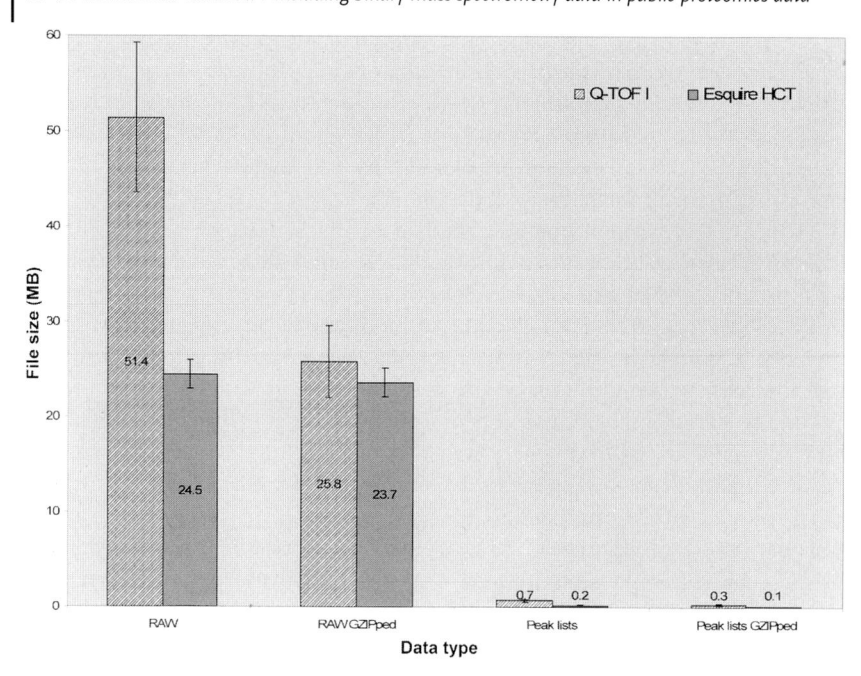

Fig. 1 Comparing compressed and uncompressed file sizes for RAW data and the corresponding peak lists. Figures for the data are based on the averages of multiple separate files for each measurement. Error bars denote one SD on the averages. For the raw data, the sizes were averaged over ten individual files. Q-TOF I (Micromass, Cheshire, UK) peak list data consist of 720 individual files, the Esquire HCT (Bruker Daltonik, Bremen, Germany) IT peak list data count 1050 distinct files. Both file sets were grouped into ten subgroups, with each subgroup corresponding to the spectra extracted from a single parent raw file. File format chosen for the peak lists was the intermediately verbose MASCOT Generic Format (http://www.matrixscience.com/help/data_file_help.html). Peak lists have been tarred by GNU tar (http://www.gnu.org) to compensate for size-bloating due to the minimal file size limit of the NTFS file system. Compression for both RAW files and peak lists was done using GZIP with default compression settings. Note the extreme difference in file sizes between raw data files and peak lists. Also notable is the difference in compression efficiency between Q-TOF I RAW files and their Esquire HCT counterparts, especially since the compressed results are highly similar, indicative of a built-in compression in the Esquire HCT files. Compressibility of the peak lists can be deduced from the data labels and is always greater than 50%.

Therefore, in the PPP, peak lists are part of the core data structure, whereas submission of raw files is considered an optional yet highly encouraged addition. The reason for this optional inclusion of raw data is purely technical in origin, as the sheer size of the files involved pushes infrastructure requirements for both storage of the data and their subsequent distribution to their limits.

Typically, funding for these infrastructure issues is evaluated using a standard cost/benefit model, yet for raw data files, the costs will surely outweigh the benefits in the short term. Storing raw files will require large amounts of disk space, which typically should be made redundant (*e.g.*, using RAID systems), thus disk space requirements will be at least twice the size of the data. Back-ups of this amount of data also present a nontrivial challenge. Due to typical low compression ratios, the amount of uncompressed tape media space (which tends to be more expensive than hard drive space) required will be roughly equivalent to the total data size. The distribution of the data after they have been successfully stored, also accounts for a large part of the cost involved since bandwidth does not come free, either. As an illustration of the data storage requirements, we consider the raw data for a single ICAT [13] or COFRADIC [14] run through a complete proteome (30–40 separate LC-MS/MS runs, with a 2 h gradient each) to have a compressed size of roughly 1.5 GB for older or less sophisticated machines, up to a massive 45 GB for newer, state-of-the-art instruments! It can be expected that future machines will generate even larger files as instrument accuracy and resolution increases. Put in perspective, a single proteome thus requires at least three times as much storage space as the NCBI nonredundant protein database (ftp://ftp.ncbi.nih.gov/blast/db/nr.tar.gz) in FASTA format, or three times as much as the full Swiss-Prot database [15] in the native text format! And although a 100 GB low end hard disk can currently be purchased for about US $100, a conservative cost estimate from the EBI averages to a total cost of US $2000 *per* 100 GB stored for data on a public high-availability FTP server, including distribution and back-up costs!

Even though a truly distributed system (every lab hosting its own raw data) maximizes cost-efficiency through distribution of both the storage and bandwidth cost, it is typically undesirable in the long run as the turn-over for availability of academic sites tends to be quite high. The installation of centralized repositories, located at dedicated institutes such as the EBI or the National Center for Biotechnology Information (NCBI), would be far more reliable in the long run, yet these organizations typically suffer from a lack of resources to host this amount of data. Compared to sequence databases, for instance, the growth in data storage requirements (and hence the rise of the cost) will be far greater for raw data, whereas the benefits (typically calculated in number of downloads or resulting publications) will most probably be less. The lack of open formats for the raw data adds to the difficulty of establishing funding for centralized repositories, which brings us to a catch-22: for a true incentive towards routine dissemination of raw data for published papers, we need open standards for the data formats used, but in order to push such open standards on the vendors, a large user community is needed that can actively define these standards as well as demand support for them from the vendors.

As a conclusion, the following recommendations can be made concerning the dissemination of MS data: (1) peak lists should be made available by default. There is no reason not to make these publicly available, and there are no real storage or distribution issues to be considered. (2) raw data have some clear benefits over peak lists, yet currently lack both standardized formats as well as the required infrastructure for centralized storage and distribution. Therefore, information on how to obtain raw data should at the very least be referenced in the published results for the time being. This can easily be done by providing links to individual lab websites from the journal websites (note that this is a version of the "truly distributed system" discussed above). (3) Efforts should be started at centralized repositories to create the necessary infrastructure so that in the mid- to long-term, source data will preferentially be submitted in the raw format. Meanwhile, (4) vendor support should be enlisted for open formats or at least open access to software tools that allow users to read and interpret the different formats of raw data. Since these latter developments are mutually dependent, the most important breakthrough to achieve seems to be the establishment of centralized repositories. Perhaps some lessons can be learned in this respect from the microarray community, as they have faced (and largely overcome) similar problems in the recent past [16].

The authors would like to thank David States for sharing his experiences from the data gathering efforts executed by his core bioinformatics unit at the University of Michigan, Ann Arbor; Ilan Beer for interesting discussions and his views on the subject matter; Peter Stoehr for the storage cost estimates; Jimmy Eng for contributing raw data file sizes for several instruments; and An Staes for her help in preparing the chart in Fig. 1. K.G. is a Postdoctoral Fellow and L.M. a Research Assistant of the Fund for Scientific Research-Flanders (Belgium) (F.W.O. Vlaanderen). Parts of the data used in this paper were generated in the context of the GBOU-research initiative (Project number 20204) of the Flanders Institute of Science and Technology (IWT).

15.1
References

[1] Hanash, S., Celis, J. E., *Mol. Cell. Proteomics* 2002, *1*, 413–414.

[2] Omenn, G. S., *Proteomics* 2004, *4*, 1235–1240.

[3] Adamski, M., Blackwell, T. W., Menon, R., Martens, L., Hermjakob, H., Taylor, C. F., Omenn, G., States, D., *Proteomics* 2005, *5*, this issue.

[4] Martens, L., Hermjakob, H., Jones, P., Adamski, M., Taylor, C. F., States, D., Gevaert, K., Vandekerckhove, J., Apweiler, R., *Proteomics* 2005, *5*, this issue.

[5] Carr, S. A., Aebersold, R., Baldwin, M., Burlingame, A., Clauser, K., Nesvizhskii, A., *Mol. Cell. Proteomics* 2004, *3*, 531–533.

[6] Prince, J. T., Carlson, M. W., Wang, R., Lu, P., Marcotte, E. M., *Nat. Biotechnol.* 2004, *4*, 471–472.

[7] Orchard, S., Hermjakob, H., Randall, K. J., Jr., Runte, K., Sherman, D., Wojcik, J., Zhu, W., Apweiler, R., *Proteomics* 2004, *4*, 490–491.

[8] Li, X. J., Zhang, H., Ranish, J. A., Aebersold, R., *Anal. Chem.* 2003, *75*, 6648–6657.

[9] Beer, I., Barnea, E., Ziv, T., Admon, A., *Proteomics* 2004, *4*, 950–960.

[10] Wiley, H. S., Michaels, G. S., *Nat. Biotechnol.* 2004, *22*, 1037–1038.

[11] Pedrioli, P. G., Eng, J. K., Hubley, R., Vogelzang, M., Deutsch, E. W., Raught, B., Pratt, B. *et al.*, *Nat. Biotechnol.* 2004, *22*, 1459–1466.

[12] Petritis, K., Kangas, L. J., Ferguson, P. L., Anderson, G. A., Pasa-Tolic, L., Lipton, M. S., Auberry, K. J. *et al.*, *Anal. Chem.* 2003, *75*, 1039–1048.

[13] Gygi, S. P., Rist, B., Gerber, S. A., Turecek, F., Gelb, M. H., Aebersold, R., *Nat. Biotechnol.* 1999, *17*, 994–999.

[14] Gevaert, K., Goethals, M., Martens, L., Van Damme, J., Staes, A., Thomas, G. R., Vanderkerckhove, J., *Nat. Biotechnol.* 2003, *21*, 566–569.

[15] Apweiler, R., Bairoch, A., Wu, C. H., Barker, W. C., Boeckmann, B., Ferro, S., Gasteiger, E. *et al.*, *Nucleic Acids Res.* 2004, *32* Database issue, D115–D119.

[16] Ball, C. A., Sherlock, G., Brazma, A., *Nat. Biotechnol.* 2004, *22*, 1179–1183.

16
A functional annotation of subproteomes in human plasma*

Peipei Ping, Thomas M. Vondriska, Chad J. Creighton, TKB Gandhi, Ziping Yang, Rajasree Menon, Min-Seok Kwon, Sang Yun Cho, Garry Drwal, Markus Kellmann, Suraj Peri, Shubha Suresh, Mads Gronborg, Henrik Molina, Raghothama Chaerkady, B. Rekha, Arun S. Shet, Robert E. Gerszten, Haifeng Wu,, Mark Raftery, Valerie Wasinger, Peter Schulz-Knappe, Samir M. Hanash, Young-ki Paik, William S. Hancock, David J. States, Gilbert S. Omenn and Akhilesh Pandey

The data collected by Human Proteome Organization's Plasma Proteome Pilot project phase was analyzed by members of our working group. Accordingly, a functional annotation of the human plasma proteome was carried out. Here, we report the findings of our analyses. First, bioinformatic analyses were undertaken to determine the likely sources of plasma proteins and to develop a protein interaction network of proteins identified in this project. Second, annotation of these proteins was performed in the context of functional subproteomes involved in the coagulation pathway, the mononuclear phagocytic system, the inflammation pathway, the cardiovascular system, and the liver; as well as the subset of proteins associated with DNA binding activities. Our analyses contributed to the Plasma Proteome Database (http://www.plasmaproteomedatabase.org), an annotated database of plasma proteins identified by HPPP as well as from other published studies. In addition, we address several methodological considerations including the selective enrichment of post-translationally modified proteins by the use of multi-lectin chromatography as well as the use of peptidomic techniques to characterize the low molecular weight proteins in plasma. Furthermore, we have performed additional analyses of peptide identification data to annotate cleavage of signal peptides, sites of intra-membrane proteolysis and post-translational modifications. The HPPP-organized, multi-laboratory effort, as described herein, resulted in much synergy and was essential to the success of this project.

---

* Originally published in Proteomics 2005, 13, 3506–3519

Exploring the Human Plasma Proteome. Edited by Gilbert S. Omenn
Copyright © 2006 WILEY-VCH Verlag GmbH & Co. KGaA, Weinheim
ISBN: 3-527-31757-0

16.1
Introduction

Proteomic technologies and applications have the potential to facilitate the development of novel diagnostic tools for clinical medicine. The proteome, representing the functional translation of the genome, affords great opportunities for the discovery of novel biomarkers/biosignatures for human diseases. More closely linked than the genome to physiologic and pathologic human conditions, proteins offer an opportunity to serve as functional biomarkers for human diseases. Despite considerable progress in this regard [1], many challenges remain, including proteomic-based diagnostic tools for routine clinical applications. One impediment to this process has been the absence of informative proteomic map(s), similar to the blueprint of the human genome. Among all possible proteomes, the Human Plasma Proteome is of particular interest because of its broad interface with other organs in the body. Many of the plasma protein constituents are considered as attractive candidates for biomarkers as they can reflect the diseased or healthy states of various tissues and organ systems. Despite the obvious challenges–as plasma contains a wide dynamic range of component concentrations (*i.e.*, nine to ten orders of magnitude) [2]–several recent investigations have discussed the potential of plasma proteome in biomarker research [3–7].

The approaches used for protein identification by HUPO's Plasma Proteome Project (HPPP) laboratories were based on tandem mass spectrometry, although the exact instruments and pre-analytical separation methods varied across laboratories. A critical step subsequent to mass spectrometry-based protein identification is careful data mining and annotation, which place the identified proteins in the context of what is known about the human biology. To address this challenge, the initial list of proteins identified by HPPP was annotated for their relevance to physiological and pathological states. This included annotation by researchers studying the coagulation pathway, the mononuclear phagocytic system, the liver, the cardiovascular system, the glycosylated proteins and DNA binding proteins. In addition, a bioinformatic analysis was conducted to discover signal peptide cleavage sites by aligning semi-tryptic peptides to the amino termini of proteins. Finally, an analysis of the interactome of plasma proteins with reference to the sites of protein localization was also performed.

16.2
Materials and methods

Eighteen participating laboratories collected data on the plasma proteome using tandem MS/MS. These findings constituted the basis of the initial dataset. A core dataset of 3020 proteins were selected from the total based upon a predetermined criterion of two or more identified peptides [9]. These were unambiguously mapped (as described ref. [10]) to 2446 distinct gene products and used in the following analyses.

16.2.1
Coagulation pathway and protein interaction network analysis

A list of proteins involved in the coagulation pathway was prepared from the catalog of plasma proteins identified by HPPP based on scientific evidence documented in the literature as well as review by experts in coagulation. The protein interaction network was analyzed by using the 2446 proteins as 'seed' proteins to extract their interacting proteins from the Human Protein Reference Database (HPRD) [11]. The subcellular component of both seed and interactor proteins was also obtained from HPRD and an interactome map was drawn using the Osprey tool [12]. Each protein was represented as a node and the nodes were arranged according to their subcellular localizations in the protein interaction map.

16.2.2
Gene ontology annotations

The proteins identified by the HPPP were assigned gene ontology (GO) terms, which rely on a controlled vocabulary for describing a protein in terms of its molecular function, biological process, or subcellular localization [13].

16.2.3
Analysis of MS-derived data for identification of proteolytic events and post-translational modifications

A bioinformatic analysis of the data generated using a quadrupole time-of-flight mass spectrometer by one of the participating laboratories (Waniger and Raftery) was carried out. We focused on those peptides that were semi-tryptic in nature (*i.e.*, cleaved by trypsin only at one end) and could, therefore, arise from *in vivo* cleavage events. All the identified semi-tryptic peptides were mapped to the corresponding proteins and overlaid with the protein domain architecture information including signal peptides and transmembrane domains. In addition, we carried out a database search using N-acetylation and hydroxylation as variable modifications to identify peptides containing these post-translational modifications. RefSeq database was searched and only peptides with a MASCOT score greater than 30 were considered for further analysis. All MS/MS spectra were also manually interpreted to validate the assignment.

16.3
Results and discussion

The primary criterion that we used for assembling the initial list for annotation in this study was that the proteins were identified on the basis of two or more peptides from tandem mass spectrometry experiments. All of these identifications were deemed 'high confidence' by the individual HUPO laboratories (for details, see ref [8]). Peptide mass fingerprinting data were not included in these analyses. In all, 3020 IPI protein entries were identified based on these criteria. These IPI proteins were unambiguously mapped as 2446 gene products. The major reasons we were unable to map the remaining identifications was because the IPI entries were deleted (203 proteins), redundant (81 proteins), corresponded to pseudogenes (32 proteins) or immunoglobulin fragments (9 proteins), did not have a complete open reading frame (33 proteins) or were not identical to any other protein entry in the non-redundant database (216 proteins). All of these proteins, in addition to proteins that have been described by previous studies to exist in plasma, have been annotated in the Plasma Proteome Database (http://www.plasmaproteomedatabase.org) (see ref [10]).

16.3.1
Bioinformatic analyses of the functional subproteomes

Projects of the magnitude such as the HPPP are likely to provide interesting and biologically significant information when protein identification is coupled to bioinformatics analyses. To this end, we carried out an analysis on protein-protein interaction "network" of the HPPP identified proteins. We assigned probable functions to the proteins according to the Gene Ontology convention. Our goal for these analyses was to use the static protein identification data to gain functional insights pertaining to the role(s) of these proteins in biological processes.

16.3.1.1 An interaction map of human plasma proteins

The access to this large plasma protein data base provided an opportunity to explore the global and systemic properties of the underlying molecular networks/pathways of the plasma proteome. The afore-mentioned 2446 gene products were further analyzed with respect to their subcellular localization and potential interacting partners. These 'seed' proteins were used to extract all documented interactors from the Human Protein Reference Database [11]. The subcellular components of both seed and interacting proteins were determined and subsequently, an interactome map was created utilizing the Osprey tool [12]. Most of the identified proteins possess multiple interacting partners. The average number of interactions per protein were 5.8. In the plasma interaction map (Fig. 1), each protein is represented as a node, and the nodes are arranged according to their subcellular localization. The lines connecting the nodes represent protein interactions.

Among the 2446 proteins studied, 652 proteins were determined to be involved in potentially 3811 interactions based on published literature. The proteins were found to be distributed evenly in the extracellular compartment, plasma membrane, cytoplasm and nucleus. Extracellular proteins, in addition to interactions among themselves and to members on the plasma membrane, were found to have an equally high number of interactors with cytoplasmic and nuclear proteins. There are several possible biological explanations for this result. First, it suggests that many extracellular proteins may exhibit heretofore unrecognized intracellular localization. Second, it is conceivable that these interactions may occur when a

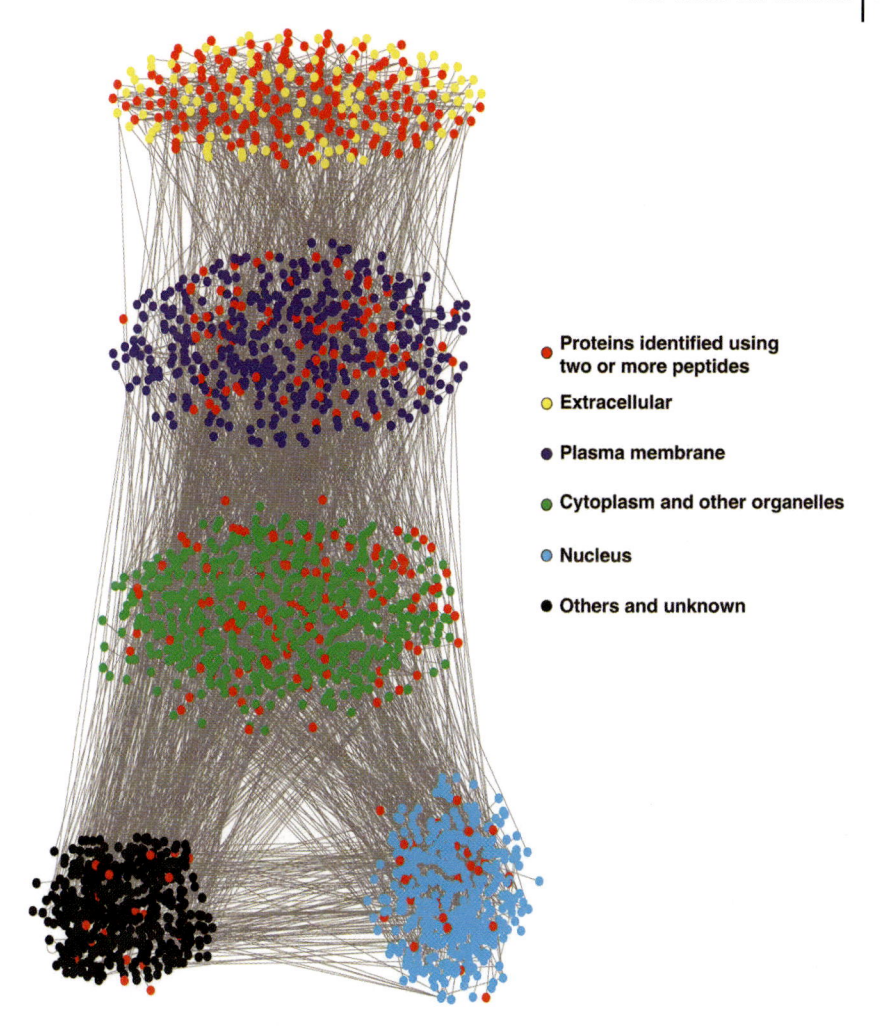

Fig. 1 Protein-interaction map of the plasma proteome. The figure shows the organization of protein interactions in human plasma. Each node represents a protein; the colors depict their localization in the cell. Red nodes represent 652 out of 2446 unique gene products that were identified by HUPO PPP that had interacting proteins listed in the HPRD. Yellow, blue, green, magenta and grey colors represent proteins that are localized to extracellular, plasma membrane, cytoplasm and other organelles, nucleus and others/unknown subcellular compartments, respectively.

large number of intracellular molecules interact with plasma membrane-bound proteins. Finally, our protein network analysis may illustrate interactions that are dependent on physiological/pathological or experimentally-induced release of intracellular proteins into the plasma.

Gene Ontology term enrichment (fraction of proteins)

Cellular Component

- nucleus
- extracellular region
- cytoskeleton
- extracellular space
- plasma membrane
- extracellular matrix (sensu Metazoa)
- extracellular matrix
- actin cytoskeleton
- microtubule cytoskeleton
- intermediate filament cytoskeleton
- intermediate filament
- myosin
- microtubule associated complex
- collagen

■ HPPP composite list
□ all proteins

Molecular Function

- protein binding
- ion binding
- nucleic acid binding
- nucleotide binding
- hydrolase
- cation binding
- purine nucleotide binding
- ATP binding
- adenyl nucleotide binding
- DNA binding
- transporter activity
- transferase activity
- structural molecule
- calcium ion binding
- transferase activity, phosphorus
- enzyme regulator activity
- kinase activity
- peptidase activity
- cytoskeletal protein binding
- protein kinase activity
- hydrolase activity, acid anhydrides
- hydrolase activity, ester bonds
- actin binding
- hydrolase activity, in phosphorus
- pyrophosphatase activity
- nucleoside-triphosphatase activity
- protein serine/threonine kinase activity
- enzyme inhibitor activity
- endopeptidase activity
- motor activity
- carbohydrate binding
- endopeptidase inhibitor activity
- protease inhibitor activity
- structural constituent of cytoskeleton
- ATPase activity
- helicase activity
- serine-type endopeptidase
- calmodulin binding
- chymotrypsin activity
- trypsin activity
- serine-type endopeptidase inhibitor
- ATP-dependent helicase activity
- extracellular matrix structural constituent
- glycosaminoglycan binding
- polysaccharide binding
- lipid transporter activity
- microtubule motor activity

Biological Process

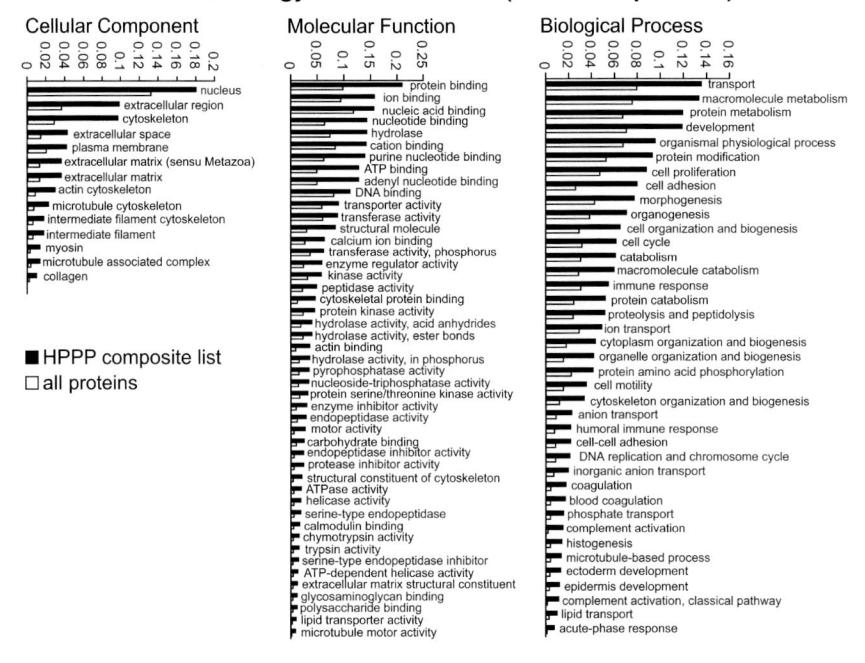

- transport
- macromolecule metabolism
- protein metabolism
- development
- organismal physiological process
- protein modification
- cell proliferation
- cell adhesion
- morphogenesis
- organogenesis
- cell organization and biogenesis
- cell cycle
- catabolism
- macromolecule catabolism
- immune response
- protein catabolism
- proteolysis and peptidolysis
- ion transport
- cytoplasm organization and biogenesis
- organelle organization and biogenesis
- protein amino acid phosphorylation
- cell motility
- cytoskeleton organization and biogenesis
- anion transport
- humoral immune response
- cell-cell adhesion
- DNA replication and chromosome cycle
- inorganic anion transport
- coagulation
- blood coagulation
- phosphate transport
- complement activation
- histogenesis
- microtubule-based process
- ectoderm development
- epidermis development
- complement activation, classical pathway
- lipid transport
- acute-phase response

Fig. 2 Significantly enriched GO terms for the set of 2446 HUPO PPP proteins. GO term annotations were obtained from LocusLink (http://www.ncbi.nlm.nih.gov/LocusLink/), 1943 of which were in the HUPO PPP composite list. GO terms that applied either to less than 20 genes or to more than 2800 genes in LocusLink were not considered. All GO terms shown are significant with $p < 0.001$ (after correcting for multiple term testing by Bonferonni, using 1000 random gene lists of 1943).

It is also noteworthy that this investigation represents one of the first studies to consider the plasma as containing a network of interacting proteins. Many recent investigations in other tissue types have emphasized the role of multiprotein complexes as building blocks of hierarchical cellular machinery, and a goal of these bioinformatic analyses was to provide an initial framework to conceptualize proteins in the plasma in a similar manner.

16.3.1.2 Gene Ontology annotation of protein function

We linked the 2,446 proteins to their Gene Ontology (GO) annotations, a procedure relying upon a controlled vocabulary for describing proteins with respect to their molecular functions, biological processes, or subcellular localization [13]. About 100 GO annotation terms showed significant enrichment in the HPPP list ($p < 0.001$, Bonferroni-adjusted), as compared to the entire list of proteins in the LocusLink database (Fig. 2). Many of these enriched GO terms represent protein classes commonly associated with plasma, such as *extracellular region* (191 proteins found), *cell adhesion* (155), *protease inhibitor activity* (47), *blood coagulation* (33), and *complement activation* (29). Other enriched terms are somewhat unexpected, such as *DNA binding* (216 proteins found) and *nucleus* (350), which observations may in fact be the result of the secretion of cellular breakdown products into the circulation. For each tissue represented in the Novartis atlas [14], we obtained an ordered list of genes that were ranked on the basis of higher mRNA expression in the given tissue as compared to that in other tissues (by rank sum). When the HUPO PPP proteins are ranked by the number of peptides used to identify them across all laboratories, we observe a very high correspondence (approaching $p < 10\text{E-}50$) between genes expressed in the liver over other tissues and HUPO PPP proteins that had the most peptide hits. This correspondence is measured by taking the top genes from each ranked list, identifying the genes in common between these lists, and determining whether this overlap significantly exceeds that expected due to random occurrence (Supplementary Fig. 1). This observation probably reflects the fact that the liver is a major contributor to proteins in plasma. Such a correspondence was not found for any other tissues represented in the atlas. It is worthy to emphasize that this does not imply that tissues other than liver do not contribute to the plasma proteome, rather, that proteins derived from these other tissues may not be the most abundant in plasma (or the most readily detectable by proteomics). This trend also does not exclude the contribution of proteins to the plasma from other tissues, as has been described elsewhere in this article with regard to organ-specific protein components of plasma (such as heart and smooth muscle-derived proteins). Another important implication from these bioinformatics analyses is that the number of times a protein is identified across laboratories might correlate with its relative abundance in the plasma. Likewise, the results of this analysis suggest that many of the proteins identified by only a single laboratory may indeed be present in plasma, perhaps in low abundance, and making their detection a rare event. This important information could not have been obtained without a combined effort of multiple laboratories to analyze a standardized sample. Future studies will provide absolute quantitation of these proteins and can be directed to unequivocally reveal the tissue(s) of origin on a protein-by-protein basis.

16.3.2
Proteins involved in the blood coagulation pathway

Coagulation of blood is a critical physiological process that prevents excessive bleeding at the site of injury. It is a proteolysis-driven pathway in which procoagulant factors and cofactors interact to constitute a clotting cascade, which ultimately leads to fibrin clot formation [15, 16]. Concomitant with this, regulatory mechanisms, such as the thrombomodulin/protein C pathway, are activated to limit the extent of clot formation and to prevent thrombosis [17]. Coagulation-related proteins identified by the HPPP (Tab. 1) can be classified into three categories from a functional standpoint:

Tab. 1 Proteins known to be involved in blood coagulation

	IPI ID	PPD ID	Gene Symbol	Name of the protein
1	IPI00019568	HPRD_01488	F2	Prothrombin
2	IPI00022937	HPRD_01964	F5	Coagulation factor V
3	IPI00017603	HPRD_02384	F8	Coagulation factor VIII
4	IPI00296176	HPRD_02385	F9	Coagulation factor IX
5	IPI00019576	HPRD_01966	F10	Coagulation factor X
6	IPI00008556	HPRD_07524	F11	Coagulation factor XI
7	IPI00019581	HPRD_01992	F12	Coagulation factor XII
8	IPI00297550	HPRD_00604	F13A1	Coagulation factor XIII, A1
9	IPI00025862	HPRD_00404	C4BPB	Complement component 4 binding protein, beta
10	IPI00009920	HPRD_01956	C6	Complement component 6
11	IPI00007240	HPRD_00605	F13B	Factor XIII, B subunit
12	IPI00021885	HPRD_00619	FGA	Fibrinogen, alpha chain
13	IPI00298497	HPRD_00620	FGB	Fibrinogen, beta chain
14	IPI00021891	HPRD_00621	FGG	Fibrinogen, gamma chain
15	IPI00022418	HPRD_00626	FN1	Fibronectin 1
16	IPI00011255	HPRD_01976	GP1BA	Glycoprotein Ib, platelet, alpha polypeptide
17	IPI00027410	HPRD_01431	GP5	Glycoprotein V platelet
18	IPI00292950	HPRD_00795	SERPIND1	Heparin cofactor II
19	IPI00008558	HPRD_01971	KLKB1	Kallikrein B
20	IPI00032328	HPRD_01970	KNG1	Kininogen
21	IPI00032256	HPRD_00072	A2M	Macroglobulin, alpha 2
22	IPI00019580	HPRD_01417	PLG	Plasminogen
23	IPI00296180	HPRD_01883	PLAU	Plasminogen activator urokinase
24	IPI00021817	HPRD_01466	PROC	Protein C
25	IPI00007221	HPRD_03503	SERPINA5	Protein C inhibitor
26	IPI00294004	HPRD_01473	PROS1	Protein S, alpha
27	IPI00027843	HPRD_07182	PROZ	Protein Z
28	IPI00296099	HPRD_01765	THBS1	Thrombospondin I
29	IPI00021834	HPRD_01064	TFPI	Tissue factor pathway inhibitor
30	IPI00023014	HPRD_01906	VWF	von Willebrand factor
31	IPI00032215	HPRD_00120	SERPINA3	Alpha 1 antichymotrypsin
32	IPI00305457	HPRD_02463	SERPINA1	Alpha 1 antitrypsin
33	IPI00032179	HPRD_00122	SERPINC1	Antithrombin III
34	IPI00293057	HPRD_04374	CPB2	Carboxypeptidase B2, plasma

The first category includes the major coagulation factors and their cofactors. The plasma concentration of the major coagulation proteins has been described previously [16] and thus it was not surprising that all of the most abundant procoagulant proteins were identified in the HPPP studies. Some coagulation proteins with concentration below 1 µg/mL, such as factor VIII, were also successfully identified in the present study. The second category of proteins includes components that limit the extent of blood coagulation. This category includes protein C, a serine protease, and its cofactor, protein S, both of which possess anticoagulant activity. Serine protease inhibitors (SERPINs) also represent another major regulatory mechanism controlling enzyme activity of activated coagulation factors. Numerous serpins such as antithrombin, heparin cofactor II, and protein C inhibitor were identified. Proteins involved in regulating fibrinolysis – of which most known members were identified – are also included in this category. One example of a protein identified in this category is carboxypeptidase B, also known as thrombin activable fibronolysis inhibitor (TAFI), which is a 58-kDa enzyme that circulates in plasma as a zymogen [18]. The third category contains other coagulation-related proteins: ADAMTS-13 is a newly identified metalloprotease enzyme, which cleaves von Willebrand factor (vWF) multimers [19]. Impairment of ADAMPTS-13 activity is caused either by a hereditary deficiency or by acquired autoantibodies that specifically inhibit ADAMTS-13 function. Both lead to an excess of ultra large vWF and thereby to thrombotic thrombocytopenic purpura [20, 21]. Interestingly, there are several well-known coagulation factors that escaped identification by all HPPP participating laboratories; these include platelet glycoprotein IX, factor VII, tissue factor and thrombomodulin.

16.3.3
Proteins potentially derived from mononuclear phagocytes

The mononuclear phagocytic system consists of bone marrow monoblasts, promonocytes, blood monocytes, and tissue macrophages. Under normal circumstances, monocytes derived from bone marrow circulate in the blood for less than 48 h and enter into various tissues to differentiate into macrophages [22, 23]. Along with polymorphonuclear neutrophils, mononuclear cells (MCs) provide a defense against microbial invasion via chemotaxis, phagocytosis and release of inflammatory cytokines. Additionally, monocyte-derived tissue macrophages such as dendritic cells are primarily responsible for processing antigen and interact with CD4 positive T cells to initiate host adaptive immunity [22]. HPPP proteins in Supplementary Tab. 1 are potentially of MC origin and the possible biological implications of these findings are discussed below.

Macrophages produce an array of cytokines at sites of inflammation and contribute to the development of overall host inflammatory responses [23]. Plasma Interleukin 1β, 6, 10, 12α, and 15 are all potentially contributed by activated MCs. Two chemokines CCL3 and CCL5 were also observed in plasma. CCL3, or macrophage inflammatory protein 1α (MIP-1α), regulates the migration of various effector cells such as monocytes, T cells, neutrophils, eosinophils, basophils, and natural killer cells [24]. CCL5 is a chemoattractant for CD4+ memory T lymphocytes, monocytes, and eosinophils. CCL5 is expressed in many cell types including MCs [25, 26]. MCs constitutively express Colony Stimulating Factor 1 receptor (CSF1R, CD115), and Colony stimulating Factor 2 receptor alpha chain (CSF2RA, CD116) and thus these plasma proteins may have arisen from MCs. Upon activation, both of these receptors mediate monocyte differentiation and prolong cell survival. Inflammatory stimulation also triggers release of a soluble isoform of CSF2RA potentially by metalloprotease action [27, 28].

CD14 is the lipopolysaccharide (LPS)-receptor protein in MCs; interaction between which triggers an inflammatory response in monocytes via the TLR4/MD-2 complex [29, 30]. CD163 is a hemoglobin-haptoglobin scavenger, a monocyte/

macrophage-restricted member of the scavenger receptor family. In addition to being expressed on the MC surface, a soluble form of CD163, resulting from proteolytically cleavage of membrane bound CD163, has been reported [31]. Plasma level of CD163 may be used as a marker for inflammatory disorders [32], or as immunophenotypic marker of monocytic lineage during the diagnosis of acute myeloid leukemia [33]. The mannose receptor (MR) is primarily expressed on macrophages. It recognizes a range of carbohydrates present on, or shed from the surface of microorganisms and can mediate endocytosis and phagocytosis of microbial pathogens [34, 35]. Mannose receptor C type 2 (CD69) is a newly described transmembrane glycoprotein with a C-type lectin binding domain that is constitutively expressed on human monocytes [37] and potentially contributes to innate immunity [36].

MCs contain numerous lysosomal thiol proteases and aspartic proteases, supporting their activities in proteolysis of pathogens and antigen processing [38–42]. Several thiol proteases (Cathepsins L, S, Z) and one aspartic protease (Cathepsin D), identified in plasma, were potentially derived from MCs. Urokinase inhibitor, also named plasminogen activator inhibitor 2 (PAI2), is primarily synthesized by monocytes [43–45]. The physiological function of PAI2 in blood coagulation is incompletely understood; however, it is known to inhibit intracellular proteases and has been implicated in coronary heart disease [46].

16.3.4
Proteins involved in inflammation

Systemic or local inflammation is a pathology associated with a number of human disease. Therefore, the inflammatory subproteomes within the plasma has the potential to be relevant for the management of a host of inflammatory human disorders. As seen in Supplementary Tab. 2, a number of inflammatory mediators were identified by the HUPO PPP, including chemokines, adhesion molecules, as well as other proinflammatory cytokines. It must also be pointed out that many classic protein markers of leukocytes were not identified including CD19, 20, 21 23, 24 and 25 (markers of B cell lineage), CD 1, 2, 3, 5, 8, and 28 (markers of T cell lineage), CD 15, 16, 33, 35 (markers of granulocytic lineages), CD 61 (marker of platelets) and CD 68 (marker of MCs). This probably reflects the low abundance of many of these molecules coupled with the lack of shedding from cell surface. Nevertheless, some of the proteins that were identified are believed to be present in relative low abundance in normal serum or plasma. For example, vascular cell adhesion molecule (VCAM)-1 has been reported in healthy individuals at a concentration of 200–300 ng/mL [47]. Surprisingly, multiple laboratories identified a far less abundant cytokine, Interleukin-6, which is present at levels as low as 1.0 pg/mL [48]. It is interesting to note that several of the proteins identified are more closely associated with ongoing inflammation, rather than classic homeostatic processes *per se*. Intriguingly, the HPPP also identified multiple viral and bacterial proteins in purportedly normal plasma [8]. Inflammatory mediators may thus reflect ongoing subclinical challenges by antigens or infectious agents. These groundwork analyses of plasma notwithstanding, the "normal" spectrum of inflammatory mediators defined in broader cohorts of human subjects remains the subject of future investigation. In summary, currently available technologies can already achieve sufficient depth so as to identify immune mediators, which has clear implications for biomarker discovery for the diagnosis and prognosis of human inflammatory diseases.

16.3.5
Analyzing the peptide subproteome of human plasma

Endogenous peptides play an important role in many physiological processes in their capacity as messengers, hormones or cytokines. Furthermore, alterations in peptide levels under pathophysiological conditions implicate this class of molecules as potential drug targets or biomarkers. Alterations in peptide levels under pathophysiological conditions indicate their usefulness as potential drug targets. In addition, peptides can be derived from turnover processes of larger blood- or tissue-borne proteins. A limited number of studies have been published that concentrate on native, circulating peptides, also referred to as the peptidome (low molecular weight proteome) in blood [3, 4, 49], and therefore, we investigated these peptide constituents of plasma in the HPPP.

In one of our labs (Kellman and Schulz-Knappe), Peptidomics-Technologies® were developed specifically for peptidome analysis. This method combines a multiplexed peptide display, differential analysis (Differential Peptide Display®, DPD) and sequence analysis on the basis of liquid chromatography and mass spectrometry, providing single markers, and marker panels or patterns. Peptidomics Technologies provides a multiplexed display of approximately 2000 endogenous peptides derived from 1 mL human plasma down into the picomolar concentration range. Further details are available in the accompanying paper by Tammen *et al.* [50]. In the near future, population-based approaches that display thousands of circulating peptides in a (semi-)quantitative manner will allow for the search of biomarkers from the blood-peptidome.

16.3.6
Liver related plasma proteins

In addition to modification and function-specific annotation of proteins described in the preceding sections, we also annotated the proteins based on whether they could be derived from the liver or the cardiovascular system. The liver is one of the largest, most functionally complex, organs in the body and plays a major role in the metabolism of carbohydrate, amino acids and lipids. Currently, the functional status of the liver is surveyed by quantifying blood components to reveal the extent and type of liver damage. These measurements include chemicals (*e.g.*, bilirubin) and enzyme activities (*e.g.*, aminotransferases) in a serum specimen [51]. Furthermore, the presence and level of certain proteins in the plasma is commonly used for diagnosing hepatic disorders. Levels of albumin, the most abundance protein in

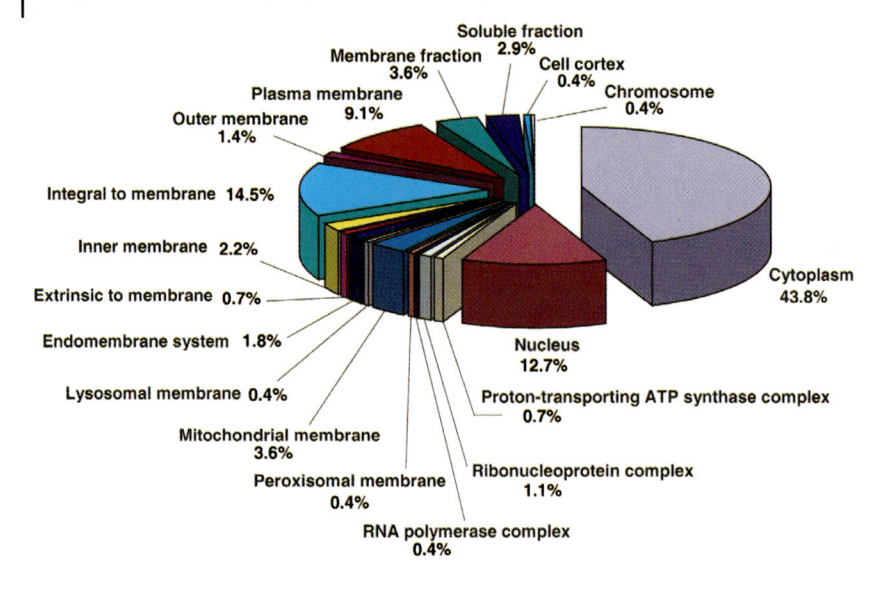

Fig. 3 Distribution of 362 proteins of liver-origin according to the sites of subcellular localization.

plasma, can aid in determining generalized liver disease, since hypoalbuminaemia is a feature of advanced chronic liver disease [52]. Serum alkaline phosphatase activity indicates cholestasis or blockage of bile flow [53]. Elevated activity of aminotransferases in serum indicates acute damage to hepatocytes irrespective of its etiology [54]. Other proteins such as α1-antitrypsin and ceruloplasmin are measured in the diagnosis of specific diseases affecting the liver. The existing plasma protein markers for liver function provide clear impetus to examine the plasma as a source of liver and liver disease-associated proteins.

We prepared a partial list of 362 proteins of liver-origin, based on HPPP master list. Annotation was performed on these confirmed lists using the information present in the published literature (using PubMed) and the online Mendelian inheritance in man (OMIM) database. Proteins of hepatic origin were classified according to their subcellular localization (Fig. 3) (for details, see [55]). The proteins are highly localized in the cytoplasmic (121), followed by integral membrane (40), nuclear (35), extracellular space (26) and plasma membrane (25). When these proteins of liver-origin were examined in the context of various liver diseases using NCBI Mesh DB, only a few represent known proteins used to assess liver function (*e.g.*, alpha-1 anti-trypsin and ceruloplasmin), which was not surprising because plasma samples were obtained from 'normal' individuals, *i.e.*, persons without overt liver disease. Alanine or aspartate aminotransferases or alkaline phosphatase enzymes, three key proteins that are elevated in liver disorders, were not identified. However, two proteins, termed insulin-like growth factor-I (IGF-I), a marker of hepatic function [56] and transforming growth factor beta (TGF β) [57, 58], a marker with known involvement in hepatocellular carcinoma, were identified. Hepatocellular carcinoma is the fifth most common malignant disorder that causes nearly 1 million deaths worldwide. The circulating IGF-I is mainly liver-derived and the development of hepatocellular carcinoma is accompanied by a significant reduction of serum IGF-I levels independent of the grade of impairment of liver function. Importantly, alteration in IGF-I levels precedes the morphologic appearance of the disease, permitting a diagnosis of the tumor [59, 60]. TGF beta levels are elevated in the sera of patients with hepatocellular carcinoma [58]. In conclusion, many proteins potentially of liver-origin, including some disease markers, were found in normal human plasma.

16.3.7
Cardiovascular system related plasma proteins

To obtain information regarding proteins of cardiac or vascular origin that were identified in the HPPP analyses, we used PubMed to individually search the proteins for relevance to cardiovascular function using the key words "heart" or "vascular." When searches of protein names returned no results, we searched abbreviations of the protein names (*e.g.*, protein kinase C would also be searched as PKC) and/or non-isoform-specific names (*e.g.*, PKCβ would also be searched as PKC). In-house expertise with cardiovascular medicine and physiology was used to examine the database search results for possible importance to CV function and/or disease. Our analyses identified cardiovascular-related functions for at least 345 of these 2446 proteins and suggested that these proteins can be divided into at least eight different categories based on function as shown in Fig. 4 (a comprehensive analysis of these cardiovascular-associated proteins can be found in ref [61]). These categories include: markers of inflammation, myocardial ischemia and/or cardiovascular disease, signaling proteins, growth- and differentiation-associated proteins, transcription factors, cytoskeletal proteins, vascular proteins, heart failure- and remodeling-related proteins and channel and receptor proteins.

There are at least two reasons why the organ-based annotative approach performed herein can yield further important information regarding plasma proteins. First, if these proteins in the plasma indeed originate from the heart or other organs, they may herald pathologic (such as cell death-associated) or physiologic (normal interplay between functioning tissues and the plasma) relationships between plasma and organ systems that were previously not appreciated. Second, if the individual proteins are of plasma or hematopoietic cell origin, insights into their function in other tissues will advance the understanding of their potential actions in the plasma. The potential roles of cardiac- or vascular-derived signaling molecules in the plasma to influence cardiovascular function, and the understanding of whether their release into the plasma is a controlled or catastrophic process, will be the pursuit of future basic science investigations.

In addition, the list of cardiovascular-related proteins itself constitutes a blueprint from which physiologically relevant information regarding cardiovascular diseases can be further mined. It is reasonable to hypothesize that changes in the levels and/or presence of these cardiovascular-related proteins may serve as a diagnostic tool, or readout, of the underlying function of the heart.

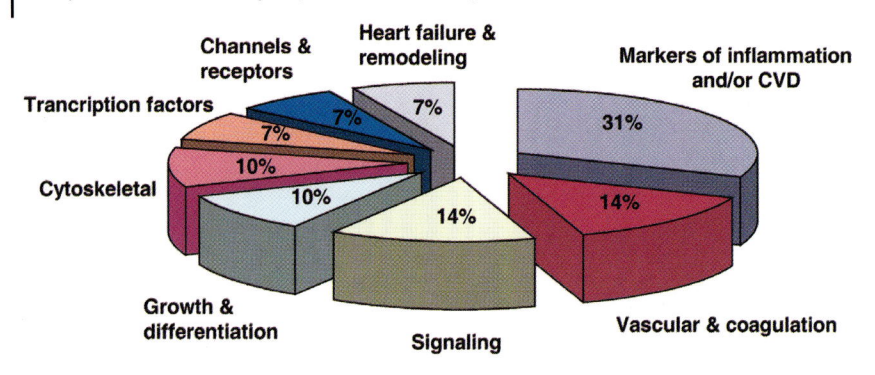

Fig. 4 Annotative breakdown based on known function of cardiovascular-related proteins identified in plasma. The initial subset of plasma proteins utilized for this study (3,020) was searched individually with the keywords "heart" and "vascular" on the PubMed search engine. Of this initial group, 345 proteins were ascribed cardiovascular-related function on the basis of previous studies, and have been divided into eight groups: markers of inflammation and/or cardiovascular disease; signaling; growth- and differentiation-associated; transcription factors; cytoskeletal; vascular; heart failure and remodeling; and channels and receptors. Percentages of total cardiovascular-related proteins represented by each group are labeled.

16.3.8
Glycoproteins

Various post-translational modifications, including glycosylation, can lead to altered protein function and activity. Selective detection of such modified proteins is a useful tool for modification-based subproteomes mining. A total of 170 proteins were identified using multi-lectin affinity chromatography out of which 84 (50%) were previously shown to be glycosylated. The exact glycosylation site was known in a majority of cases (Supplementary Tab. 3). Further, almost 66 of 170 (40%) of the proteins identified had previously been shown to be present in serum, plasma or whole blood. The details of the experimental protocol are available in reference [62]. We also carried out a functional annotation of the glycoproteins. Fig. 5 and Supplementary Tab. 3 show the variety of tissues where the proteins identified by multi-lectin affinity chromatography are expressed. The existence of glycosylation-based subproteomes in plasma demonstrate the utility of this approach and suggest that it could be used to identify modifications of potential physiological relevance.

16.3.9
DNA-binding proteins

The next component of our annotative effort included the characterization of DNA-binding proteins, a functional subproteome within the plasma. DNA-binding proteins were classified as such based on analysis of the published literature and a bioinformatics analysis of the protein domains. DNA-binding proteins bind to

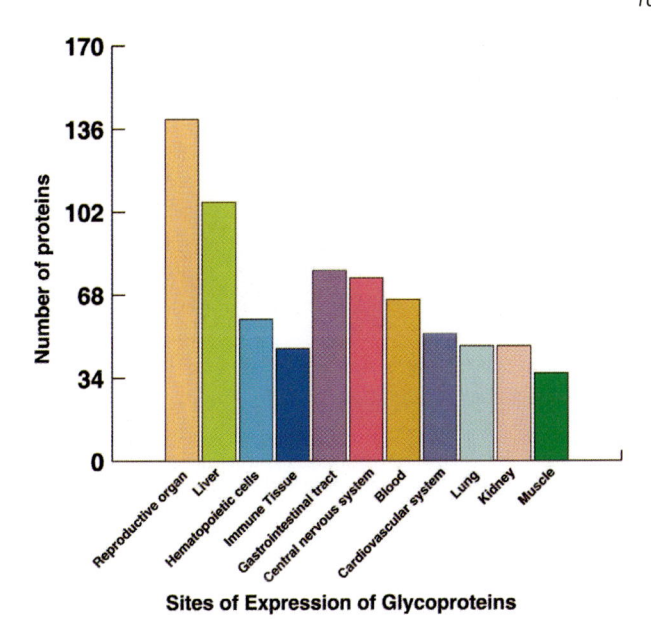

Fig. 5 Sites of expression of proteins enriched by multi-lectin affinity chromatography. The histogram shows the sites of expression of 170 proteins that were identified using multi-lectin affinity chromatography. The sites of expression are from the published literature based on mRNA and/or protein expression as annotated in the Human Protein Reference Database.

double-stranded as well as ssDNA. Because DNA is normally localized to the nucleus, DNA-binding proteins are expected to localize to the nucleus or to shuttle between the cytoplasm and nucleus. Thus, it was of particular interest that normal human plasma proteins were found to include a number of proteins, which have previously been attributed DNA-binding functions. Several independent classes of DNA-binding proteins were seen on the HPPP list, including histones, helicases, zinc finger proteins and proteins involved in cell division (Supplementary Tab. 4). In addition to the HPPP identifications discussed herein, previous studies [63] have also reported the presence of members of these protein families in human serum. Another possible explanation for the presence of nuclear proteins in plasma or serum could be due to the contamination by broken/dead cells caused by the isolation procedure.

16.3.9.1 Histones

Histones are small basic proteins that form the structural core of chromatin nucleosomes, the macromolecular conformation of DNA. Each nucleosome is composed of two H2A and H2B subunits and two H3 and H4 subunits, however, many variants and histone related proteins have been identified in the genome. Histone H1 and related proteins bind to DNA in the linker region between nucleosomes. Histone proteins, which were identified in HPPP samples, include H2AFJ, HIST1H4A, and HIST1H2AL. H2AFJ and HIST1H2AL belong to the H2A histone family. The H2A subunit performs essential roles in maintaining structural integrity of the nucleosome, chromatin condensation, and binding of specific chromatin-associated proteins. There are previous reports which support the presence of histones in the serum. Waga *et al.* [64] identified and isolated soluble histones from bovine milk and serum. Using Western blot analysis, they detected H4 and partially degraded H2A in bovine serum. Holdenrieder *et al.* [65] reported circulating nucleosomes in serum. During cell death, particularly apoptosis, endonucleases are activated and cleave chromatin into multiple oligo and mononucleosomes. Subsequently, these nucleosomes are packed into apoptotic bodies and engulfed by macrophages or neighboring cells. Under conditions of high cellular turnover and cell death, nucleosomes can also be released into the circulation and can be detected in serum or plasma. As enhanced cell death occurs under a number of pathologic conditions, the presence of elevated amounts of circulating nucleosomes is not specific for any specific benign or malignant disorder.

Serum amyloid P component (SAP) is a highly conserved plasma protein named for its universal presence in amyloid deposits. SAP is the single normal circulating protein that shows specific calcium-dependent binding to DNA and chromatin under physiological conditions. Binding of SAP displaces H1-type histones and thereby solubilizes native chromatin, which is otherwise insoluble at the physiological ionic strength of extracellular fluids. Also, SAP binds *in vivo* both to apoptotic cells, the surface blebs of which bear chromatin fragments, and to nuclear debris released by necrosis. SAP may therefore participate in handling of chromatin exposed by cell death [66]. The role played by nucleosomes in the pathogenesis of diseases is not well studied. In autoimmune diseases such as SLE (Systemic Lupus Erythematosus), circulating nucleosomes might be one of the elements that initiate and maintain the activated state of the immune system [67]. In malignant diseases, it has been proposed that the presence of large amounts of circulating nucleosomes might be a part of the tumor counterattack that overwhelms the immune system [68].

16.3.9.2 Helicases

Helicases are proteins that promote unwinding of duplex DNA during replication by binding cooperatively to single-stranded regions of DNA or to short regions of duplex DNA that are undergoing transient opening. Helicases, identified in HPPP samples include HELLS (helicase, lymphoid specific or Lsh), RECQL5, CHD1, 5 and 7 (chromodomain helicase DNA binding proteins). Lsh is thought to be involved with cellular proliferation and may play a role in leukemogenesis. A study by Yan *et al.* demonstrated that Lsh is an exclusively nuclear protein, and they defined a nuclear localization domain within the N-terminal portion of Lsh [69]. Lsh strongly associates with chromatin and requires the internal and C-terminal regions of the protein for this interaction. Interestingly, many of the HPPP participating labs have confirmed the presence of Lsh in plasma. Thus, tissue leakage or release of nuclear debris into the extracellular fluid could account for its presence in the plasma. The CHD family of proteins is characterized by the presence of chromo (chromatin organization modifier) domains and SNF2-related helicase/ATPase domains.

16.3.9.3 Zinc finger proteins

Zinc fingers include multiple distinct structures with the commonality that they require zinc for correct formation of their tertiary and/or quaternary structure. These structural motifs are involved in a broad range of biological activities including double and single stranded DNA binding, RNA recognition, and coordination of protein-protein interactions [70]. The zinc finger proteins that were identified in HPPP samples include RAG 1, MGC26914, FLJ30791, ZNF21, ZNF291, ZNF22 and KAISO-L1. A reasonable explanation for the presence of high number of zinc finger proteins in the plasma could be that some of these domains are involved in the formation of multiprotein complexes and thereby in regions of the cells exposed to exocytosis. Likewise, these proteins may have been released into the plasma by protein transport or cell death.

A number of proteins either directly shown to bind DNA or containing DNA domains thought to confer DNA binding capacity are present in normal human blood [7, 71]. A nonredundant list of 1175 gene products was developed by combination of four separate sources [7]. The methodologies used in this study were 1) literature search for proteins reported to occur in plasma or serum; 2) multidimensional chromatography of proteins followed by two-dimensional electrophoresis and mass spectrometric (MS) identification of resolved proteins; 3) tryptic digestion and multidimensional chromatography of peptides followed by MS identification; and 4) tryptic digestion and multidimensional chromatography of peptides from low-molecular-mass plasma components followed by MS identification. At least 10 transcription factors were observed in the experimental sets, and none of these were found in the literature accession set. In some cases, such as SAP, these proteins may have a physiologic role in binding and solubilizing DNA and chromatin that leak into the plasma. In other cases, these proteins most likely represent breakdown products released in the process of cell death and disintegration. Concrete determination of the roles of these individual DNA-binding-like molecules in the plasma to bind to nucleotides or other functions will require additional studies.

16.3.10
Annotation through reanalysis of mass spectrometry data

Although the primary endpoint in the HPPP was to identify as many proteins as possible in normal human plasma, an interesting offshoot is that the raw data from such mass spectrometric experiments are also a potential goldmine of information regarding post-translational modifications. To cultivate this resource, we carried out a bioinformatic analysis of the data generated using a quadrupole time-of-flight mass spectrometer by the one of the laboratories participating in HPPP (Raftery and Waniger) to specifically investigate signal peptides, post-translationally modified (e.g. acetylated) proteins or proteins that potentially underwent regulated intramembranous proteolysis.

16.3.10.1
Cleavage of signal peptides and transmembrane domains

The plasma proteome is rich in proteins that have undergone at least one major proteolytic cleavage event-the removal of the signal peptide. These signal peptides are found at the N termini of proteins that are either extracellular or generally bound to the plasma membrane. The exact location of these signal peptides is usually assigned on the basis of prediction by algorithms that recognize general features of signal peptides including the presence of a hydrophobic stretch flanked on one end by basic amino acids. Although this is currently the most popular method of annotating the location of a signal peptide, such assignments can be inaccurate. Precise identification of signal peptides thus requires experimental determination of the location of the cleavage event that results in the formation of mature proteins. After this cleavage has occurred, the mature proteins lack a signal peptide. In many instances, the mature proteins undergo further cleavage and the resultant cleavage products have different biological activities. If these proteins have to be produced recombinantly (as is the case with cytokines and growth factors administered in humans such as erythropoietin, IL-2, insulin etc.), addition or deletion of a few amino acids because of incorrect assignment of signal peptides can have disastrous consequences ranging from immunological reactions to lack of biological activity.

We focused our attention on those peptides that were semitryptic in nature (*i.e.*, cleaved by trypsin only at one end) and could, therefore, arise from *in vivo* cleavage events. All the semi-tryptic peptides identified through this analysis were mapped to the corresponding proteins and overlaid with the protein domain architecture information including signal peptides and transmembrane domains. As shown in Tab. 2, we were able to validate the correct signal peptide assignments in the case of

Tab. 2 A list of semi-tryptic peptides indicating signal peptide cleavage sites

Accession #	Protein Name	Gene Symbol	Peptide Sequence
1. NP_000005	Macroglobulin, Alpha-2	A2M	MGKNKLLHPSLVLLLLVLLPTDASVSGKPQYMVLVPSLLHTETTEKGCVL
2. NP_068657	Fibrinogen, alpha chain	FGA	MFSMRIVCLVLSVVGTAWTADSGEGDFLAEGGGVRGPRVVERHQSACKDS
3. NP_001076	Alpha-1 antichymotrypsin	SERPINA3	MERMLPLLALGLLAAGFCPAVLCHPNSPLDEENLTQENQDRGTHVDLGLA

Gray color signifies predicted signal peptide sequence and underline indicated the semitryptic peptide that was identified.

macroglobulin alpha-2 and fibrinogen, alpha chain as the semi-tryptic peptide that we identified matched the assigned site of signal peptide cleavage. However, our analysis showed the presence of one instance where the identified peptide indicated the site of cleavage of signal peptide but did not coincide with the predicted site. The signal peptide in the case of alpha-1-antichymotrypsin (SERPINA3) was predicted to be cleaved after residue 25 whereas the semi-tryptic peptide that we identified indicated that the cleavage occurred after residue 23. Thus a shorter mature protein was predicted in the databases for alpha-1-antitrypsin because of this incorrect prediction of signal peptide. Similar validation and correction of assignment of signal peptide cleavage sites has recently been shown for proteins identified from hemodialysis fluid [72].

Regulated intramembrane proteolysis (RIP) is a recently discovered phenomenon that refers to the cleavage of the hydrophobic stretches that comprise the transmembrane domain(s) of membrane-bound proteins [73]. The cleaved product is rendered soluble and might possess a biological activity not contained within the protein itself. An example of RIP is SREBP, an ER membrane protein whose cytoplasmic domain is cleaved in response to low sterol concentrations and this soluble protein fragment then translocates to the nucleus and facilitates the transcription of genes involved in lipid synthesis or uptake [73]. We found 2 previously undescribed candidates for RIP – the first protein is a syntaxin in which the cleavage event would lead to release of SynN and t-SNARE domains into the cytoplasm, which are known to mediate binding to v-SNARES to facilitate membrane fusion. Thus, the loss of these domains by a cleavage event could represent a possible mechanism to regulate membrane targeting. The second protein is the programmed cell death 1 ligand 2. A cleavage event in this case would lead to the release of the extracellular immunoglobulin domain, which is involved in interactions with other proteins and its release would probably be a point of regulation of the immune function [74]. Tab. 3 shows the peptides sequences that were identified from these proteins.

16.3.10.2 Identification of PTMs

MS/MS spectra obtained during the project by the above-mentioned laboratory were also carefully reanalyzed with the specific intent of identifying post-translational modification(s). We specifically searched the RefSeq database for the presence of modified peptides and found one N-terminally acetylated peptide as well as a peptide containing a hydroxyproline residue. Neither of these post-translational modifications had previously been reported regarding the proteins in question (Tab. 4). A detailed analysis of the entire dataset is likely to reveal many additional examples as illustrated by a recent proteomic analysis of hemodialysis fluid [72]. Supplementary Fig. 2 shows the MS/MS spectra for all semitryptic and post-translationally modified peptides that were identified in this analysis.

Tab. 3 A list of proteins with TM domain and a semi-tryptic peptide corresponding to intramembrane cleavage

	Accession #	Protein Name	Gene Symbol	Peptide Sequence
1.	NP_003754	Syntaxin 16	STX16	IVLIVVLVGVKSR
2.	NP_079515	Programmed cell death 1 ligand 2	PDCD1LG2	IATVIALR

Tab. 4 A list of peptides with post-translational modifications (PTMs)

	Accession #	Protein Name	Gene Symbol	Type of PTM	Peptide Sequence
1.	NP_068657	Fibrinogen, alpha chain	FGA	Hydroxylation	TFPGFFSP* MLGEFVSETESR
2.	NP_005521	Isocitrate dehydrogenase 3 alpha Subunit	IDH3A	N-acetylation	Ac-AGPAWISK

* Denotes hydroxylation and Ac denotes an acetyl group

16.4
Concluding remarks

Large scale proteomics investigations across numerous laboratories, notwithstanding the advantages of using standardized samples, face challenges for data analyses. The HPPP approach presented in this study capitalized on an international array of expertise from across a spectrum of human physiology and pathophysiology toward the common goal of analysis and annotation of human plasma. It is essential to reiterate that the investigators involved in the annotation process, as presented in this manuscript, were also centrally involved in the plasma analysis/protein identification stage. This integration of analysis and annotation provides for an accurate assessment of physiological context of proteins and peptides by investigators with direct knowledge of the experimental parameters under which the data were obtained.

Through this annotative effort, we have characterized functional subproteomes in the human plasma with potential relevance to cardiovascular and liver disease, DNA binding, coagulation and mononuclear phagocytosis. We have also performed a detailed bioinformatic analysis of the peptides used to identify these proteins, both with regard to the potential independent roles of these peptides in plasma, and as their unique properties can provide insights into the proteins from which they originate. These results suggest that the proteins that were rarely identified may either be lower abundance plasma components, and/or may arise from non-hematopoietic cell types. Furthermore, these findings suggest that the *groups* of proteins enriched in the lower confidence dataset are classes that are either under-represented in plasma or that have been released into the plasma through cellular breakdown. Resolution of this dichotomy will clearly be one of the most exciting challenges of the future of plasma proteomics. Specifically, the field now faces the challenge to decipher the physiological/pathological roles of these proteins in the plasma while simultaneously working from these annotated datasets to discover and develop novel biomarkers/biosignatures of disease. The model of subproteome analysis employed in this study can be applied to numerous other disease states and in other bodily fluids and tissues. Ultimately, it is hoped that subproteome analysis will couple changes in subsets of proteins to alterations in human phenotype.

We would like to thank all the laboratories that participated in the HPPP. We thank Dr. David Ginsburg for helpful discussions regarding the coagulation pathway. This work is supported in part by the UCLA Theodore C. Laubisch Endowment to Dr. Ping. Dr. Pandey serves as Chief Scientific Advisor to the Institute of Bioinformatics. The terms of this arrangement are being managed by the Johns Hopkins University in accordance with its conflict of interest policies.

16.5
References

[1] Granger, C. B., Van Eyk, J. E., Mockrin, S. C., Anderson, N. L., *Circulation* 2004, *109*, 1697–1703.

[2] Anderson, N. L., Anderson, N. G., *Mol. Cell. Proteomics* 2002, *1*, 845–867.

[3] Tirumalai, R. S., Chan, K. C., Prieto, D. A., Issaq, H. J., et al., *Mol. Cell. Proteomics* 2003, *2*, 1096–1103.

[4] Schulz-Knappe, P., Raida, M., Meyer, M., Quellhorst, E. A., Forssmann, W. G., *Eur. J. Med. Res.* 1996, *1*, 223–236.

[5] Pieper, R., Gatlin, C. L., Makusky, A. J., Russo, P. S., et al., *Proteomics* 2003, *3*, 1345–1364.

[6] Adkins, J. N., Varnum, S. M., Auberry, K. J., Moore, R. J., et al., *Mol. Cell. Proteomics* 2002, *1*, 947–955.

[7] Anderson, N. L., Polanski, M., Pieper, R., Gatlin, T., et al., *Mol. Cell. Proteomics* 2004, *3*, 311–326.

[8] Omenn, G. S., States, D. J., Adamski, M., Blackwell, T. W., et al., *Proteomics* 2005, 5. DOI 10.1002/pmic.200500358

[9] Adamski, M., Blackwell, T., Menon, R., Martens, L., et al., *Proteomics* 2005, 5. DOI 10.1002/pmic.200500186

[10] Muthusamy, B., Hanumanthu, G., Suresh, S., Rekha, B., et al., *Proteomics* 2005, 5. DOI 10.1002/pmic.200400588

[11] Peri, S., Navarro, J. D., Amanchy, R., Kristiansen, T. Z., et al., *Genome Res.* 2003, *13*, 2363–2371.

[12] Breitkreutz, B. J., Stark, C., Tyers, M., *Genome Biol.* 2003, *4*, R22.

[13] Ashburner, M., Ball, C. A., Blake, J. A., Botstein, D., et al., *Nat. Genet.* 2000, *25*, 25–29.

[14] Su, A. I., Wiltshire, T., Batalov, S., Lapp, H., et al., *Proc. Natl. Acad. Sci. USA* 2004, *101*, 6062–6067.

[15] Butenas, S., Mann, K. G., *Biochemistry (Mosc)* 2002, *67*, 3–12.

[16] Mann, K. G., *Thromb. Haemost.* 1999, *82*, 165–174.

[17] Bourin, M. C., Lindahl, U., *Biochem. J.* 1993, *289 (Pt. 2)*, 313–330.

[18] Sakharov, D. V., Plow, E. F., Rijken, D. C., *J. Biol. Chem.* 1997, *272*, 14477–14482.

[19] Furlan, M., Robles, R., Galbusera, M., Remuzzi, G., et al., *N. Engl. J. Med.* 1998, *339*, 1578–1584.

[20] Tsai, H. M., Lian, E. C., *N. Engl. J. Med.* 1998, *339*, 1585–1594.

[21] Triplett, D. A., *Arch. Pathol. Lab. Med.* 2002, *126*, 1424–1429.

[30] Furie, B., Bouchard, B. A., Furie, B. C., *Blood* 1999, *93*, 1798–1808.

[22] Cerundolo, V., Hermans, I. F., Salio, M., *Nat. Immunol.* 2004, *5*, 7–10.

[23] van Furth, R., *Res. Immunol.* 1998, *149*, 719–720.

[24] Rollins, B. J., *Blood* 1997, *90*, 909–928.

[25] Melchjorsen, J., Pedersen, F. S., Mogensen, S. C., Paludan, S. R., *J. Virol.* 2002, *76*, 2780–2788.

[26] Fessele, S., Boehlk, S., Mojaat, A., Miyamoto, N. G., et al., *Faseb. J.* 2001, *15*, 577–579.

[27] Flamant, S., Lebastard, M., Pescher, P., Besmond, C., et al., *Microbes. Infect.* 2003, *5*, 1064–1069.

[28] Prevost, J. M., Pelley, J. L., Zhu, W., D'Egidio, G. E., et al., *J. Immunol.* 2002, *169*, 5679–5688.

[29] Guha, M., Mackman, N., *Cell. Signal* 2001, *13*, 85–94.

[30] O'Neill, L. A., *Trends Immunol.* 2002, *23*, 296–300.

[31] Sulahian, T. H., Pioli, P. A., Wardwell, K., Guyre, P. M., *J. Leukoc. Biol.* 2004, *76*, 271–277.

[32] Moller, H. J., Aerts, H., Gronbaek, H., Peterslund, N. A., et al., *Scand. J. Clin. Lab Invest. Suppl.* 2002, *237*, 29–33.

[33] Walter, R. B., Bachli, E. B., Schaer, D. J., Ruegg, R., Schoedon, G., *Blood* 2003, *101*, 3755–3756.

[34] Apostolopoulos, V., McKenzie, I. F., *Curr. Mol. Med.* 2001, *1*, 469–474.

[35] Stahl, P. D., Ezekowitz, R. A., *Curr. Opin. Immunol.* 1998, *10*, 50–55.

[36] Natarajan, K., Sawicki, M. W., Margulies, D. H., Mariuzza, R. A., *Biochemistry* 2000, *39*, 14779–14786.

[37] Marzio, R., Mauel, J., Betz-Corradin, S., *Immunopharmacol. Immunotoxicol.* 1999, *21*, 565–582.

[38] Menard, R., Nagler, D. K., Zhang, R., Tam, W., et al., *Adv. Exp. Med. Biol.* 2000, *477*, 317–322.

[39] Sakai, K., Nii, Y., Ueyama, A., Kishino, Y., *Cell. Mol. Biol.* 1991, *37*, 353–358.

[40] Chapman, H. A., Jr., Munger, J. S., Shi, G. P., *Am. J. Respir. Crit. Care Med.* 1994, *150*, S155–159.

[41] Rossman, M. D., Maida, B. T., Douglas, S. D., *Cell. Immunol.* 1990, *126*, 268–277.

[42] Riese, R. J., Chapman, H. A., *Curr. Opin. Immunol.* 2000, *12*, 107–113.

[43] Ritchie, H., Jamieson, A., Booth, N. A., *Thromb. Haemost.* 1997, *77*, 1168–1173.

[44] Ritchie, H., Booth, N. A., *Thromb. Haemost.* 1998, *79*, 813–817.

[45] Ritchie, H., Booth, N. A., *Exp. Cell Res.* 1998, *242*, 439–450.

[46] Risse, B. C., Chung, N. M., Baker, M. S., Jensen, P. J., *J. Cell. Physiol.* 2000, *182*, 281–289.

[47] Sampson, M., Davies, I., Gavrilovic, J., Sussams, B., et al., *Cardiovasc. Diabetol.* 2004, *3*, 7.

[48] Yun, K. A., Lee, W., Min, W. K., Chun, S., et al., *Scand. J. Clin. Lab Invest.* 2004, *64*, 223–228.

[49] Harper, R. G., Workman, S. R., Schuetzner, S., Timperman, A. T., Sutton, J. N., *Electrophoresis* 2004, *25*, 1299–1306.

[50] Tammen, H., Schulte, I., Hess, R., Menzel, C., et al., *Proteomics* 2005, *5*. DOI 10.1002/pmic.200400419

[51] Rajkumar, S. V., Richardson, R. L., Goellner, J. R., *Mayo Clin. Proc.* 1998, *73*, 533–536.

[52] Hanif, M., Raza, J., Qureshi, H., Issani Z., *J. Pak. Med. Assoc.* 2004, *54*, 119–122.

[53] Sapey, T., Mendler, M. H., Guyader, D., Morio, O., et al., *J. Clin. Gastroenterol.* 2000, *30*, 259–263.

[54] Goris, R. J., Boekhorst, T. P., Nuytinck, J. K., Gimbrere, J. S., *Arch. Surg.* 1985, *120*, 1109–1115.

[55] Cho, S. Y., Lee, E.-Y., Chun, Y.-W., Lee, J. S., et al., *Proteomics* 2005, *5*. DOI 10.1002/pmic.200400497

[56] Vyzantiadis, T., Theodoridou, S., Giouleme, O., Harsoulis, P., et al., *Hepatogastroenterology* 2003, *50*, 814–816.

[57] Murawaki, Y., Ikuta, Y., Nishimura, Y., Koda, M., Kawasaki, H., *J. Gastroenterol. Hepatol.* 1996, *11*, 443–450.

[58] Song, B. C., Chung, Y. H., Kim, J. A., Choi, W. B., et al., *Cancer* 2002, *94*, 175–180.

[59] Mazziotti, G., Sorvillo, F., Morisco, F., Carbone, A., et al., *Cancer* 2002, *95*, 2539–2545.

[60] Stuver, S. O., Kuper, H., Tzonou, A., Lagiou, P., et al., *Int. J. Cancer* 2000, *87*, 118–121.

[61] Berhane, B., Zong, C., Liem, D. A., Huang, A., et al., *Proteomics* 2005, *5*. DOI 10.1002/pmic.200401084

[62] Yang, Z., Hancock, W. S., Richmond-Chew, T., Bonilla, L., *Proteomics* 2005, *5*. DOI 10.1002/pmic.200400411

[63] King, C., Chan, D. A. L., Hise, D. M., Schaefer, C. F., Xiao, Z., Janini, G. M., Buetow, K. H., et al., *Clin. Proteomics* 2004, *1*, 101–226.

[64] Waga, S., Tan, E. M., Rubin, R. L., *Biochem. J.* 1987, *244*, 675–682.

[65] Holdenrieder, S., Stieber, P., Bodenmuller, H., Busch, M., et al., *Ann. NY Acad. Sci.* 2001, *945*, 93–102.

[66] Bickerstaff, M. C., Botto, M., Hutchinson, W. L., Herbert, J., et al., *Nat. Med.* 1999, *5*, 694–697.

[67] Amoura, Z., Chabre, H., Koutouzov, S., Lotton, C., et al., *Arthritis Rheum.* 1994, *37*, 1684–1688.

[68] Igney, F. H., Behrens, C. K., Krammer, P. H., *Eur. J. Immunol.* 2000, *30*, 725–731.

[69] Yan, Q., Cho, E., Lockett, S., Muegge, K., *Mol. Cell. Biol.* 2003, *23*, 8416–8428.

[70] Leon, O., Roth, M., *Biol. Res.* 2000, *33*, 21–30.

[71] Girschick, H. J., Grammer, A. C., Nanki, T., Mayo, M., Lipsky, P. E., *J. Immunol.* 2001, *166*, 377–386.

[72] Molina, H., Bunkenborg, J., Reddy, G. H., Muthusamy, B., et al., *Mol. Cell. Proteomics* 2005, *4*, 637–650.

[73] Brown, M. S., Ye, J., Rawson, R. B., Goldstein, J. L., *Cell* 2000, *100*, 391–398.

[74] Liu, X., Gao, J. X., Wen, J., Yin, L., et al., *J. Exp. Med.* 2003, *197*, 1721–1730.

17
Cardiovascular-related proteins identified in human plasma by the HUPO Plasma Proteome Project Pilot Phase*

Beniam T. Berhane, Chenggong Zong, David A. Liem, Aaron Huang, Steven Le, Ricky D. Edmondson, Richard C. Jones, Xin Qiao, Julian P. Whitelegge, Peipei Ping and Thomas M. Vondriska

Proteomic profiling of accessible bodily fluids, such as plasma, has the potential to accelerate biomarker/biosignature development for human diseases. The HUPO Plasma Proteome Project pilot phase examined human plasma with distinct proteomic approaches across multiple laboratories worldwide. Through this effort, we confidently identified 3020 proteins, each requiring a minimum of two high-scoring MS/MS spectra. A critical step subsequent to protein identification is functional annotation, in particular with regard to organ systems and disease. Performing exhaustive literature searches, we have manually annotated a subset of these 3020 proteins that have cardiovascular-related functions on the basis of an existing body of published information. These cardiovascular-related proteins can be organized into eight groups: markers of inflammation and/or cardiovascular disease, vascular and coagulation, signaling, growth and differentiation, cytoskeletal, transcription factors, channels/receptors and heart failure and remodeling. In addition, analysis of the peptide *per* protein ratio for MS/MS identification reveals group-specific trends. These findings serve as a resource to interrogate the functions of plasma proteins, and moreover, the list of cardiovascular-related proteins in plasma constitutes a baseline proteomic blueprint for the future development of biosignatures for diseases such as myocardial ischemia and atherosclerosis.

17.1
Introduction

The potential of proteomics to deliver improved health care has catalyzed a surge in diagnostic and/or "clinical" proteomic studies over the past decade. Foremost

---

* Originally published in Proteomics 2005, 13, 3520–3530

Exploring the Human Plasma Proteome. Edited by Gilbert S. Omenn
Copyright © 2006 WILEY-VCH Verlag GmbH & Co. KGaA, Weinheim
ISBN: 3-527-31757-0

among goals pursued by these studies is the identification of biomarkers or biosignatures of human diseases, more closely linked than genomic information to physiologic and pathologic human conditions. Despite some considerable progress in this regard [1], proteomics has yet to deliver concrete diagnostic/therapeutic tools to advance medicine.

One impediment to this process has been the absence of a clear proteomic map, or starting point, similar to the initial blueprints of the human genome, for a clinically relevant human tissue. One tissue of overt clinical significance is plasma, which is easily sampled *via* standardized and noninvasive procedures in virtually every outpatient clinic in the world. Notwithstanding multiple proteomic studies on human plasma and serum to date [*e.g.*, 2, 3], an organized and systematic analysis of human plasma using distinct proteomic techniques has never been reported. An international team of investigators recently undertook the HUPO Plasma Proteome Project pilot phase (PPP) specifically to address this paucity of information.

17.1.1
HUPO Plasma Proteome Project pilot phase

The collaborative effort of the HUPO PPP has analyzed standardized human plasma samples using various proteomic platforms. Human blood samples were collected and pooled from fasting healthy adult males and post-menopausal females, with appropriate anticoagulant concentrations of K-EDTA, lithium heparin or sodium citrate for plasma, or clot activator for serum as described by Omenn *et al.*, in this issue [4]. Donors were deemed healthy at the blood collection site on the basis of a screening questionnaire and each sample tested negative for human immunodeficiency virus (HIV-1 and HIV-2 antibodies, and HIV-1 antigen), hepatitis B virus (hepatitis B core antigen), human T-lymphotropic virus type I and II, (anti-HTLV-I/-II), hepatitis C virus (anti-HCV antibodies), and syphilis. The method for protein identification used for all proteins was peptide-sequencing MS/MS. However, the instruments and pre-analytical separation varied across individual laboratories. For a comprehensive description of the collection techniques and different platforms utilized please see [4].

A critical step subsequent to protein identification is annotation, or the placing of the identified proteins in context with what is known of the organism's biology. This is a manifold process in the case of plasma which, as the fluid that provides nourishment to the entire body, is impinged upon by virtually every organ. To address this challenge, a subset of proteins from the HUPO PPP was annotated for relevance to cardiovascular disease.

17.1.2
Need for novel insights into cardiovascular disease

Cardiovascular disease (CVD) remains the leading cause of death in the industrialized world and the number one non-communicable condition worldwide [5, 6]. Despite intense research in this field over the last 50 years, the CVD epidemic in developed nations continues to worsen. As a result, there are at least two aspects of cardiovascular physiology and medicine that will benefit from the mapping of the human plasma proteome. First, a comprehensive list of proteins that reside in plasma in normal humans will significantly advance our understanding of the physiological interactions among plasma, heart, and vasculature. That is, a list of cardiac- or vascular-derived proteins in plasma will catalyze future basic science investigations to unravel the functional importance of these plasma constituents. Second, the identification of novel biomarkers and/or biosignatures of health and disease residing in an accessible bodily fluid, such as plasma, has the potential to greatly impact human health.

17.2
Materials and methods

Although gene ontology analyses and other automated annotation methods can provide useful information about gene classification, organ-specific annotation of proteins requires a discerning evaluation of protein function as supported by scientific literature. To reap information regarding proteins of cardiac or vascular origin that were identified in the PPP analyses, we researched an initial 3020-protein subset of the total proteins identified by the PPP. Members of this group of proteins, as described in detail elsewhere in this issue [4], were identified in human plasma by the HUPO PPP investigators on the basis of two or more peptides from MS/MS experiments (trypsin digest of proteins was performed in all of the studies discussed herein). For annotation, we used the PubMed search engine (National Library of Medicine, USA; www.ncbi.nlm.nih.gov) to individually search each of these 3020 proteins for relevance to cardiovascular function and disease using the key words "heart", "cardiac", "vasculature" or "vascular". When searches of protein names returned no results, we searched abbreviations of the protein names (*e.g.*, protein kinase C would also be searched as PKC) and/or non-isoform-specific names (*e.g.*, PKCΘ would also be searched as PKC). It is important to emphasize that no part of this process was automated: abstracts and manuscripts were individually read and relevance of specific proteins to cardiovascular physiology/pathology extracted.

Our analyses identified cardiovascular-related functions for at least 345 of these 3020 proteins and suggested that these proteins can be divided into eight different categories based on their known functions (Fig. 1). Each of these categories is discussed in detail below, and individual members-of-interest from these categories addressed in turn. A complete list of potentially cardiovascular-related proteins can be found in the Supplemental Tables, and Tables 1–8 list example members from each group with clear relevance to cardiovascular function.

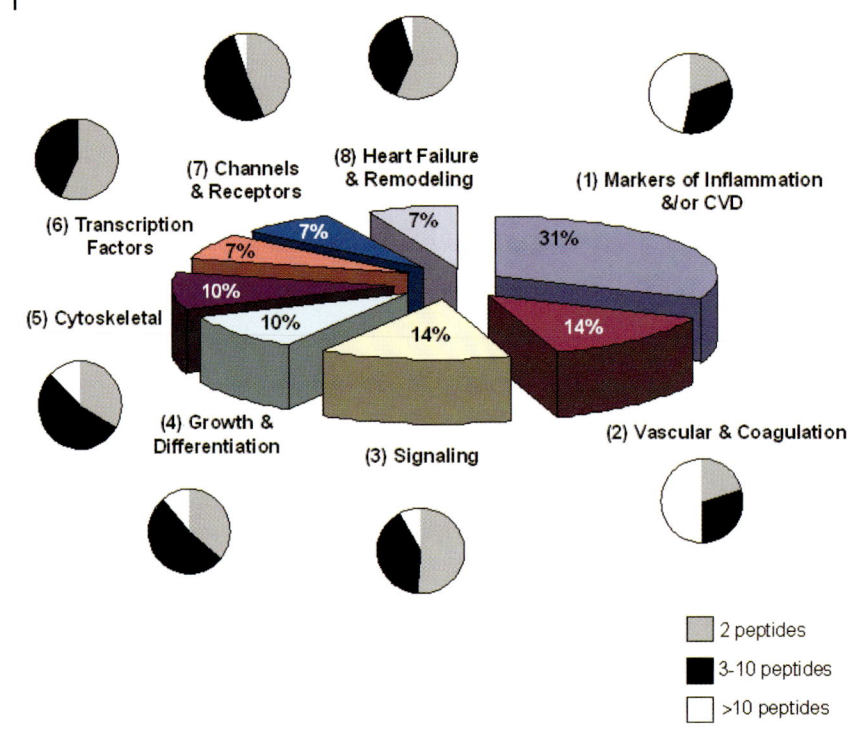

Fig. 1 Annotative break-down based on known function of cardiovascular-related proteins identified in plasma. 345 of the 3020 proteins identified with ≥2 peptides were ascribed cardiovascular-related function on the basis of previous studies, and have been divided into eight groups (central, colored chart): Markers of inflammation and/or CV disease (CVD); vascular and coagulation; signaling; growth- and differentiation-associated; cytoskeletal; transcription factors; channels and receptors; and heart failure and remodeling. Percentages of total 345 cardiovascular-related proteins represented by each group are labeled. Peripheral charts (black and white) indicate the respective numbers of LC/MS/MS-sequenced peptides used to identify the individual proteins in each group (2 peptides; 3–10 peptides; and, >10 peptides).

17.3
Groups of cardiovascular-related proteins

17.3.1
Markers of inflammation and CVD

CVD has long been associated with a dynamic interplay between vasoactive agents (proteins, lipids, and other factors) that influence the patency of the vasculature and thereby the function of the heart. This is especially pertinent in the case of acute myocardial infarction, or heart attack, which involves the blockage of blood flow to the heart and is thought to be influenced directly by inflammatory mediators in the blood [7, 8]. The identification of these proteins was therefore not surprising, and moreover, these findings provided an important "positive control" for the proteomic analyses.

Tab. 1 Example markers of inflammation and/or cardiovascular disease in plasma

Protein name	IPI identifier	Known function
CD 14 Antigen	IPI00029260	Mediates inflammatory response
C-reactive protein	IPI00022389	Marker of inflammation and heart disease
Glutathione peroxidase 3	IPI00026199	Associated with arterial thrombosis
Lactotransferrin	IPI00298860	Associated with ischemic stroke
Glutathione S transferase	IPI00019755	Related to Huntington's and Alzheimer's disease
Osteoglycin (mimecan)	IPI00025465	Proteoglycan in atherosclerotic plaques
Serine (cysteine) proteinase inhibitor	IPI00305457	Prevents ischemia/reperfusion damage during heart attack
Transferrin	IPI00022463	Associated with hemochromatosis, iron levels
Thrombospondin 1	IPI00296009	Inflammation and atherosclerosis related
Serum amyloid A2	IPI00022368	Biomarker for acute myocardial infarction

Please see Supplemental Tables for a complete listing of markers of inflammation and cardiovascular disease identified in plasma by the HUPO PPP (including the number of peptides used to identify each protein)

C-reactive protein (CRP) is a clinically-established marker for acute phase reactions and inflammation [7, 9], which plays a significant role in the development and progression of coronary artery disease. CRP has been shown to be elevated in blood/plasma during diseases that directly or indirectly involve inflammation, and thus is a clear marker of cardiovascular status (Tab. 1). Transferrin is a serum protein that binds and transports iron for delivery to cells [10]. The levels of serum transferrin depend upon the iron need of the cells, and although necessary for multiple cellular functions, free iron also has toxic and insoluble properties. During anemia, serum transferrin levels tend to increase, whereas during hemochromatosis, transferrin levels decrease [10]. Since most chronic diseases are accompanied by anemia, the levels of transferrin together with ferritin (not identified in these studies) which binds to iron in tissue, similar to the manner in which transferrin binds to iron in plasma, serve as general harbingers of chronic disease.

17.3.2
Vascular and coagulation proteins

The vasoactivity of coronary vessels directly regulates cardiac function, and in addition, the independent roles of systemic atherosclerosis and hypertension as risk factors for CVD have been established [7]. The plasma serves as an ideal sample from which to evaluate the proteomic status of the vasculature as well as the heart. In addition to proteins from other groups that are of vascular origin (e.g., cytoskeletal proteins from vascular tissues), we also annotated a group of proteins with primarily vascular and/or coagulation functions that did not classify in the other categories. As expected, our analyses of the plasma identified

Tab. 2 Example vascular and coagulation proteins in plasma

Protein name	IPI identifier	Known function
Angiopoietin-like 3	IPI00004957	Regulates angiogenesis
Angiotensin I converting enzyme	IPI00025852	Central regulator of vascular tone; role in hypertension (ACE inhibitors)
Angiotensinogen proteinase inhibitor	IPI00032220	Central regulator of vascular tone; regulates actions of angiotensinogen
Apolipoprotein A-I	IPI00021841	Involved in atherosclerosis and heart disease
Cadherin, vascular endothelial	IPI00012792	Regulates endothelial cell adhesion, growth
Carbonic anhydrase 2	IPI00218414	Regulates acid/base levels; vasoactive roles
Fibrinogen A alpha polypeptide	IPI00021885	Involved in platelet aggregation and atherosclerosis
Fibronectin	IPI00022418	Regulates platelet adhesion
Heparin sulfate proteoglycan	IPI00024284	Regulates angiogenesis, peripheral resistance and blood pressure
Vascular cell adhesion molecule	IPI00018136	Associated with atherosclerosis

Please see Supplemental Tables for a complete listing of vascular proteins identified in plasma by the HUPO PPP (including the number of peptides used to identify each protein)

many proteins from vascular cells (such as smooth muscle cells) as well as known components of the blood itself (including numerous coagulation factors) (Tab. 2).

A primary vasoregulatory mechanism in mammals is the renin-angiotensin system. Renin is released by the kidneys upon a decrease in blood pressure and cleaves angiotensinogen into angiotensin I. Subsequently, angiotensin I is cleaved by angiotensin converting enzyme (ACE; a protein identified by these analyses), primarily in the lung capillaries, to angiotensin II. Angiotensin II plays a major role in increasing peripheral resistance, and therefore blood pressure, in normal and pathological states, and thus ACE inhibitors are some of the most widely prescribed drugs for congestive heart failure and hypertension [11].

Another known plasma constituent identified by these studies is plasminogen, which plays a critical role in thrombotic disease [12]. The major reaction of the fibrinolytic system involves the conversion of the inactive proenzyme plasminogen into the active enzyme plasmin. Subsequently, plasmin can degrade fibrinogen, fibrin monomers and cross-linked fibrin. Thus, plasmin can cleave fibrin clots associated with atherosclerosis and coronary artery disease and is an ideal therapeutic target. Numerous thrombolytic drugs, often used in the wake of myocardial infarction, directly or indirectly aid the conversion of plasminogen into plasmin.

Tab. 3 Example signaling proteins in plasma

Protein name	IPI identifier	Known function
Adenylate cyclase 6	IPI00011938	Downstream regulator of adrenergic cascade
Calcium/calmodulin dependent kinase	IPI00000026	Regulates calcium-dependent signaling in hypertrophy and other settings
Caspase 6, cysteine protease	IPI00023876	Central mediator of apoptotic cell death
Heat shock 70k Da protein 8	IPI00003865	Involved in hypertrophy and cardiac protection
Janus kinase 1 (tyrosine kinase)	IPI00011633	Signals to STAT transcription factors, involved in cardioprotection
Mitogen-activated protein kinase kinase 2 (MEK2)	IPI00003783	Involved in growth signaling and ischemia/reperfusion injury in cardiac cells
Phosphodiesterase 4B, c-GMP-specific	IPI00220621	Degrades the second messenger cGMP, related to NO signaling
Phospholipase A2, group IVC	IPI00003166	Receptor-activated, upstream of PKC signaling in heart and vasculature
Pleckstrin homology domain-containing protein, family C	IPI00000856	PH domains regulate interaction of signaling proteins with lipids in cardiac and other cells
SH3 binding protein 4	IPI00171093	SH2 and SH3 domains play important roles in protein-proteins interactions

Please see Supplemental Tables for a complete listing of signaling proteins identified in plasma by the HUPO PPP (including the number of peptides used to identify each protein).

17.3.3
Signaling proteins

A group of signaling proteins (Tab. 3) containing known mediators of cardiac and/or vascular signal transduction pathways was identified by these analyses [13]. While many of these "signaling" proteins have been implicated in the cardiovascular system, it is important to note that the proteomic experimentation herein provides no evidence to unequivocally indicate that these proteins identified in the plasma indeed originate from heart or vascular cells. Consideration of this caveat is pertinent for all groups of proteins, but especially for signaling proteins, which may have disparate roles in distinct cell types [14]. Nevertheless, numerous signaling proteins known to directly modulate protective and deleterious cardiac phenotypes, such as PKC, adenylate cyclase 6, and the MAPK activator MEK2, were observed in plasma. The enzyme phosphodiesterase was also identified, which degrades cGMP, the second messenger of nitric oxide, a ubiquitous vasoactive substance.

Also found in this group were caspases (including caspase 6) and phospholipase A2. Caspases are a group of cysteine proteases essential for apoptosis, the primary mechanism of programmed cell death [15]. Recent evidence suggests a role for apoptosis in myocardial ischemia/reperfusion injury (*i.e.*, myocardial infarction), making the role of caspases and other canonical "apoptotic signaling proteins" an area of intense research recently. The human genome encodes 11 caspases that can be divided into subgroups depending on inherent substrate specificity. Caspases are synthesized as a single-chain zymogen and carry out cell death in response to apoptotic stimuli, although the activation state of the caspase(s) cannot be ascertained from the present studies.

Phospholipase A2, known as platelet-activating factor acetyl hydrolase, comprises a family of enzymes that hydrolyze phospholipid ester bonds to yield fatty acids and lysophospholipids-signaling molecules that mediate a multitude of cellular processes. During inflammation, the main function of phospholipase A2 is to convert membrane lipids into arachidonic acid. Subsequently, arachidonic acid is further converted into thromboxane, prostaglandins, prostacyclin and leukotrienes, all central inflammatory signaling molecules. Accordingly, corticosteroids (known inhibitors of phospholipase A2) are clinically used to counteract inflammation [16].

17.3.4
Growth- and differentiation-associated proteins

The heart displays a limited ability for self-renewal following injury and thus the role of growth and differentiation factors in cardiovascular health remains a hotbed of research interest [17]. The identification of such factors in the circulating plasma of a normal human has important implications for our understanding of the physiological importance of normal cardiac and vascular growth in adults (Tab. 4).

These findings also support the paracrine/endocrine actions of many growth factor molecules (such as IGF-1) that may use the plasma as a vehicle to reach adjacent cells. Insulin-like growth factors (IGFs) are polypeptides with high sequence similarity to insulin [18]. They can trigger the same cellular responses as insulin, including cell division and proliferation. Most IGF-1 is secreted in the liver and transported to other tissues, acting as an endocrine/paracrine hormone. More recently, IGF-1 and IGF-related proteins (identified in this study) have been found to have a role in cardiac growth and potentially regeneration [17].

17.3.5
Cytoskeletal proteins

As noted above, many of the salient "biomarkers" of cardiac injury were not identified by these approaches (e.g., troponins), suggesting that cardiomyocyte rupture and death in the setting of infarction was not robustly occurring in the donor patients. The finding that heretofore unrecognized cytoskeletal components, potentially of cardiac and/or vascular cell origin, reside in the plasma, suggests that release of cytoskeletal proteins from muscle cells may represent an area of research warranting further investigation. In addition, the cardiac- and vascular-specific proteins identified in these studies are discussed in detail below.

Actin is a major component of the cytoskeleton in all eukaryotic cells and is comprised of several isoforms whose distribution in vertebrates is tissue-specific (interestingly, the cardiac and skeletal isoforms were detected in these analyses).

17.3 Groups of cardiovascular-related proteins

Tab. 4 Example growth- and differentiation-associated proteins in plasma

Protein name	IPI identifier	Known function
Alpha-macroglobulin	IPI00032256	Acute phase protein involved in cardiac hypertrophy
Bone morphogenic protein 10	IPI00030115	Centrally involved in cardiogenesis and growth
Endotehlin converting enzyme	IPI00002478	Regulates endothelin levels, involved in cardiac hypertrophy and vascular tone
Growth arrest-specific 6	IPI00032532	Associated with normal heart development in mammals
IGF binding protein	IPI00020996	Retains IGF in blood, alters ability of IGF to stimulate growth in cardiac and other cells
P300/CBP-associated factor	IPI00022055	Acetylase that modulates transcription factor involved in growth
Prodynorphin	IPI00000832	Opioid signaling and cardiac gene expression
Phosphoglucomutase 5	IPI00014852	Vascular and cardiac development and differentiation
Polo-like kinase 2	IPI00302787	Related to permanent withdrawal of cardiomyocytes from cell cycle
Wingless-type MMTV	IPI00011031	Role in normal cardiac development

Please see Supplemental Tables for a complete listing of growth- and differentiation-associated proteins identified in plasma by the HUPO PPP (including the number of peptides used to identify each protein).

The thin filaments of actin intertwine in a helical pattern. Each actin filament is arranged on a serpentine backbone of tropomyosin molecules. Crucial to the interaction between actin and myosin, both found in these studies, are the troponin-complexes, themselves markers of cardiac cell rupture and necrosis as mentioned above. The interaction between actin and myosin constitutes the functional basis for force generation (Tab. 5).

17.3.6
Transcription factors

Several of the identified transcription factors (Tab. 6) have known roles in normal cardiac function, as well as in innate cardiac protective responses, such as ischemic preconditioning (*e.g.*, STAT; [19]). Likewise, transcription factors involved in vascular function were identified in the analyses.

The plasma localization of transcription factors with known relevance to cardiac and smooth muscle function has intriguing implications for analysis of cardiovascular health. Little is known about transcriptional processes in the plasma, but its hosting of several factors which are linked to particular cardiac or vascular phenotypes raises the possibility that these transcriptional agents are either released from

Tab. 5 Example cytoskeletal proteins in plasma

Protein name	IPI identifier	Known function
Actin beta	IPI00021439	Central component of contractile apparatus
Laminin, alpha 2	IPI00218725	Major component of muscle cell basement membrane
Myosin, heavy polypeptide 6, cardiac isoform	IPI00302328	Central component of contractile apparatus
Nebulin	IPI00303335	Aids specification of actin filament length
Smoothelin	IPI00024007	Smooth muscle actin-associated protein
Supervillin	IPI00018370	Actin-associated protein
Talin 1	IPI00298994	Cardiac cytoskeletal protein
Tropomodulin 3	IPI00005087	Regulates actin structure/assembly
Tubulin beta polypeptide	IPI00013475	Heart and vascular cytoskeletal protein
Villin 2 (ezrin)	IPI00216311	Actin-associated protein

Please see Supplemental Tables for a complete listing of cytoskeletal proteins identified in plasma by the HUPO PPP (including the number of peptides used to identify each protein).

Tab. 6 Example transcription factors in plasma

Protein name	IPI identifier	Known function
E2F transcription factor	IPI00180615	Cell cycle regulator protein
Forkhead	IPI00294826	Regulates genes involved in cardiac development
Mitochondrial translation initiation factor 2	IPI00005039	Regulates mitochondrial protein expression in heart and other tissues
SIN3 homolog B, transcriptional regulator	IPI00027351	HDAC-associated transcription factor
Signal transducer and activator of transcription 1	IPI00030781	Contributes to protection of ischemic myocardium
Topoisomerase II beta 180kDa	IPI00027280	Target of doxorubicin that causes cardiomyopathy
Zinc finger, BTB domain containing 4	IPI00001838	Zn finger transcription factors involved in development in heart and other tissues

Please see Supplemental Tables for a complete listing of transcription factors identified in plasma by the HUPO PPP (including the number of peptides used to identify each protein).

cardiac/vascular cells, or that they may be directly regulating similar processes in the plasma. It is also possible that, like cytoskeletal proteins, these transcription factors have arisen in plasma due to cardiac and/or muscle cell rupture or exocytosis, not associated with cell death. This possibility is discussed in detail below.

Tab. 7 Example channel and receptor proteins in plasma

Protein name	IPI identifier	Known function
Epidermal growth factor receptor	IPI00018274	Central role in numerous cardiac signaling processes
Glutamate receptor, ionotropic, N-methyl D-aspartate 2B	IPI00297933	Role in cardiovascular response to NMDA agonists
Mannose receptor C, type 1	IPI00027848	Involved with parasite infection of heart and other tissues
Nucleoporin 133 kDa	IPI00291200	Involved in nuclear trafficking in cardiomyocytes and other cells
Opioid growth factor receptor	IPI00021537	Involved in hyperplasia after angioplasty and in cardiac protection against ischemia
Poliovirus receptor	IPI00299158	Gene polymorphisms associated with cardiovascular disease
Ryanodine receptor (cardiac isoform)	IPI0002317	Calcium release channel in sarcoplasmic reticulum of cardiac muscle
Voltage-gated sodium channel type X	IPI00008522	Regulates cellular ionic homeostasis
Tight junction protein (zona occludens) 2	IPI00003843	Gap junction-associated protein
Toll-like receptor 10	IPI00008887	Interleukin receptors, regulate some forms of growth/hypertrophy

Please see Supplemental Tables for a complete listing of channel and receptor proteins identified in plasma by the HUPO PPP (including the number of peptides used to identify each protein).

17.3.7
Channel and receptor proteins

Many of the proteins in this category are imbedded within cell membranes for the purpose of transporting molecules/ions, maintaining homeostasis and eliciting contraction [20] (Tab. 7). The finding of these integral membrane proteins in plasma highlights a concern with proteomic investigations employing enzymatic (usually trypsin) digestion of proteins prior to identification (as is common practice and was performed prior to collection of all the HUPO PPP data analyzed herein). Other analyses within the PPP are mapping peptides from identified proteins to the regions of the intact molecules, and this information will be especially enlightening for the membrane proteins identified in plasma. Specifically, it will shed light onto whether these peptides/proteins made their way into plasma *via* controlled cleavage events (as would be supported if the peptides used for identification mapped to exposed, hydrophilic regions of the intact proteins) or whether their presence in plasma may be indicative of muscle cell exocytosis or membrane rupture (as would be supported if the peptides mapped to membrane-buried, hydrophobic regions).

Tab. 8 Example heart failure- and remodeling-associated proteins in plasma

Protein name	IPI identifier	Known function
Acyl-Coenzyme A dehydrogenase	IPI00028031	Down-regulated in cardiac hypertrophy and failure
ATP-binding cassette, subfamily C	IPI00024278	Mutated in some idiopathic cardiomyopathies
Integrin, alpha 1	IPI00328531	Potential role in remodeling after myocardial infarction
Leucine aminopeptidase 3	IPI00220067	Involved in cardiac hypertrophy signaling
Lysosomal-associated membrane protein	IPI00009030	Potential role in hypertrophic cardiomyopathy
Matrix metalloproteinase 9	IPI00027509	Matrix remodeling in cardiac valvular and other diseases
Neuropeptide Y receptor	IPI00019491	Induces cardiac hypertrophy
Regulator of G-protein signaling 12	IPI00024714	G protein signaling has clear role in cardiac hypertrophy
Transforming growth factor beta	IPI00018219	Participates in blood vessel growth, cytokine-induced inflammation and heart failure
Tumor necrosis factor super-family member 6	IPI00007577	Inducer of cardiac hypertrophy, inflammation and atherosclerosis; protective in some settings

Please see Supplemental Tables for a complete listing of heart failure- and remodeling-associated proteins identified in plasma by the HUPO PPP (including the number of peptides used to identify each protein).

Endogenous opioid peptides are known to be potent regulators of neuromodulators/neurotransmitters and growth [21], and their receptors were observed in the plasma in these investigations. Stimulation of delta-opioid receptors has been shown to be involved in protection against myocardial infarction [22]. The poliovirus receptor which is used by all three viral serotypes to initiate infection of cells was also found in plasma. More recently, polymorphisms in human poliovirus receptor, along with those in apolipoprotein (also identified in these analyses, listed in Markers of Inflammation and CVD) have been associated with coronary artery disease [23]. Identification of receptors, or receptor-derived peptides, in the plasma, has the potential to yield a biomarker of the activation status of these receptors in the heart and vasculature, and/or to identify a physiological cleavage event, potentially altering receptor structure/function.

17.3.8
Heart failure- and remodeling-related proteins

Heart failure remains the number one cause of morbidity in the elderly and significant research is targeted to discern the cellular mechanisms responsible for this debilitating phenomenon [17]. Cardiac remodeling following injury (*e.g.*, infarction) is an initially favorable process that ultimately leads to heart failure. Therefore, as with cardiac growth, the identification of heart failure/remodeling factors in the plasma will potentially aid basic science investigations into the mechanisms of this adaptive process (Tab. 8).

As an example of a clear mediator of heart failure, cytokines are soluble proteins secreted by lymphocytes, monocytes-macrophages and natural killer cells [24]. Among the cytokines, tumor necrosis factor (TNF), is known to stimulate T-cell proliferation and interleukin-2 production by T-helper cells. Furthermore, TNF initiates B-cell proliferation and attracts neutrophils. The role of TNF as a pro-inflammatory cytokine in the cardiovascular system has been suggested, and studies indicate a clear role for TNF in detrimental cardiac remodeling and heart failure [25].

17.4
Functional analyses and implications

The goal of these analyses was to discover proteins in the plasma with relevance to cardiovascular disease through directed, manual annotation of a large list of plasma proteins. Nevertheless, the proteomic approaches employed in these PPP investigations preclude assertion of an organ-specific origin of the vast majority of the proteins, regardless of the extent of literature describing their importance in these systems. As a result, cardiovascular system-directed annotation of plasma proteins affords two distinct results. First, the list of cardiovascular-related plasma proteins, and the putative functional groups, serve as a basis for biomarker/biosignature development for CVD. Second, the centralized information regarding the known roles of these proteins in the cardiovascular system provides a resource to aid investigators in pursuit of their actions in plasma. In both cases, focused basic science investigations are clearly required to unravel the function of the protein as a biomarker and/or as a physiological mediator in the plasma. Notwithstanding these considerations, there were proteins identified by these studies that are specific to cardiac and/or vascular cells, a brief discussion of which is warranted.

17.4.1
Organ specific cardiovascular-related proteins in plasma

In the Channels and Receptors group, the cardiac isoform of the ryanodine receptor was identified representing, to our knowledge, the first report of this protein in normal human plasma. The ryanodine receptor is an intracellular calcium channel in the sarcoplasmic reticulum membrane of contractile muscle (including skeletal and cardiac; incidentally, isoforms from both muscle types were identified). This protein plays a fundamental role in excitation-contraction coupling in the heart [26], a process that facilitates the conversion of electrical stimuli into the mechanical force to pump blood. Previous studies have demonstrated antibodies against the ryanodine receptor in sera from myasthenia gravis patients [27]. However, the

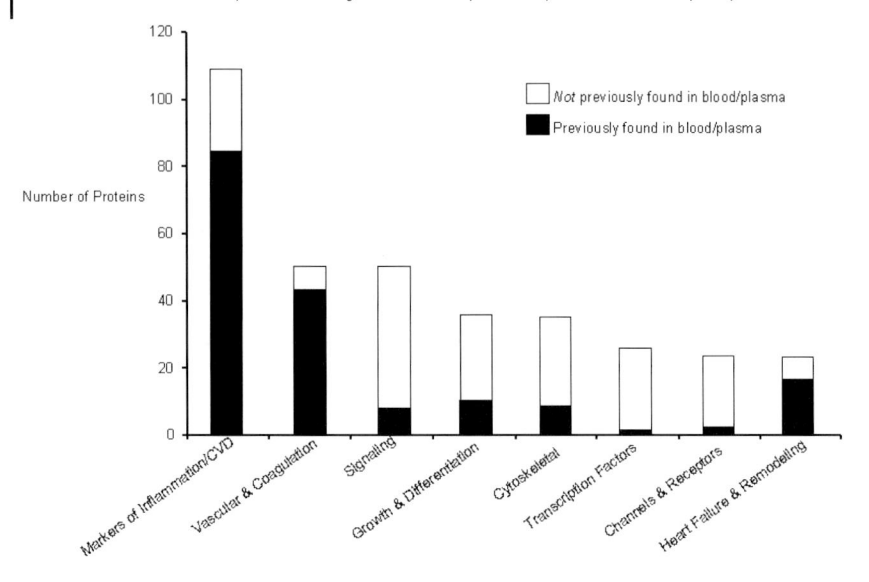

Fig. 2 Analysis of proteins with previously recognized presence and/or function in plasma. Following initial query to identify 345 proteins with cardiovascular-related functions, these proteins were re-searched on PubMed to determine whether they had been identified in plasma/blood by previous studies. The Vascular and Coagulation and Markers of Inflammation and CVD groups had the highest percentage of proteins that had been shown by other investigations to take residence in the plasma/blood.

identification of this cardiac-specific protein in the plasma of a healthy adult, on the basis of a high number (19) of peptides, is of particular importance because as an integral membrane protein within a subcellular organelle, ryanodine receptor presence in the plasma suggests a clear cell rupture or exocytosis event (*i.e.*, contamination from myocardial cells during the plasma isolation process is unlikely).

In the Cytoskeletal group, smoothelin is a structural protein with a distribution restricted to smooth muscle cells [28]. It is exclusively expressed in fully differentiated (contractile) smooth muscle cells, though its function and regulation are largely unknown. Interestingly, one study has shown that smoothelin colocalizes with smooth muscle actin in humans, and a potential role for this protein in modulating atherosclerotic lesions has been suggested [28]. These proteins serve as prime examples of organ-specific molecules identified with high confidence in the plasma that could potentially be developed into selective biomarkers of cardiovascular function.

17.4.2
Novel cardiovascular-related proteins identified in plasma

A secondary annotative analysis of the cardiovascular-related proteins was conducted to determine which of the proteins had previously been identified in the plasma/blood (Fig. 2). Not surprisingly, most of the proteins in the Vascular and Coagulation and Markers of Inflammation and CVD groups had been shown by other investigators to localize to the plasma, whereas the majority of the other groups hosted a larger number of novel plasma components. This underscores an advantage of large proteomic analyses of systems like the plasma: the anthropomorphic bias associated with studying a protein of interest is reduced and potential participants in normal and diseased human health can be identified and pursued by future studies. The analyses provided in this manuscript serve as a conceptual framework for future studies of cardiovascular-related plasma proteins, and for the mining of the plasma proteome for biosignatures of cardiovascular health and disease. Furthermore, knowledge from the cardiovascular system regarding the actions of these novel plasma constituents can provide insights into their roles in the plasma.

As noted in Section 2, the criteria for protein identification in these studies was two high-scoring tandem mass spectra. After generating a list of cardiovascular-related proteins, we examined the exact numbers of peptides used to identify these proteins in the original MS/MS experiments (Fig. 1). As compared to the other groups, a significant portion of the proteins identified in the Vascular and Coagulation (50%) and Markers of Inflammation and CVD (47%) groups were done so on the basis of greater than 10 peptides. This is an intriguing finding, in context with the aforementioned point that these same two groups harbored the greatest number of previously known plasma proteins. In the case of Transcription Factors, on the other hand, no proteins were identified on the basis of more than 10 peptides and 56% were identified with only two peptides. As another example, known abundant plasma proteins such as alpha-macroglobulin (associated with cardiac hypertrophy) and transferrin (associated with iron handling) were identified on the basis of 207 and 242 peptides, respectively. Although the peptide number is an inexact estimation of protein abundance, correlation between these two factors is often observed. In the context of the annotative efforts in the present study, there are several potential interpretations of the peptide number data. First, increased numbers of peptides identified for a given protein could indicate a higher copy level of that protein in the plasma due to altered abundance, stability or both. Second, if cell lysis or controlled exocytosis must occur to release a given protein, than the rate of this secondary process in "normal" humans will inherently determine the amount of these proteins (and hence their peptides) in the plasma, regardless of their abundance in the tissue of origin. Third, a given analytical technique may selectively enrich or deplete a given species of peptide, artificially altering its level in a manner uncorrelated with that of the intact protein.

It is important to highlight that the identification of proteins with known cardiovascular relevance in the plasma does not indicate that these proteins originate from the heart or vasculature, nor that they serve as biomarkers for CVD. The goal of these initial studies, performed in plasma from health individuals, was to establish an annotative baseline for proteins of potential relevance to cardiovascular function in the normal patient. It is only with this baseline information in hand that one can now pursue additional analyses in which biomarkers and biosignatures of CVD may be revealed (that is, by examining plasma from the diseased patient). Indeed, understanding of the time course changes in cardiovascular proteins in the plasma as humans progress through the early stages of CVD and into more acute conditions like myocardial infarction and stroke will be essential for use of plasma information for diagnoses. An ideal biomarker of CVD would be (i) specific for a certain stage of the disease, (ii) readily detectable in an accessible bodily fluid (such as plasma) or tissue, (iii) reproducibly observed across patients and correlated to other established clinical measurements of cardiovascular function, and (iv) amenable to simple detection in the clinical setting by standardized methods. This information can only be obtained through rigorous additional experimentation in plasma from healthy and diseased individuals.

17.5
Methodology considerations

It should be noted that proteins were assigned to groups in an exclusive manner, that is, no protein appears in more than one group. Nonetheless, it is certainly true that many proteins have disparate functions in the cardiovascular system, and indeed could be classified in multiple groups (for example, forkhead is a transcription factor with multiple functions and appears in the Transcription Factor group in our annotation but has been associated with cardiac development by some studies). For clarity, we listed proteins in a single annotative group; however, all proteins are listed for all groups in the Supplemental Tables such that the reader can garner information about specific cardiovascular-related proteins identified regardless of whether he/she agrees with the authors' classification.

Another important caveat of these studies is that protein identification is inferred on the basis of peptide identification by MS. Because peptide-sequencing MS/MS is currently the state-of-the-art for protein identification in proteomics, and this technique is very often preceded with enzymatic digestion of proteins (as was the case for all of the protein identifications discussed herein), confirming the presence in plasma of the protein from which these peptides presumably originate will require intact protein analyses by MS, antibody-based, or other methods. Moreover, the antithetic analyses must also be performed, to examine whether peptides themselves, independent of experimental digestion, are resident and perhaps biologically active in the plasma, as implicated by recent studies of low molecular weight plasma components [3].

Many of the proteins identified by these analyses (*e.g.*, transcription factors, signaling molecules and other low abundance proteins) would not be expected, *a priori*, to make their way into the plasma from cardiac or vascular cell origin in the absence of cell rupture (*i.e.*, oncosis/necrosis) or controlled exocytosis. However, the observation that several cardiac contractile proteins, far more abundant than transcription factors or signaling proteins, that are known to be released into plasma upon cell necrosis (*e.g.*, troponins) were *not* identified in these studies, suggests that the proteins that were observed in plasma may have localized there by a mechanism other than necrotic cell death and uncontrolled cell rupture. Because the current investigations examined only plasma from healthy individuals, it is impossible to know at this juncture the exact mechanism by which these proteins arose in the plasma. In future investigations, the proteins must be examined on an individual basis to determine whether physiological cell rupture or exocytosis, unaccounted for experimental error, or some other mechanism leads to this protein's identification in plasma.

17.6
Conclusions and future directions

In collaboration with the HUPO PPP, these studies report a group of 345 proteins with cardiovascular-related functions. This represents ~11% of the 3020 proteins identified thus far on the basis of two or more high-confidence MS/MS spectra. These proteins can be classified into at least eight functional groups: markers of inflammation and/or cardiovascular disease, vascular, signaling, growth and differentiation, cytoskeletal, transcription factors, channels/receptors and heart failure and remodeling.

There are at least two reasons why the organ-based annotative approach performed herein can yield important information regarding plasma proteins. First, if these proteins in the plasma indeed originate from the heart or vasculature, they may herald pathologic (such as CVD-associated) or physiologic (regarding normal interplay between functioning tissues and the plasma) relationships between plasma and organ systems that were previously not appreciated. Second, if the individual proteins are of plasma or blood cell origin, insights into their function in other tissues will advance the understanding of their potential actions in the plasma. That is, one can formulate hypotheses regarding plasma protein function on the basis of known roles for individual proteins in other tissues. This type of information could not be obtained without large-scale proteomic analyses (like the HUPO PPP) in conjunction with annotation efforts of equivalent magnitude by individual laboratories with the respective organ systems expertise. The potential roles of cardiac- or vascular-derived signaling molecules in the plasma to influence cardiovascular function, and the understanding of whether their release into the plasma is a controlled or catastrophic process, will be the pursuit of future basic science investigations.

In addition to the groups defined above, it should not escape observation that the list of cardiovascular-related proteins itself constitutes a phenotypic blueprint of baseline cardiovascular physiology. It is reasonable to hypothesize that changes in the levels and/or presence of these cardiovascular-related proteins may serve as a diagnostic tool, or readout, of the underlying function of the heart and vasculature under normal conditions and during common CVDs such as atherosclerosis, myocardial ischemic and heart failure. Although the specific approaches will differ between laboratories, we envision a necessary series of further investigations that will bring the present groundwork studies to fruition with regard to human physiology and disease. First, the presence of these proteins in plasma must be examined across age, gender and ethnic group. Quantitation of protein abundance must be established with regard to these same parameters, and new assays should be developed (e.g., immunoassays) to aid with the detection of the proteins across multiple samples and establish reproducible conditions for measurement. These endpoints will soundly establish the "quantitative baseline" of the plasma proteome with current technologies and allow us to move forward to characterize abnormalities in this proteome associated with cardiovascular disease. A multimarker [8] or biosignature strategy, using data from multiple proteins within a subproteome, can then be employed to distinguish phenotypic differences associated with patient health.

This study was supported by a HUPO Small Projects Grant and the Laubisch Endowment at UCLA.

17.7 References

[1] Granger, C. B., Van Eyk, J. E., Mockrin, S. C., Anderson, N. L., *Circulation* 2004, *109*, 1697–703.

[2] Anderson, N. L., Polanski, M., Pieper, R., Gatlin, T., Tirumalai, R. S., Conrads, T. P., Veenstra, T. D et al., *Mol. Cell. Proteomics* 2004, *3*, 311–326.

[3] Petricoin, E. F., Liotta, L. A., *Curr. Opin. Biotechnol.* 2004, *15*, 24–30.

[4] Omenn, G. S.,States, J. D., Adamski, M., Blackwell, T. W., Menon, R., Hermjakob, H., Apweiler, R. et al., *Proteomics* 2005, *5*, this issue.

[5] American Heart Association. *Heart Disease and Stroke Statistics: 2004 Update*, American Heart Association, Dallas, Texas, USA.

[6] World Health Organization. The World Health Report, 2004. www.who.int.

[7] Libby, P., *Nature* 2002, *420*, 868–874.

[8] Morrow, D. A., Braunwald, E., *Circulation* 2003, *108*, 250–252.

[9] Yeh, E. T. H., *Circulation* 2004, *109*, I11–II14.

[10] Baker, H. M., Anderson, B. F., Baker, E. N., *Proc. Natl. Acad. Sci. USA* 2003, *100*, 3579–3583.

[11] Franke, F. E., Pauls, K., Metzger, R., Danilov, S. M., *APMIS* 2003, *111*, 234–244.

[12] Deitcher, S. R., Jaff, M. R., *Rev. Cardiovasc. Med.* 2002, *3*, S25–S33.

[13] Olson, E. N., *Nat. Med.* 2004, *10*, 467–474.

[14] He, W., Miao, F. J., Lin, D. C., Schwandner, R. T., Wang, Z., Gao, J., Chen, J. L. et al., *Nature* 2004, *429*, 188–193.

[15] Salvesen, G. S., Abrams, J. M., *Oncogene* 2004, *23*, 2774–2784.

[16] Barnes, P. J., Adcock, I. M., *Ann. Intern. Med.* 2003, *139*, 359–370.

[17] Nadal-Ginard, B., Kajstura, J., Leri, A., Anversa, P., *Circ. Res.* 2003, *92*, 139–150.

[18] Laron, Z., *J. Clin. Pathol: Med. Pathol.* 2001, *54*, 311–316.

[19] Bolli, R., Dawn, B., Xuan, Y. T., *Trends Cardiovasc. Med.* 2003, *13*, 72–79.

[20] Rockman, H. A., Koch, W. J., Lefkowitz, R. J., *Nature* 2002, *415*, 206–212.

[21] Zagon, I. S., Verderame, M. F., McLaughlin, P. J., *Brain Res. Rev.* 2002, *38*, 351–376.

[22] Gross, G. J., *J. Mol. Cell Cardiol.* 2003, *35*, 709–718.

[23] Freitas, E. M., Phan, T. C., Herbison, C. E., Christiansen, F. T., Taylor, R. R., Van Bockxmeer, F. M., *J. Cardiovasc. Risk* 2002, *9*, 59–65.

[24] Dayer, J.-M., *Best Prac. Res. Clin. Rheumatol.* 2004, *18*, 31–45.

[25] Bradham, W. S., Bozkurt, B., Gunasinghe, H., Mann, D., Spinale, F. G., *Cardiovasc. Res.* 2002, *53*, 822–830.

[26] Wehrens, X. H., Marks, A. R., *Nat. Rev. Drug Discov.* 2004, *3*, 565–573.

[27] Romi, F., Gilhus, N. E., Varhaug, J. E., Myking, A., Skei, G. O., Aarli, J. A., *Ann. NY Acad. Sci.* 2003, *998*, 481–490.

[28] Niessen, P., Clement, S., Fontao, L., Chaponnier, C., Teunissen, B., Rensen, S., van Eys, G. et al., *Exp. Cell Res.* 2004, *292*, 170–178.

Index

a
albumin 115
annotation 329
antibody microarrays 91
anticoagulants 91
atherosclerosis 353

b
bioinformatics 323, 329
biomaker 353
biosignature 353
blood 249, 353

c
cLC-FT-ICR MS 249
cluster analysis 249
CV 273

d
databases 2, 317, 323
2-DE 201

f
fractionation 115
free flow electrophoresis 201

g
gel pixelation 115
glycoprotein 159
glycoproteome 159

h
haptoglobin 273
human serum proteome 221
HUPO 37, 159, 249, 317
HUPO Plasma Proteome Project 2, 37, 185, 201

i
immunoaffinity column 201
immunoassays 91
ischemic heart disease 353

l
LCQ 159
lectin 159
liquid chromatography-tandem mass spectrometry 135
LTQ 159

m
major protein depletion 115
MASCOT 289
mass spectrometry 221, 249, 289, 323
MS 329
multidimensional liquid chromatography 221
multilectin affinity chromatography 159
myocardial infraction 353

p
peptideProphet 289
plasma 2, 63, 201, 249, 273
plasma proteome 91, 135, 317
prefractionation 185

Exploring the Human Plasma Proteome. Edited by Gilbert S. Omenn
Copyright © 2006 WILEY-VCH Verlag GmbH & Co. KGaA, Weinheim
ISBN: 3-527-31757-0

protein arrays   115, 135
protein fractionation   221
protein identification   37
proteomics   329
proteomics database   37
proteomics data integration   37
proteomics methods   135

s
SELDI   273
SEQUEST   289
serum   2, 185, 249, 329
serum proteome   135
sonar   289
Specimen collection and handling   63
Spectrum Mill   289

t
tandem mass spectrometry   317
two-dimensional gel electrophoresis   221

x
X!Tandem   289